Quantum Computing: From Alice to Bob

Quantum Computing: From Alice to Bob

Alice Flarend

Physics Teacher

Bellwood-Antis High School

Bob Hilborn

Associate Executive Officer

American Association of Physics Teachers

OXFORD

UNIVERSITY PRESS

OXFORD
UNIVERSITY PRESS

Great Clarendon Street, Oxford, OX2 6DP,
United Kingdom

Oxford University Press is a department of the University of Oxford.
It furthers the University's objective of excellence in research, scholarship,
and education by publishing worldwide. Oxford is a registered trade mark of
Oxford University Press in the UK and in certain other countries

Impression: 2

Published in the United States of America by Oxford University Press
198 Madison Avenue, New York, NY 10016, United States of America

British Library Cataloguing in Publication Data
Data available

Library of Congress Control Number: 2021951706

ISBN 978–0–19–285797–2 (hbk)
ISBN 978–0–19–285798–9 (pbk)

DOI: 10.1093/oso/9780192857972.001.0001

Printed and bound by
CPI Group (UK) Ltd, Croydon, CR0 4YY

Cover image: Buntoon Rodseng/Shutterstock.com.

Contents

Preface

You have probably picked up this book because you are intrigued or puzzled by what you have heard about quantum computing (QC) and quantum information science (QIS) and you want to learn more. Why you should you keep reading? What is different in this book? We have tried to position this book between the highly technical books aimed at professional scientists and engineers with expertise in formal quantum mechanics and advanced mathematics and those general audience books that use almost no math, though some are quite clever in finding pictorial substitutes for the math.

Our presentation is aimed at readers who want an introduction to quantum computing that gives them a strong basic understanding and prepares them to talk intelligently with the "experts." If readers are so inclined, they will be ready to dig into the more technical aspects of the field after working through this book. The material should be accessible to a typical undergraduate student (or a senior-level high school student), whose mathematics background includes secondary school algebra and a passing acquaintance with sines and cosines. No background in physics is required, but if you are fortunate enough to have had a reasonable introductory physics course in high school or college, what you learned there will provide a broader perspective on quantum computing. High school mathematics and physics teachers as well as college and university faculty members who are not experts in quantum information science and quantum computing should also enjoy, and benefit from, reading the book.

The following distinguish our presentation from others:

- We make the treatment accessible to those who do not have much background in physics, computer science, and mathematics by providing just enough information about those areas to enable you to understand the fundamental issues. We have avoided unnecessary mathematics and mathematical jargon, but we treat carefully what we do use with examples and explanatory comments.

- Discussions are spread out and we take time to build an understanding of the critical elements of quantum computing. Inspired by Galileo's *Dialogues Concerning Two New Sciences*, we have structured the presentation as a conversation between the two of us, Alice and Bob, and Cardy, a typical undergraduate student, without much background in physics, computer science, or mathematics, who wants to learn about quantum computing. The tone of the writing is more conversational than didactic.

- Applications of quantum computing are developed in some detail to show where and how quantum computing has advantages over traditional computational methods. We use those applications to ground the mathematics in specific and accessible problems to be solved.

- We provide brief surveys of current quantum information and quantum computing devices, advice about what you should do next if you want to learn more about those fields, and an overview of career opportunities in those areas.

Although we have worked hard to make the book accessible to a wide audience, we also want to give you the concepts and vocabulary to read intelligently and knowledgeably about quantum computing. We have employed the following methods to accomplish this:

- We introduce and thoroughly explore Dirac quantum state vector notation since it is the most widely used notation in quantum computing and at the same time the notation that you in our target audience have probably not seen before, even if you have taken introductory courses in physics, mathematics, or computer science.

- There is a focus on multiple representations of quantum states and operations on quantum states, using abstract symbols, graphical (state space) diagrams, and mathematical column vectors and matrices. We emphasize multiple representations because quantum states are the bread-and-butter of quantum computing. No prior background in vectors and matrices is needed to understand the arguments in this book.

- Clear ways of thinking about the role of quantum measurements in quantum computing and how you represent those measurements formally are presented. In Chapter 14, we show you how to make a "quantum sandwich," which encapsulates everything we can learn from quantum measurements.

- We provide in-text exercises (Try Its) that you can use to build your fluency in writing descriptions of quantum states and to check your understanding of the concepts.

- A summary is given of quantum formulas and matrix terminology in the Appendix: Quantum Toolkit.

The quantum information science and quantum computing revolutions are just getting underway. To give you a sense of the diversity in both people and their educational backgrounds in this rapidly growing field, we have included at the end of Chapter 16 links to short biographical sketches of those involved who work in those fields in academia, in government labs, and in private business and industry.

Although the fundamental principles of quantum information science and quantum computing are reasonably well established, the details of the applications of those principles are likely to change dramatically over the next few decades. In the last chapter of the book, we write about what we believe are likely to be the most exciting developments in these fields. We also discuss moral and ethical concerns raised by the possibility of dramatic changes in computational power that are likely to be the product of the second quantum revolution.

Perhaps we should also tell you about what is *not* in the book—things that you might have expected to see. The following are avoided:

- Formal linear algebra. Although much of the math we use is based on linear algebra, we don't, at our level of discussion, need to use linear algebra jargon and theorems. We are confident that our treatment will prepare you to learn more about those topics if you want to go further in quantum computing.

- Complex numbers. Standard quantum mechanics uses complex numbers and complex variables. But we demonstrate that almost all the basic concepts of quantum computing and quantum information science can be treated without them. In Chapter 15, we do show you why complex numbers are needed for more general situations in quantum mechanics and explain the additional resources they provide for quantum computing.

- Stern–Gerlach measurements using magnetic field gradients. To appreciate those measurements, which were important in the historical development of quantum mechanics, you need a deeper grounding in physics: atomic and molecular beams, magnetic dipole moments, and forces exerted on dipoles due to magnetic field gradients. More importantly, those kinds of measurement are unlikely to play a role in quantum computing and quantum information even though they show up in many books on quantum computing. We believe that polarized light provides a much more accessible physical system to illustrate the basics of quantum states. We do give an introduction to so-called spin-1/2 quantum states but without the details of how measurements are carried out on those systems. We refer the reader who is interested in those details to many fine textbooks and online articles.

- Computational complexity. Many computer scientists and quantum computer people spend a lot of time building classifications of the difficulty of the problems addressed by computational algorithms. You may have read about P and NP problems. Those categories have proliferated into an alphabet soup of acronyms. Although these classifications help you to understand the kinds of problems quantum computers can advantageously solve (compared to traditional computers), we believe that knowledge of these categories is not necessary to understand the fundamentals of quantum computing and are best left aside until the basics are mastered. In a few specific examples, we do provide arguments that show that quantum computation often requires fewer (and, in some cases, far fewer) computational steps compared to the corresponding classical algorithms, but we avoid connecting these to the traditional computational complexity categories.

For those of you who are educators, this book provides more than enough content for a semester-long course. The core material is in Chapters 1–10. Chapters 11–13 provide detailed discussions of several quantum algorithms. The algorithms are relatively independent. Pick those that your students will find most interesting and discuss the details at a level appropriate for your students. If you have time to do only one algorithm in detail, we suggest the Grover algorithm in Chapter 12. It requires the least mathematical background to understand the goal and we have presented several different and complementary ways of understanding how the algorithm works. Chapters 14 and 15 treat some fundamental issues in quantum mechanics. Those chapters are important for readers who want a deeper dive into QIS and QC in preparation for further coursework or reading. Students interested in current technology and careers in QC and QIS can read Chapter 16 on their own.

Throughout the book, as we have mentioned, we make use of three ways of working with quantum states: symbolic state vectors, graphical representations (state space), and column vectors and matrices. We have found that the matrix methods are relatively unfamiliar to many students. Even though the matrix methods involve only addition and multiplication, the lack of familiarity means that many students have difficulties figuring out what goes where. You should encourage your students to make use of online "matrix calculators" until they are more comfortable with the matrix manipulations. (Just search online for "matrix calculator.") The book is also well suited to self-study. We encourage students to form collaborative groups to read the book together.

Our introduction to quantum computing is built around a fundamental theme: Quantum states, quantum entanglement (we will explain what that means), and quantum measurements provide "resources" for computation that in principle far exceed what can be done with traditional computers. We will play this leitmotif again and again throughout the book. However, quantum states are far from our everyday intuitions about how the world works. In fact, when it comes to quantum states and quantum computing, we are all "novices," and it takes many hours of work to move from being a novice to being an expert. So, it would be presumptuous for us to say that working through this book will make you an expert in quantum computing. Our goals are to provide you with the basic concepts and formalism of quantum computing, to show you how quantum states can be manipulated using quantum algorithms to carry out computationally useful processes, and to look to the future of where quantum computing is likely to go.

We treat many of the "classic" quantum algorithms as case studies to show how our fundamental theme has been applied to solve computational problems that are difficult if not impossible for traditional computers. We don't expect that most of you will be developing new quantum algorithms, just as the productive use of traditional computers does not require that you develop new numerical algorithms. In practice, you simply make use of what others have created and assemble the computer code to tackle the problem at hand. But we do want you to have the tools needed to be able to read and talk intelligently about quantum computing.

As mentioned previously, we have, by design, limited the math background required to benefit from this book. That said, we do recognize that some experience with symbolic (algebraic) reasoning, with graphical and geometric representations, and, more importantly, with persistence and perseverance will be important to grasp the fundamental concepts and ideas. Our advice to all readers: Give yourself plenty of time to work through and digest the material and to reread the arguments to deepen your understanding. Keep in mind that it took physicists and mathematicians several decades to complete the development of quantum mechanics and another 50–60 years to recognize the revolutionary potential of quantum mechanics to rebuild completely the foundations of information science and computing. So, as another Bob (Bobby McFerrin) sang, "Don't worry, be happy" if some concepts seem rather puzzling at first. You are in good company. As dedicated teachers, we need to remind you, as we remind our students, if you are not confused, you are not learning. Or as a skiing instructor once told Bob, you are never going to learn to ski unless you fall down.

Let us close by mixing metaphors: we hope you will persevere through your confusion as you navigate the moguls on the slopes of quantum mechanics, quantum computing, and quantum information science.

The book's website QuantumComputingA2B.org will include solutions to the Try Its, additional problems for exams and group problem solving, resources for teachers, links to other websites for concept assessments tools, and "breaking news" about quantum computing and quantum information science.

Your devoted authors,
Alice and Bob

Acknowledgments

We express our deep appreciation to Steve Arendt, John Donohue, and Diandra Leslie-Pelecky for providing feedback on the manuscript, encouraging us to clarify our thoughts and to focus on the central message. Many students and physics education colleagues over the years helped us to understand and learn to explain quantum mechanics more effectively, though that is a lifelong journey. Oxford University Press's Sonke Adlung provided both encouragement and thoughtful comments from several anonymous reviewers. Giulia Lipparini, our OUP Project Editor, skillfully and gently led us through the intricate journey of getting the manuscript into its final form. We also thank Roopa Vineetha Nelson, our senior production manager, and Henry MacKeith, our copy editor, for their thorough, but gentle guidance in refining and polishing the manuscript.

1 Introduction

In Nature's infinite book of secrecy,
A little I can read.

Shakespeare, *Antony and Cleopatra*

1.1 Meet Alice, Bob, and Cardy

ALICE: A nuclear engineer and educator with a strong theoretical background, she is very experienced with translating quantum computing and quantum cryptography for high school students and teachers.

BOB: An experimental physicist who likes to think about how quantum computers will actually be built and how you can use lasers and microwaves to prepare, manipulate, and analyze quantum states. But he also has a philosophical bent: What does quantum computing tell us about fundamental issues in quantum mechanics and information theory? What does it teach us about the fundamental nature of reality?

CARDY: A first-year university student with a double major in business and English—wants to know more about quantum computing both because it sounds cool, with many philosophical puzzles, and because it may be the future of the information technology (IT) business world. CARDY hasn't had physics since high school and math ended with algebra II and trig.

1.2 What's the Big Deal about Quantum Computing?

ALICE: Welcome, Cardy.

CARDY: Hi, Alice and Bob. Thanks for inviting me over. I heard that you two had figured out some simple ways of understanding why quantum computing (QC) and quantum information science (QIS) are important and can explain all of that to a non-expert like me. I saw your note in the student newspaper about wanting someone to critique your new book manuscript. That's why I called. As a business and English major, I want to find out more about these subjects. I keep reading that they are going to be the next "big thing" in the business world and that means opportunities for building new businesses and developing new innovations. Even our School of Business has had a few talks about quantum computing. Besides, it would be cool to wow my friends with my knowledge of quantum computing, entanglement, and all the stuff I've read about online.

Quantum Computing: From Alice to Bob. Alice Flarend and Bob Hilborn, Oxford University Press.
© Alice Flarend and Robert C. Hilborn (2022). DOI: 10.1093/oso/9780192857972.003.0001

In fact, I read about you and Bob in a blog post about quantum computing, and it is a great pleasure to meet you in person.

BOB: I hate to disappoint you, Cardy, but Alice and I are not the actual characters whose message-sending is described in both popular books and technical papers about quantum information, cryptography, and quantum computing. In fact, "Alice and Bob" first appeared in print in 1978 in a cryptography article in the technical journal *Communications of the Association for Computing Machinery*.

CARDY: I remember my parents talking about a 1969 movie *Bob & Carol & Ted & Alice*. Was that you guys?

BOB: Alas, no again. But the common mention of our names in quantum information and quantum computing books was part of the motivation for our putting together these materials. What better way to learn about quantum computing than from Alice and Bob!

ALICE: Bob and I decided that we ought to find a way to introduce people who aren't experts in quantum mechanics, linear algebra, and computer science to the key ideas and vocabulary of QC and QIS. There are lots of books that take a fairly high-level approach, which is fine for math and physics majors but leaves most folks baffled. There are also several books that avoid math entirely and focus on general concepts, often with idiosyncratic models and weird terminology that don't make contact with how most scientists, engineers, and mathematicians talk about quantum computers. We worked hard to find the "sweet spot" between insider jargon wrapped in unfamiliar mathematics and fluffy treatments that seem strange and inadequate to both experts and novices. Since both of us struggled to penetrate the many layers of jargon that expert authors have brought to the field, we felt we were in a good position to find that sweet spot. Jargon is fine of course for the experts, but many authors use it mainly to show that they are part of the "tribe" and to separate the tribe from outsiders.

CARDY: Speaking of confusing jargon, what's the difference between quantum computing and quantum information science? To me, they sound the same.

ALICE: Thanks for calling us out! We need to practice what we preach. Please do that whenever we get carried away like that. "Quantum information science" has become the general term for the use of quantum principles and technology for the transfer of information (what you might call quantum communications), the processing of information (QC), and for the sensitive detection of various kinds of physical properties (quantum sensing). People are already talking about building a quantum internet to enhance the security of information exchange. In this book we focus on the first two–QIS and QC. To appreciate quantum sensing, you need some background in physics and how quantum systems interact with their environment.

BOB: As we said in our email messages, we'll be delighted to spend several days with you. We hope that your background in English will help you tell us where our presentation is muddled or confusing.

Of course, quantum computing is built on the basic principles of quantum mechanics and there are many parts of quantum mechanics that are conceptually quite challenging for people new to the field and to experts. Even though scientists and mathematicians know how to do quantum calculations, there is still a lot of controversy about what those calculations tell us about the nature of reality. We have tried to develop quantum computing ideas in a way that avoids, at least for a while, those conceptual challenges, but we will need to deal with them eventually. And, in their own way, they are rather cool challenges.

CARDY: I am a bit worried about my math preparation for quantum computing. I keep hearing about linear algebra, Hilbert spaces, operators, state vectors, complex numbers, and the like—none of which I have studied in my math courses.

BOB: I hope you will find that Alice and I have been careful to minimize the formal mathematics in our story of quantum computing by choosing examples carefully and avoiding unnecessary mathematical jargon. Of course, the math is important. Every community has its own language and for the community of physicists, engineers, and computer scientists, mathematics is the lingua franca.

To understand how quantum computing works, we are going to need some math to help us with the reasoning. Fortunately, all you will need is some simple algebra, basic concepts of vectors (which we will help you with pictorially), and a few sines and cosines—the latter in fact appear only in only a few places in this book. However, since we want you to learn to "read" the equations of quantum computing, we have used the symbols that are commonly deployed in the field. Those symbols were borrowed from quantum mechanics, and they will look strange when you first meet them. But if you are patient and practice writing and reading those symbols, you will get used to them.

ALICE: Let me warn you that the experts may be annoyed by our approach because our examples do simplify the math. For example, ordinary quantum mechanics makes use of complex numbers and we have found a way to avoid them for most of this book. Our approach is to help you understand the basic ideas and prepare you to then move into the more complex mathematics if you so choose.

CARDY: Now that you mention it, I do recall one of my math teachers in high school saying something about real and imaginary numbers. I never did figure out what an imaginary number means.

ALICE: It turns out that we can introduce the basics of quantum computing with only ordinary ("real") numbers. In Chapter 15, the next-to-last chapter, we will show you *why* we need complex numbers in some aspects of quantum information science, but they are not essential for the main ideas. There is also some mathematical jargon we will mention only in passing in case you read about those ideas in the quantum computing literature. We'll do our best to avoid excess jargon as much as possible.

1.3 A Brief Preview

BOB: Let me give you a quick preview of the story we tell in this book. To help you develop an understanding of quantum computing, we will need the key concept of a quantum state. We say more about quantum states below. Like many concepts in science and math, it will be important to think about states using multiple ways of representing quantum states. We will introduce you to (1) abstract state "vectors," (2) graphical pictures of quantum state "spaces," and (3) numerical forms of mathematical objects called column vectors and row vectors, which are straightforward generalizations of ordinary vectors.

CARDY: I'm confused already. I don't know anything about vectors, state spaces, or column vectors.

ALICE: We don't expect you to be familiar with the concepts. We will introduce them one by one and give you several ways of thinking about their role in QC and QIS.

We have also found that many introductions to quantum computing and quantum information science use confusing language about quantum measurements, collapse of quantum states, jumps between quantum states, and the results of measurements on entangled states. Keeping in mind how actual quantum measurements are carried out, we will show you a way of talking about those concepts that will help sweep away the fog surrounding many of those discussions.

CARDY: Before we get too far into the weeds, can you explain how quantum theory is different from quantum mechanics? I've heard both terms used in reading about quantum computing.

BOB: They are essentially the same. When quantum theory emerged in the first third of the 20th century, its way of thinking about how the world works was significantly different from what physicists had used before. To distinguish the new from the old, which physicists called "Newtonian mechanics" or "classical mechanics," they invented the term quantum mechanics. In physics, mechanics means theories of motion, forces, and the like.

CARDY: So, "classical" is like classical music? You know, Joseph Boulogne (Chevalier de Saint George), Wolfgang Amadeus Mozart, Clara Schumann, Richard Wagner, Nadia Boulanger, Marin Alsop, William Grant Still, and the like.

BOB: Well, not quite. What "classical" means is physics that does not need quantum mechanics to explain what is going on. Classical physics can be just as mind-bending as quantum physics, particularly when it comes to things like chaos theory.

To give you more context about what we are going to do, let me explain a bit about the differences between the Newtonian worldview and that of quantum mechanics. In Newtonian mechanics—before quantum mechanics came along—we thought of the world as being made of objects whose positions and velocities we can track and predict if we know the interactions among those objects. With the addition of theories of electricity and magnetism (now unified as electromagnetism), physicists thought they had a reasonably comprehensive picture of how the world works. Interactions among material entities (such as atoms, protons, electrons) and fields like the electromagnetic field and the gravitational field could explain everything. Of course, there were many details to be worked out, but the basic conceptual structure was in place, or so they thought. The world was essentially deterministic. What does that mean? Here is an example: If we knew the positions and velocities of all these entities that make up a system at some time, according to the Newtonian worldview, we could predict the future behavior of those entities. If you could apply that method to the bouncing numbered balls in a lottery machine, you could get rich very quickly!

CARDY: That's scary. You mean you could predict everything that might happen?

BOB: That indeed is the Newtonian worldview. But there is a catch. In real life, trying to do that becomes practically impossible if there are more than a few objects in the system. To make matters worse, even for simple systems with only a few objects, the precision with which we can make those predictions diminishes as the systems evolve in time because we can never really know the current state absolutely perfectly. Furthermore, with chaotic systems (like the weather), the precision of the predictions diminishes surprisingly rapidly because chaotic behavior changes dramatically. Such systems are hyper-sensitive to even small changes in their initial conditions. You may have heard of the butterfly effect: The flapping of a butterfly's wings in Brazil can cause a tornado in Texas.

All of that said, science has moved beyond the Newtonian worldview. With the discovery of quantum mechanics in the early 20th century, we now know that the world at its most fundamental is not deterministic. As far as we know today, there is randomness and probability at the core of nature.

We will see that the strange aspects of quantum states—combinations of states called superposition states, a property of those states called entanglement, and the role of randomness and probability when we make observations on ("measure") a quantum system—will turn out to be of crucial importance in quantum information science and quantum computing. In fact, sets of procedures—algorithms—in QIS and QC make use of all these strange aspects.

We will start off with a look at traditional computers, which are often called "classical computers" since they don't directly involve quantum mechanics. The concepts of classical computers will form a good platform from which to launch ourselves into quantum mechanics and ultimately quantum computing. We hope you will be patient because it will be awhile before we get into the meat of quantum computing. We have found that if we don't spend some time on basic quantum concepts, then quantum computing and quantum information processing more generally will seem even more strange. Well armed with those concepts, you will find quantum computing, though still strange, much more comprehensible.

Once we have laid the foundation with the crucial quantum concepts, we will show you several algorithms that demonstrate the advantages of quantum computing over classical computing. We will also introduce some key issues in quantum cryptography and error correction. That may sound rather boring but they are of great importance in any kind of computing and their quantum versions raise many intriguing issues about what is information and what is "noise."

ALICE: We will end our tour of QC and QIS by visiting several quantum issues that are not directly part of quantum computing but that, given the tools we will have learned about, we will be able to understand. Those topics reinforce the notion that the quantum world is conceptually and experimentally far removed from the Newtonian world. For example, there is some deep quantum weirdness in a famous result called Bell's theorem (which makes us question the nature of reality) and in the relationship between classical computing and quantum computing.

BOB: We'll conclude our voyage with a look at the future—both your personal future, Cardy, what you ought to do next if you want to learn more about quantum computing, and also the future of quantum computing itself.

CARDY: That was a big help. All that sounds exciting. I'm ready to get going with the real stuff. But could you give me a quick preview of why we should worry about quantum computing, besides its being cool?

ALICE: Fair question, but one that is hard to answer because the field of quantum computing in some ways is still in its infancy. But we do know that quantum computing allows us to carry out calculations that would be completely impractical with classical computers. More importantly, quantum computing allows us to think about problems in a completely different way from the way we think about classical computation. As an analogy, I might point to the discovery of microbes (bacteria and viruses) as the carriers of diseases. After that discovery, we had entirely new ways of preventing and treating disease. Similarly, I believe that ultimately quantum computing will open our eyes to whole new ways of thinking and issues that we don't even recognize today. Quantum computing is not just doing traditional computing more efficiently or faster, though that in itself would be worthwhile, but it is doing computation in entirely new ways. Even though we don't know exactly how that is going to work out, the possibilities are just mind-blowing. Already we know that QCs can find energies and configurations of molecules more efficiently than classical computers. That opens the door to new methods of drug discovery, for example. We also know that quantum computers can crack many of the encryption methods used to keep data secure. But quantum methods can be used to develop encryption systems that are yet more secure.

1.4 How to Use This Book: Encouragement and Coaching

BOB: To get you immersed into quantum computing as smoothly as possible, we have broken up our story into chapters and the chapters into sections to provide a gradual introduction to new information and concepts. Along the way, we suggest various activities labeled "Try It" to encourage you to reflect on what you are learning and to examine the details that are important in understanding the concepts. By doing those Try Its, you will build your ability to think "quantumly" and you will increase your confidence that you can tackle cutting-edge concepts in quantum information science.

Both Alice and I believe that you can master all parts of this book because we firmly believe that the human mind is a "muscle" that can become stronger with exercise. Your abilities, both mental and physical, are not fixed but can grow over time. But we also recognize what you have all experienced: learning requires effort, practice, and coaching. Just think of what you went through to learn to speak, read, and understand your native language. But you persisted and with help from your family, friends, and teachers you now do those things naturally and usually without much effort. We have written this book to help direct your QC and QIS efforts in productive ways, to provide you with practice, and some coaching. QC and QIS, like all fields of human knowledge, have a new vocabulary unfamiliar to beginners, new concepts, and new ways of thinking about and reasoning with those concepts. We urge you to work with friends as you learn about QC and QIS to coach and to encourage each other when things seem difficult.

ALICE: When Bob and I teach physics classes, we often hear students say, "I understand the concepts but I'm not a math person." That statement does express students' feelings about math, but it is not grounded in the reality of how one learns. While learning about QC and QIS, it is important to realize that there are no "QC and QIS people." In fact, we all belong to the large group of people trying to understand how QC and QIS work and what those fields are good for. The fields are brand new and are still in a state of growth and flux. Furthermore, the quantum world does not come naturally or intuitively to anyone—don't let the experts get away with saying "it is obvious that. . . ." It only becomes obvious after you have spent hours, days, weeks, and longer figuring out the details and making connections among concepts.

We do acknowledge that people pick up some things more easily than others. You might be great at drawing, but you have a tough time learning a foreign language. It all depends on what you already know and how big a stretch it is to understand and master the new activity. It also critically depends on how much effort you are willing to put in. We have written this book by breaking up an extremely rich meal into chewable and digestible bites and providing lots of encouragement and feedback.

If you don't know any algebra, learning what is going on in QIS and QC, which is the goal of this book, will be difficult because the language of the quantum world is mathematics. However, your algebraic fluency should increase with practice, and we provide a gentle on-ramp for all the math that we need in this book. If you feel that the math in a particular section is too dense, feel free to skim through that section, if so inclined, and return to the details later. We will point out specific sections that are mathematically challenging and others than can be skimmed without damaging the overall understanding of the book.

Since your thoughts and attitudes about learning are critical to the benefits you will get from this book, we ask you to take a few minutes to reflect on what it takes to learn. This may seem like an unusual exercise for a book on QC and QIS, but research and our experience have shown that this kind of reflection is important in your journey to become an effective learner. We, as learners, need to pay attention to what we do when we learn because there is more to it than just taking notes. Often we are working hard but not noticing it as being an integral part of learning.

Try It 1.1

Write a few sentences about an experience you had learning something new (a sport, knitting, a game, cooking, woodworking, drawing, playing a musical instrument . . .) and how you used persistence to become better over time at that activity. Did you do it all in one go or did you get better after repeating it? Did you learn the activity just by reading about it or did you have to pick up the ball, for example, and practice throwing it?

Can we learn QC and QIS? We urge you to join Ceasar Chavez, educator and civil rights leader, and former President Barack Obama, in shouting "Yes, we can!"

CARDY: All of this sounds great. Let's get started!

     FURTHER READING

A history of the characters "Alice and Bob" is available at https://en.wikipedia.org/wiki/ Alice_and_Bob.

2 Traditional Computing

And those who were seen dancing were thought to be insane by those who could not hear the music.

Friedrich Nietzsche

2.1 Traditional Computing versus Quantum Computing

ALICE: Before we get into quantum computing, we need to say a few words about traditional computing and traditional computers. Some physicists like to call this "classical" computing, because as we mentioned in Chapter 1, the history of physics is often divided into a pre-quantum era called "classical physics" and the contemporary era of "quantum physics." Classical physics is the realm of Isaac Newton and his study of motion and forces; James Clerk Maxwell, working on electricity and magnetism; and Albert Einstein, who extended those studies to objects traveling close to the speed of light. Of course, there were hundreds of others contributing to these studies, but scientists, like people in many fields, prefer focusing on a small hagiography to honor.

Quantum physics, not surprisingly, focuses on those physical phenomena where the concepts and theories of quantum mechanics are needed to understand what is going on. When talking about quantum physics, people mean the structure of atoms and molecules, nuclear physics, elementary particle physics, semiconductors, lasers, and so on.

The important point is that the classical physics world was turned on its head in the 1920s with the development of quantum theory—the theory describing the world at the atomic and subatomic levels. That theory became known as quantum mechanics and you can think of QC and QIS more generally as applications of quantum mechanics to computational and information processing situations.

CARDY: I read that today's computers and computer chips are built from semiconductors. Don't we need quantum mechanics to understand the properties of semiconductors?

ALICE: Classical (traditional) computing (prior to the development of quantum computers) is actually based on a mixture of classical (pre-quantum physics) and quantum physics. Quantum physics helped scientists and engineers develop the semiconductor devices that power almost all contemporary computers. Even though the understanding of the material properties of those semiconductors grew out quantum theory, the devices, as used in classical computers, operate in regimes in which classical physics (mostly electricity and magnetism)

Quantum Computing: From Alice to Bob. Alice Flarend and Bob Hilborn, Oxford University Press.
© Alice Flarend and Robert C. Hilborn (2022). DOI: 10.1093/oso/9780192857972.003.0002

are perfectly adequate to describe their behavior. The details of quantum physics are not needed.

BOB: Let's look under the hood of a classical computer. Most of us who use computers these days know that you have to tell computers what to do. Those instructions are computer programs. Writing them is called coding. An app for your smart phone is just a computer program that tells the computer inside the smart phone what to do. So, writing a best-selling app is "just" computer programming. Fortunately, decades of work in computer science have produced what are called high-level computer programming languages that are close enough to everyday language and everyday mathematics to make coding relatively straightforward. But, a big part of using a computer is first figuring out what you want the computer to do. That part is called developing an algorithm. And then you need to work out how to implement that algorithm in the programming language you decide to use.

CARDY: My high school tried to teach us some computer coding, but I remember spending most of my time getting semicolons and capital letters in the right places and even then, the code hardly even did what I wanted it to do. I hope quantum computing is not going to be like that.

ALICE: Your experience, unfortunately, is not that different from what most people find when they get started with coding. I have to warn you that some aspects of quantum computing require that same level of attention to detail. But we will deal with that when we have to. Right now, we want to focus on the basics that underpin all computation.

2.2   Binary Digits

ALICE: To understand what quantum computing is all about, we need to dig a bit deeper into the inner workings of a computer. The code you write in a high-level language is translated by another set of instructions called a compiler into code that the machinery of the computer can understand directly: This "machine language" consists of strings of 1s and 0s known as bits ("bit" is an abbreviation of "binary digit"). The binary number system has just two symbols in it, traditionally labeled 0 and 1. Our everyday number system is a decimal system, with 10 symbols: 0, 1, 2, 3, 4, 5, 6, 7, 8, and 9. It turns out that using just 0s and 1s is sufficient for almost all computational work and it greatly simplifies the design and construction of computers.

So, if you looked at the machine language version of a computer program, you would see something like 10011110001110011111100. . .. Not very informative to the typical human. Fortunately, the ordinary computer user (that includes you when you use your smart phone) never has to worry about the binary digits, though you often see them scrolling across a computer screen in tech movies. You only need the binary digits if you are trying to save the world from evil hackers or if you want to understand the fundamental systems that govern all traditional computers.

BOB: Figure 2.1 shows a photograph of the electrical circuitry of a traditional computer. Each of the "wires" (the long, narrow, light-colored lines in the figure) can be either at a high electrical voltage (usually 3–5 volts, the equivalent of a few AA batteries stacked end to end) or at a low voltage (usually 0). The electronic circuits inside the black rectangular devices ("integrated circuits") sense those voltages and carry out computation actions based on those

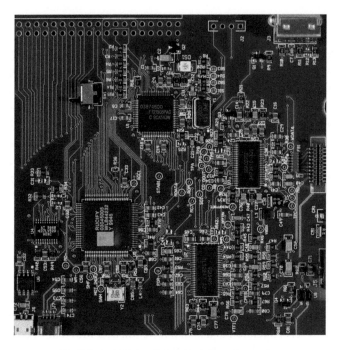

Fig. 2.1 A circuit board of a typical classical computer.
Source: Miguel-á-Padriñán-343457 pexels.com.

voltages. In principle, you could be sensing other physical quantities (the presence or absence of light, the presence or absence of sound, a switch in an "on" or "off" position, etc.), so it is useful to make a more abstract representation of what is happening in the wires.

I like to distinguish between the physical state of the wire (high voltage or low voltage) and the abstract representation, the "computational state," using the binary digits 0 and 1. The voltage state of each wire is associated with a binary digit—a bit—whose value can be either 0 or 1.

Most often, a 0 is associated with the low-voltage state and a 1 is associated with the high-voltage state, though there is no fundamental reason you couldn't make the opposite association.

If we measure the electrical voltage of a particular wire in an operating computer, we would record a series of low voltages and high voltages, which could be translated into the abstract representation of a series of 0s and 1s.

CARDY: I've heard about bits, but what does a series of 1s and 0s mean? When I see them in the movies, they are always in a long string.

ALICE: The series might represent many things. For example, it could be a coded message. Think of Morse code or a flashing light signal ("one if by land; two if by sea"). Several wires (bits) looked at together might represent a numerical value (the 8-bit sequence 00000011 could represent the regular (decimal) number 3) or an address within the computer where information is to be stored or retrieved. Or they could be a code for the instructions of what an integrated circuit in the computer is to do based on the input bits.

Going into the details of these representations would take us too far afield. More important for our understanding of quantum computing are some simple computer operations, which use one or two bits as input and produce an output of other bits.

2.3 Logic Gates and Truth Tables

ALICE: These operations are similar to binary logic (sometimes called Boolean logic after George Boole, a mathematician active in the 1800s), which was originally developed to represent the logic of combinations of statements (sometimes called "propositions") that can be either true (T) or false (F). In the 1930s, Claude Shannon and others realized that binary logic could be used to describe the operations of digital computing devices.

Let's start with a few Boolean logic examples. Proposition A ("I am hungry") can be either True (I haven't had anything to eat since breakfast) or False (because I just consumed a full bag of chips). A Boolean logic operation called NOT can act on the proposition. For example NOT A is NOT ("I am hungry") = "I am not hungry." A horizontal bar over the proposition symbol is used to label the result of NOT: for example, NOT A $= \bar{A}$. One of the quirky (but important) features of logic is that in most cases we don't care about the meaning of the proposition; we focus just on its logical structure. That drives most people nuts.

In the language of binary logic we say that if P is True, then NOT P is False and vice versa. This assumes, of course, that the proposition is either True or False. In *binary* logic there are only *two* possibilities. Other situations are not binary systems. For example, I could be hungry, ravenous, a little hungry, stuffed, not hungry, . . . the list goes on. In general, hunger is not a binary (two-state) system.

For binary digits, if a bit has the value 0, then NOT produces a bit with the value 1 and vice versa. Whenever something acts on a bit or a set of bits, we call that an "operation." We pictorially represent the devices that carry out the operations as "gates" (not related to Bill Gates). An operation can be a simple mathematical function like adding or multiplying. It can also be one of many less-familiar functions, which we will get to later.

BOB: Figure 2.2 is a pictorial representation of a NOT gate, indicated by the triangle with the small circle on the right. The input bit (or proposition in logic) is indicated by A on the left. The "output" that results from applying the gate operation to A is on the right. The horizontal line over the symbol A is used to indicate the bit "opposite" that of A (0 if A is 1 and 1 if A is 0) or the opposite "truth value" (False if A is True and True if A is False).

BOB: Let's think about the NOT operation in a slightly different way: a "truth table." The left column represents the possible values of the input (0 or 1) and the right column represents

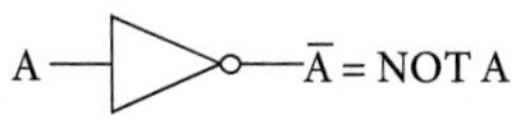

$$A \longrightarrow\!\!\!\!\triangleright\!\!\circ\!\longrightarrow \bar{A} = \text{NOT A}$$

Fig. 2.2 The triangle with the small circle on the right is a pictorial representation of a NOT gate. The gate acts on the bit A to produce "NOT A."

Table 2.1 The truth table for NOT.

A	NOT A	A	NOT A
False	True	0	1
True	False	1	0

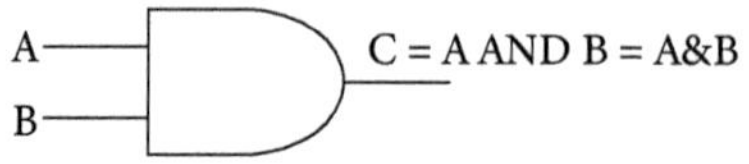

Fig. 2.3 The symbolic representation of an AND gate.

the corresponding output values. Truth tables (and their generalizations) will be helpful when we get to quantum computing operations.

BOB: In Table 2.1 the two columns on the left give the True/False values for A and NOT A. The two columns on the right give the binary 0 and 1 equivalents.

CARDY: Unless I am missing something, I don't think you can do much with a gate with just one input.

ALICE: Great point, Cardy; so, let's look at some gates that have two inputs (two bits). These are called two-bit gates.

CARDY: Hmm, two-bit sounds rather derogatory.

ALICE: I had never thought of it that way. Yes, jargon in many fields sounds absurd to newcomers. But the "experts" have gotten used to the jargon and they expect you to understand what they mean by two bits.

Our first example of a two-bit gate is an AND gate, as shown in Figure 2.3. The AND operation has two inputs A and B and one output C.

CARDY: A is for Alice and B is for Bob, right? And of course, C is for Cardy!

ALICE: I wish! Most computer scientists don't share our sense of humor. In fact, they like to leave the meaning of the symbols as unspecified as possible.

ALICE: Table 2.2 shows you the output of the AND gate for the four possible "states" of the two input bits. In terms of binary logic, we say that the AND combination of two propositions is true if and only if both propositions are true. Otherwise, the combined proposition is false. Switching to bits, you should notice a pattern that the output of an AND gate is 1 only if A and B are both 1s. Notice that this table is larger than the one for the NOT gate. As we add bits to our repertoire, the number of possible combinations grows quickly. The larger repertoire enhances our capabilities for dealing with more complex algorithms.

Try It 2.1

Write two simple sentences in everyday language and convince yourself the logic operation AND is close to what we mean by "and" in everyday language.

Table 2.2 Truth table for an AND gate.

A	B	A AND B	A	B	A AND B
False	False	False	0	0	0
False	True	False	0	1	0
True	False	False	1	0	0
True	True	True	1	1	1

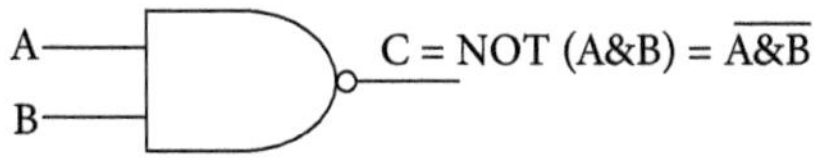

$$A \quad C = \text{NOT } (A\&B) = \overline{A\&B}$$
$$B$$

Fig. 2.4 A pictorial representation of a NAND gate—an AND gate followed by a NOT operation.

BOB: Another important two-bit gate is called a NAND gate. Its output is the negative (opposite) of the AND operation, hence the N. You can think of the NAND gate as an AND gate, followed by a NOT gate acting on C, the output of the AND gate. To simplify notation, we will often use A AND B = A&B.

BOB: Figure 2.4 displays a symbolic representation of a NAND gate. Again, the horizontal bar over A&B means NOT. The output of the NAND gate is NOT applied to the output of the AND gate as shown in the table below.

Note the right-most column in the NAND truth table (Table 2.3) has the 1s and 0s swapped compared to the AND truth table.

Try It 2.2

Using T and F (true and false) fill in the missing elements in the following NAND gate truth table.

A	B	$\overline{A\&B}$
F	F	
F		T
	F	T
T	T	

CARDY: The gates as represented by symbols or truth tables look fairly simple. How would you build a device that carries out those operations?

BOB: That's a great question, but we won't go into the details of the electronic circuitry that implements those gates. If you are interested, please take a look at Further Reading at the end of the chapter. We won't need to know anything about the actual devices in anything that follows. You can buy an integrated circuit that has four NAND gates built in for less than a dollar, and they do exactly what the table specifies without your knowing anything about the circuitry.

Table 2.3 Truth table for a NAND gate.

A	B	NOT A&B
0	0	1
0	1	1
1	0	1
1	1	0

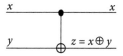

Fig. 2.5 The CNOT circuit. The symbol $\oplus$ is explained in the text. x is the 'control bit' and y is the 'target bit.'

Table 2.4 The CNOT gate.

Input		Output $z = x \oplus y$	
x	y	x	z
0	0	0	0
0	1	0	1
1	0	1	1
1	1	1	0

ALICE: Here is something that is truly amazing: *Any* binary logical operation can be implemented as a combination of NAND gates. These logic gates can be combined to add, subtract, multiply, and divide binary numbers and to carry out almost any computational task you can think of. How do you prove that? It is a theorem in binary logic. The theorem is moderately complicated, and of course, no one in their right mind would build a computer that way. However, it is a statement that is very useful in theoretical aspects of computer science.

BOB: Let's look at one final example of a two-input gate. This one, called the Controlled-NOT gate (CNOT), has two inputs and two outputs. Again, there are four possible states for the input bits. The pictorial representation of the CNOT gate is shown in Figure 2.5. Note that we switched notation, replacing A, B, and C with x, y, and z to match the notation often found in computer science and, as we shall see, in quantum computing.

The action of the gate is summarized in Table 2.4, which lists all possible two-bit inputs and the corresponding CNOT outputs.

ALICE: Cardy, what do you notice about the table?

CARDY: Hmm? It seems like x, the control bit, does not change between input and output. And y changes sometimes but not others.

ALICE: Good observation. Can you find a pattern that tells us when y changes and when it doesn't? Finding patterns is what science is all about.

CARDY: Aha! It seems to have to do with whether $x = 1$ or $x = 0$.

ALICE: Exactly! The CNOT gate is called a controlled-NOT because what happens to y is controlled by the input x. If the control bit (x) is 1, the NOT operation changes $y = 0$ to 1 or 1 to 0. If the control bit is 0 (the top two entries in the table), then the output z is the same as the input y: The NOT operation is not invoked.

CARDY: I get it, and I can guess that being able to control when something is applied or not will be very useful in a computer.

ALICE: Yes, that's right. If you replace 0 with False and 1 with True, then $z = x \oplus y$ is the "Exclusive Or" (XOR) combination of x and y. Exclusive Or means that the combination is True only if one of x and y is True. If both are True, the output of the gate is False. The standard Exclusive Or gate has two inputs but only one output (z). The CNOT gate, by design, has two inputs and two outputs.

The $+$ in $\oplus$ reminds us that you can think of the CNOT gate as a two-bit "half adder": To see what that means, let's have a quick review of binary addition. The results we need are

$$
\begin{array}{cccc}
0 & 1 & 0 & 1 \\
+0 & +0 & +\,1 & +\,1 \\
\hline
0 & 1 & 1 & 10
\end{array} \tag{2.1}
$$

CARDY: Oh, yeah. I remember that from 10th-grade math. In binary 10 means 2 in decimals. I remember my math teacher's joke: There are 10 types of people in the world: those who understand binary and those who do not.

ALICE: I love it! We won't need binary addition very much, but $\oplus$ and CNOT will show up fairly frequently in quantum algorithms. The XOR results are

$$
0 \oplus 0 = 0 \quad 1 \oplus 0 = 1 \quad 0 \oplus 1 = 1 \quad 1 \oplus 1 = 0 \tag{2.2}
$$

CARDY: It looks like the XOR is the same as regular binary adding except it drops the carry bit 1 when we add 1 to 1.

ALICE: In the next chapter we will introduce the basics of quantum computing. Then we will return to the question of the connection between quantum computing and classical computing. You will see that classical computing is a special case of quantum computing, just like, in many cases, classical physics is a special case of quantum physics.

 CHAPTER SUMMARY

- Classical computers use binary digits (bits) to represent information, instructions, addresses, etc. which the computer circuitry manipulates to carry out computational actions, store and retrieve data, produce images, communicate with your printer, and to show cute kitten videos.

- Those bits can take on values represented by the symbols 0 and 1.

- The basic operations carried out by the computer can be represented by logic gates (or just gates, for short) which use one or more bits as "input" and produce one or more bits of "output."

 FURTHER READING

Charles Petzold, *CODE: The Hidden Language of Computer Hardware and Software* (Microsoft Press, Redmond, WA, 2000) provides a nice introduction to the basics of classical computing.

3 Traditional Bits in New Clothing

Beware lest you lose the substance by grasping at the shadow.

Aesop, 'The Dog and the Shadow', *Fables*.

3.1 Quick Review

ALICE: In Chapter 2 we introduced bits as the representations of the states of circuit elements in classical computers. The states of those elements may be (and usually are) associated with the numerical binary digits 0 and 1.

Before we jump into quantum computing, let's look at several other ways of describing classical computing. In quantum computing we use symbolic representations that are based on the way physicists describe quantum mechanics, blended with some notation and concepts from computer science. At first, those representations look rather strange because you have not experienced them in your math, physics, or computer science courses, even though the underlying mathematics (mostly algebra, as we shall see) is not all that complicated. Bob and I thought it would be helpful to introduce this notation first within the framework of classical computing. This is not often done (one exception is [Mermin, 2007]) because the resulting representations are not useful in traditional computing. But in quantum computing, this notation is important and almost universally used. Thus, we think it is good to get used to it in a familiar context first. This approach will help us understand what appears to be a deep gulf between classical and quantum computing.

Cardy, you should know that this way of talking about classical computers is hardly ever used by anyone else; so, if you mention this language to a computer science major, you will probably get a puzzled expression in return.

3.2 State Symbols

BOB: Let's start by thinking about two wires in a traditional computer. The voltage differences between wires in the computer are used for several purposes: to bring information to gates for processing and to indicate where data should come from and go. Each wire has two states (or conditions): The voltage is "high" V_{high} or the voltage is "low" V_{low}. Because there are just two states, this is a *binary* system.

Quantum Computing: From Alice to Bob. Alice Flarend and Bob Hilborn, Oxford University Press.
© Alice Flarend and Robert C. Hilborn (2022). DOI: 10.1093/oso/9780192857972.003.0003

We can represent those two voltage states by the symbols $|V_{high}\rangle$ and $|V_{low}\rangle$. Each symbol consists of a vertical line on the left and an angle bracket on the right. You should think of the combination $|\rangle$ as a container. The contents of the container are *names* that label the two possible states. So, we could have $|On\rangle$ and $|Off\rangle$ states or $|True\rangle$ and $|False\rangle$ states or $|Awake\rangle$ and $|Asleep\rangle$ states, and so on. These $|\rangle$ symbols are widely used in quantum computing to represent quantum states. You might think of state symbols as a form of emoji ☺. In and of themselves, the labels or names have no numerical or even direct physical meaning.

This symbol is part of what is called Dirac notation, invented by Paul Dirac, an English physicist who lived from 1902 to 1984, who was one of the pioneers in quantum mechanics. He developed this notation in 1939, about 13 years after the birth of quantum mechanics, to clarify the procedures of quantum mechanics. QC and QIS make wide use of Dirac notation.

Try It 3.1

Write out by hand the state symbols given in Bob's previous paragraph. This exercise will get you used to writing the symbols and using them as containers. Make up some binary states of your own (for example, win-lose, dead-alive, or score-no-score) and write the corresponding state symbols.

ALICE: Anticipating the terminology used in quantum computing, we might call these classical states Cbits (C for classical). Cbit is pronounced see-bit. The 0s and 1s are called classical bits, or just bits; most people use the latter term.

CARDY: But what happens if there are more than two possible states? I would love to be able to say that my computer has more states than your computer!

ALICE: Indeed, we could develop computational systems with more than two states, but those are hardly ever used. We have already seen that we can use clusters of bits to represent numbers, alphabetical symbols, and so on. So, binary systems and bits are all we will ever need to understand commonly used computers.

3.3   State Vectors

BOB: In quantum computing, these state symbols are called "state vectors" because they are given mathematical properties that are similar to the mathematics of vector quantities like velocity, force, and acceleration, represented pictorially by arrows that point in a particular direction. The length of the vector indicates the "strength" of the quantity. You've probably seen vectors on weather maps, where the arrows (vectors) are used to indicate wind direction and speed.

CARDY: But what does it mean to say that a computing state vector points in a certain direction? Are you trying to say that in my computer the state vectors point say north or east?

BOB: Neat idea, but the "directions" for state vectors are in most cases more abstract. In fact, the state vectors "live" in an abstract space called "state space" whose "directions" have nothing

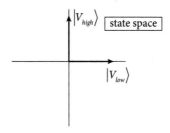

Fig. 3.1 A state space diagram for a two-state system. $|V_{low}\rangle$ and $|V_{high}\rangle$ are represented by the "vectors" along the two perpendicular axes in state space.

to do with east or north or up or down in the physical world. The role of east or north is played by the vectors representing "basis states," which we introduce in the next chapter. Though this abstract space may sound like a crazy idea, it is crazy enough to be useful. In fact, the usual mathematics of quantum theory is entirely a "state space" formalism.

Since this state space idea is weird, I find it helpful to draw pictures. These pictures are widely used in quantum mechanics and some aspects of quantum computing.

Let's see how this works for the two states $|V_{low}\rangle$ and $|V_{high}\rangle$. In Figure 3.1, I put the state vector $|V_{low}\rangle$ along the horizontal axis and $|V_{high}\rangle$ along the vertical axis. This is an arbitrary choice (like the usual choice of x along the horizontal direction and y along the vertical direction in graphs of mathematical functions). You don't have to do it this way, but most people in the field do. Reversing the roles of x and y in a graph will drive your friends nuts even though it is perfectly legitimate. All that is important, though, is that the two state vectors are perpendicular to each other.

CARDY: One of my math teachers said that perpendicular vectors are orthogonal. Is that the same as perpendicular?

ALICE: Yes, that's the same, but saying "orthogonal" gets you extra points in mathematics. In practical terms, orthogonal means that if the computer wire is in one Cbit state, then it can't be in the other state. They are mutually exclusive. If the wire is in the $|V_{low}\rangle$ state, it cannot be in the $|V_{high}\rangle$ state, and vice versa. So, for a two-state classical computing system, the actual states of a wire could be represented by arrows along either the horizontal axis or the vertical axis in state space.

In Figure 3.2, the red arrow in state space represents the state of a particular wire in a classical computer. When computations are being carried out, the wire will switch back and forth between those two states. If the system were described by a state with a "tilted" state vector pointing in some direction other than horizontal or vertical in Figure 3.2, the vector would quickly move to be either horizontal or vertical in the state space diagram. Of course, wires need to switch from one state to another as the computer carries out operations, but the circuitry is timed so the other parts of the computer "look" at the wires only after the wires have settled into the appropriate state. If the timing is not set correctly, all sorts of bad things can happen to the computer's calculations.

CARDY: What does the length of an arrow mean for a state vector? In regular space, I can imagine measuring lengths with rulers, or if I am talking about the wind speed at a particular location and associating that speed—so many miles per hour or meters per second—with the length of the wind vector. But I can't see how that would work in state space.

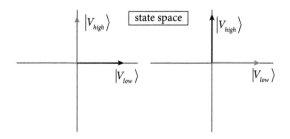

Fig. 3.2 On the left, the state of a traditional computer wire in the V_{high} state is represented by the vertical red arrow (vector) in state space. On the right, the wire is in the V_{low} state.

ALICE: Good question. The length of a state vector plays no role in classical computing, but once we get into quantum computing, we'll see that it is important to associate a "length" with a state vector.

State vector lengths lead to another helpful representation of computing states. Assume for the moment that the state vectors are one unit long. Let's also use the horizontal axis in the state space diagram as the x axis in typical mathematical graph; the vertical axis corresponds to y.

We draw the state space vectors so that their tails are at $x = 0$ and $y = 0$. We can then ask: What are the (x,y) coordinates of the tips of the arrows? For example, the state vector $|V_{high}\rangle$ has its tip at $x = 0$ and $y = 1$ in Figure 3.2.

CARDY: I get it. Then for the $|V_{low}\rangle$ vector in Figure 3.2, the tip is at $x = 1$ and $y = 0$.

ALICE: Exactly right. We may group those tip coordinates in columns:

$$|V_{high}\rangle \quad \Rightarrow \quad \begin{pmatrix} x_{tip} \\ y_{tip} \end{pmatrix} = \begin{pmatrix} 0 \\ 1 \end{pmatrix}$$

$$|V_{low}\rangle \quad \Rightarrow \quad \begin{pmatrix} x_{tip} \\ y_{tip} \end{pmatrix} = \begin{pmatrix} 1 \\ 0 \end{pmatrix}.$$

$$(3.1)$$

The x coordinate of the tip is listed at the top of the column and the y coordinate at the bottom. The large parenthesis symbols are called "column vectors." They are akin to the way you may have labeled xy coordinates of a point in a graph as (x, y).

CARDY: What do you mean by the arrow?

ALICE: The arrow $\Rightarrow$ will be used to mean "is mathematically represented by." The terminology reminds us that a Dirac state symbol is not the same as a column vector. In fact, a state vector can be represented by many different column vectors depending on how you orient your x and y axes in the diagram.

By the way, some authors use square brackets rather than parentheses. In that case, Eq. (3.1) would look like

$$|V_{high}\rangle \quad \Leftrightarrow \quad \begin{bmatrix} 0 \\ 1 \end{bmatrix}$$

$$(3.2)$$

$$|V_{low}\rangle \quad \Leftrightarrow \quad \begin{bmatrix} 1 \\ 0 \end{bmatrix}.$$

CARDY: I get it! The 1s are bits equal to 1 and the 0s are bits equal to 0.

ALICE: Sorry, Cardy. Not quite right. We haven't yet made any connection to the bits 1 and 0. Let's see how that connection works out. As we talked about in Chapter 2, in classical computing we usually associate V_{high} with the bit value 1 and V_{low} with the bit value 0. So, following that choice, we associate the state vector $|V_{high}\rangle$ with the bit 1 and the state vector $|V_{low}\rangle$ with the bit 0. It is customary to write this association as

$$|V_{high}\rangle \Rightarrow 1 \text{ and } |V_{low}\rangle \Rightarrow 0. \qquad (3.3)$$

CARDY: OK, I think I see what is going on, but I need to mull this over. I'm getting confused about what is a number, what is a coordinate, and what is a label.

BOB: I agree that the notation is a *bit* confusing (ha ha!). The way Alice and I have worked this out is to assume that what is inside the state vector "container" is always a label—just a name.

ALICE: Another piece of terminology will turn out to be important for quantum computing: *basis states*. In traditional computing, the basis states are just $|V_{low}\rangle$ and $|V_{high}\rangle$ (the voltage basis) or $|False\rangle$ and $|True\rangle$ (the True/False basis) or $|0\rangle$ and $|1\rangle$ (the classical bit basis). In all three cases, the basis state vectors point along the horizontal and vertical axes in Figure 3.1.

The physical devices in traditional computing are constructed so that only the basis states actually occur. As we mentioned before, there are rapid switches from one to the other, but the computer is constructed to ignore those switching periods. In the next chapter, we will find that the notion of basis states is more complicated for quantum states than for classical states.

For quantum states, we will see in Chapter 4 that the state need not lie along one of the basis axes. It can lie in a combination or superposition of the states. Drawing such a state vector is just like drawing any other vector based on coordinates. Start with the tail at the origin ($x = 0$ and $y = 0$). The head will be at the coordinates given by the column vector entries. For example, Figure 3.3 shows the state space vector for the state $\begin{pmatrix} 3 \\ 1 \end{pmatrix}$ with the V_{low} and V_{high} basis states. This is *not* an allowed state for a classical computer.

Most of the time we want to be more general and not have to worry about the particular physical construction of the traditional computer. Then we use the vectors $|0\rangle$ and $|1\rangle$, the so-called computational basis states. In our way of talking about state vectors, the "0" and "1"

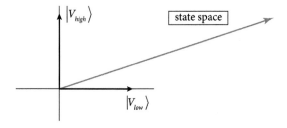

Fig. 3.3 A state space diagram with V_{low} and V_{high} basis states. The long arrow represents a state vector $3\,|V_{low}\rangle + |V_{high}\rangle$.

inside the state vector containers are just the names of the states. But given the usual conventions of computer science, we may associate the state labels with the binary numbers 0 and 1. Generally, we don't need to worry about the distinction between labels and numbers. However, it would be a big error to write $0.5\,|1\rangle = |0.5\rangle$ or $|1\rangle + |1\rangle = |2\rangle$.

Let's pause and summarize what we have done. We've looked at three representations of the states of components (wires) in a classical computer:

1. Abstract states: for example, $|True\rangle$ or $|1\rangle$ or $|V_{high}\rangle$.

2. State space vectors: arrow-like entities that "live" in the abstract state space.

3. Column vectors: a vertical array of numerical values giving the coordinates of the tips of the state vectors, assuming that the tails are at $x = 0$ and $y = 0$ in state space.

In almost every area of science, math, computer science, and engineering, a critical professional skill is being fluent in translating among different representations of concepts. We will get more fluent in those translations once we get into qubits and quantum state vectors in Chapter 4.

Try It 3.2

For the following column vectors $\begin{pmatrix} 1 \\ 1 \end{pmatrix}$ $\begin{pmatrix} 2 \\ 1 \end{pmatrix}$ $\begin{pmatrix} 0.5 \\ 0.5 \end{pmatrix}$ $\begin{pmatrix} 0.2 \\ 0.5 \end{pmatrix}$ $\begin{pmatrix} -1 \\ 1 \end{pmatrix}$, draw the corresponding state space vectors.

3.4   Gates as Matrices

BOB: We saw in Chapter 2 that computation can be carried out by using logic gates to act on the Cbits carried by the wires in the computer's electrical circuits. We described the results of those operations using logic tables (or truth tables).

There is another neat way of representing what gates do to classical computing bits. To illustrate this, let's consider the NOT gate. Remember that a NOT gate changes a bit 0 into a bit 1 and a bit 1 into a bit 0. How would we write those results in state vector language?

CARDY: I would write something like

$$\begin{aligned} \text{NOT}\,|1\rangle &= |0\rangle \\ \text{NOT}\,|0\rangle &= |1\rangle. \end{aligned} \tag{3.4}$$

Is that correct?

ALICE: Yes. We read these left to right: In the first line of Eq. (3.4), the NOT operates on the symbolic state vector $|1\rangle$ to produce the symbolic state vector $|0\rangle$. In the second line, NOT acts on $|0\rangle$ to produce $|1\rangle$.

In terms of column vectors, the representation in Eq. (3.4) looks like

$$(\text{NOT}) \begin{pmatrix} 1 \\ 0 \end{pmatrix} = \begin{pmatrix} 0 \\ 1 \end{pmatrix}$$

$$(\text{NOT}) \begin{pmatrix} 0 \\ 1 \end{pmatrix} = \begin{pmatrix} 1 \\ 0 \end{pmatrix}. \tag{3.5}$$

CARDY: I see that NOT switches between the two column vectors, like it did between True and False or 0 and 1. I am assuming the matrices with the 0s and 1s represent the computational basis?

BOB: That's right. So, the question is: What is the mathematical operation indicated by (NOT) that makes this work; the 1s change to 0s and vice versa? The answer is an array of 1s and 0s, which when multiplied times the column vectors gives the result stated in Eq. (3.5). Here is what the array looks like:

$$\begin{pmatrix} 0 & 1 \\ 1 & 0 \end{pmatrix} \begin{pmatrix} 1 \\ 0 \end{pmatrix} = \begin{pmatrix} 0 \\ 1 \end{pmatrix}. \tag{3.6}$$

ALICE: The array is called a *two-by-two matrix*: "two-by-two" because it has two rows and two columns, and "matrix" because it is an array of numbers (or symbols) that has certain mathematical properties. For now, we just need to know what happens when you multiply a column vector by a matrix. The general multiplication rule is

$$\begin{pmatrix} a & b \\ c & d \end{pmatrix} \begin{pmatrix} e \\ f \end{pmatrix} = \begin{pmatrix} ae + bf \\ ce + df \end{pmatrix}. \tag{3.7}$$

One way to remember this rule is that multiplying the array times the column vector gives another column vector. The *top element* of the resulting column vector is the sum of the products of the elements in the top row of the array times the corresponding elements in the column vector. The *bottom element* of resulting column vector is the sum of the products of the bottom row elements of the array times the corresponding column vector elements.

CARDY: Whew! That is rather weird. But it does seem to work for the NOT array.

Try It 3.3

Check that Cardy is right.

CARDY: Is that matrix array the same as *The Matrix* in the movie?

ALICE. No, but I hear from a friend who teaches math that students have been asking that question in linear algebra classes every year since the movie came out.

We have learned that the gate operations introduced in Chapter 2 are equivalent to a matrix multiplying the corresponding column vector. I remind you that this is an unusual way of talking about classical computing, but it is at the heart of quantum computing, which we will take up in Chapter 4.

Try It 3.4

Use the multiplication rule in Eq. (3.7) to check the following results:

$$\begin{pmatrix} 0 & 1 \\ 1 & 0 \end{pmatrix} \begin{pmatrix} 1 \\ 0 \end{pmatrix} = \begin{pmatrix} 0 \\ 1 \end{pmatrix} \quad \begin{pmatrix} 1 & 0 \\ 0 & 0 \end{pmatrix} \begin{pmatrix} 0 \\ 1 \end{pmatrix} = \begin{pmatrix} 0 \\ 0 \end{pmatrix} \quad \begin{pmatrix} 1 & 1 \\ 1 & 0 \end{pmatrix} \begin{pmatrix} 1 \\ 0 \end{pmatrix} = \begin{pmatrix} 1 \\ 1 \end{pmatrix}.$$

 CHAPTER SUMMARY

This chapter has introduced three state representations which we will be using throughout the remainder of this book. This summary can serve as a quick reference guide to that notation.

- In classical computing, the "state" of a wire can be at a low or a high voltage and these states can be represented by state vectors $|V_{low}\rangle$ and $|V_{high}\rangle$. In quantum computing, you will see similar use of states.

- Basis state vectors are orthogonal to each other. This has important mathematical ramifications for quantum computing algorithms.

- A state vector can be represented by

  (1) an arrow in state space,

  (2) a column vector of the x, y coordinates of the tip of the state arrow in state space, or

  (3) the Dirac state $|\rangle$ notation.

- These states live in an abstract state space, whose directions do not represent any physical space direction such as left, right, east, or west. Instead, the basis state vectors represent the mutually exclusive orthogonal states, which could be labeled as True/False, high and low voltage, or other binary properties.

- A logic gate operation can be represented by a matrix multiplying column vectors or as symbolic operations acting on states in the $|\rangle$ notation.

 FURTHER READING

N. David Mermin, *Quantum Computer Science: An Introduction* (Cambridge University Press, Cambridge, 2007). Chapter 1 uses Dirac notation to describe classical computing states.

The Khan Academy has some nice short videos and articles about matrices: https://www.khanacademy.org.

"Matrix (mathematics)," Wikipedia. Provides more information about matrices and column vectors.

4 Qubits and Quantum States

L'État, c'est moi (I am the State).

Attributed to King Louis XIV

4.1 What Is a Qubit?

ALICE: This a momentous occasion; we are now ready to tackle the formal description of quantum states—the essential ingredient of quantum computing.

BOB: Cardy, you should be sure to ask questions as we go along. But before Alice gets into the details of quantum states, I want to point out that there are two ways to think about quantum computing: First, building computers out of quantum objects. Examples of such objects are polarized photons, trapped ions, electrons trapped in "vacancies" in materials (for example, in diamonds), or superconducting electrical circuits. To understand the physics of those objects requires quantum mechanics.

The second way is developing quantum computing algorithms that make use of the mathematics of quantum mechanics. In either case, we will need to understand how the states of such objects are described in quantum mechanics, how we can manipulate those states to carry out computational tasks, and how using measurements of the objects tells us about the quantum states.

ALICE: We will use the term "qubit" (an abbreviation of "quantum bit") to mean any of these objects that have a physical property with just two possible values. Sometimes the two values arise naturally. For example, the so-called spin of an electron, neutron, or proton has just two possible values: the spin component along a particular spatial direction can be either "up" or "down."

In other cases, the quantum system is more complex: an atom or ion made up of protons, neutrons, and electrons, or a superconducting device built from millions or billions of atoms. In those cases, we manipulate the quantum system, so that only two possible values of some property (such as energy) are relevant. The key element in both cases is that there are only two possible quantum states: one associated with one value of the property and the other state with the other value.

CARDY: So, quantum computing is just like classical computing, but with qubits instead of classical bits?

ALICE: Yes. And no. We will see that there are some similarities, but also some fundamental differences. Those differences allow quantum computers to do things that are simply impossible for classical computers.

Quantum Computing: From Alice to Bob. Alice Flarend and Bob Hilborn, Oxford University Press.
© Alice Flarend and Robert C. Hilborn (2022). DOI: 10.1093/oso/9780192857972.003.0004

CARDY: Why do we limit ourselves to just two states? I think it would be better to have more than two states so you could convey more information.

ALICE: That's a great question, Cardy. In fact, there has been some QIS and QC work with quantum systems with more than two states. But so far, most QIS and QC theories and experiments have focused on qubits.

We have to be careful with the word "qubit." Some authors use the term qubit as a unit of quantum information—in analogy to a classical bit (a 0 or a 1), which itself can be viewed as a unit of information. Others use qubit to refer to the symbolic representation of the object's quantum state. As we mentioned, we will use qubit to refer to a physical object. Those multiple usages are a major cause of confusion for QC novices. Experts develop enough understanding to sort out which meaning is appropriate and when, but it is confusing to figure that out when you're just starting in QC and QIS.

The ambiguity becomes apparent in phrases such as "Alice sends Bob her qubit." Does that mean I send him the physical object (a trapped ion, for example), or do I send information that specifies the quantum state of that object? Commonly, it means sending the physical object. So, we will use qubit to refer only to the physical object and employ the phrase "quantum state" (or "quantum state vector") to refer to the mathematical representation of the state of the qubit.

4.2   What Is a Quantum State?

CARDY: Okay, I will keep that in mind, but I still need help understanding what you mean by "state."

BOB: Let's start with states in classical physics. There the state of a system is the collection of information (such as mass, position, velocity, angular momentum, electric charge . . .) sufficient to predict the behavior of the system when it is acted upon by various forces. For a classical particle, knowing its velocity and its position at some instant of time and knowing the forces acting on the particle are sufficient to predict its future position and velocity, using Newton's laws of motion. We say that the state of the particle at a particular time is given by its position and velocity at that time. Of course, for more complicated systems, we may need to specify and keep track of other properties as well.

In the world of quantum mechanics, the term "state" also means the collection of information that allows us to predict the future. But in quantum mechanics, the future is restricted to the *probabilities* of the outcomes of measurements made on that system. That statement points to a critical philosophical difference between quantum states and classical states. Quantum states tell us only the probabilities of getting various results when we measure the quantum state, whereas in classical physics, the states allow us to make exact predictions, at least in principle, for the behavior of the system.

Although the meaning of quantum state is similar to the notion of state in other areas of physics, mathematics, and engineering, the mathematics used to describe quantum states and their behavior is significantly different from that used for classical states even though most physicists believe that the classical states are just the large-scale limit of the quantum states.

ALICE: In other words, when dealing with large-size objects such as mites, grand pianos, or gadzillions of quantum particles, the quantum states of those large objects are, for all practical purposes, the same as the classical states. In particular, those probabilities Bob mentioned become certainties. However, both the conceptual connections and the mathematical connections between quantum states and classical states are less obvious. Using ideas from one realm in the other can lead to all sorts of confusion.

In what follows, we will introduce the concepts associated with quantum states in a way that gives you a good understanding of how they work and how they are different from the classical concepts. In Chapter 14, we will say more about the connection between the quantum world and the classical world. Just a warning: Thinking about that connection gives me a feeling of vertigo, particularly when I try to imagine what is "actually" going on in the world as described by quantum mechanics.

4.3 Quantum States and Quantum Computers

ALICE: Before we discuss the details of quantum states, it might be helpful to have an overview of how a quantum computer works. As shown in Figure 4.1, reading from left to right, we start with a system of qubits (objects) in some specified quantum state. The qubits then interact with a collection of quantum gates, and then each qubit interacts with a measurement device, permitting the analysis of results. That is, in most QC and QIS situations, we "prepare" a system in some quantum state, act on it with quantum gates, and "measure" the resulting state. The quotation marks remind us that we will need to be more specific about what we mean by "preparing" a quantum state and "measuring" a quantum state. The measurement results give us the "answer" to the computation. We choose the quantum gates to implement a quantum algorithm, much like a classical computer program tells a classical computer the set of calculations to be carried out on the initial set of bits.

As we saw in classical computing, we almost always use objects (classical bits) that are constructed to be in one of two possible states—V_{high} or V_{low}, for example. These are then represented more abstractly as the binary digits 1 and 0. We say that a particular object (most often a wire in a circuit) can be in either a 1 state or a 0 state. Other circuit elements (gates) sense the object's state (usually with inputs of two or more bits) and produce outputs according to the design of the particular gate.

Classical computers are built from devices with large numbers of particles and those large numbers allow the devices to be put into a state we can measure (with a voltmeter, for example) without significantly changing the state. It also means that the device (say a wire and its branches) can be connected simultaneously to many measuring or sensing devices (gates)

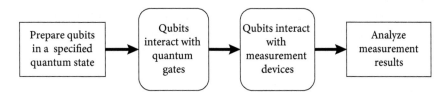

Fig. 4.1 Schematic of a quantum gate-array computer. The diagram is read from left to right.

without significantly affecting the state. It turns out that we can't do either of those things with quantum states.

BOB: In terms of symbolic representations, we know from Chapter 3 that the state of a classical bit (device) could be represented abstractly as a Dirac notation state space vector: $|0\rangle$ or $|1\rangle$, or equivalently, by column vectors $\begin{pmatrix} 1 \\ 0 \end{pmatrix}$ or $\begin{pmatrix} 0 \\ 1 \end{pmatrix}$, though for most traditional computing purposes these abstractions are, as we noted, not particularly useful.

But when we come to quantum computing and the description of quantum states, life gets more complicated, but also much more interesting. The state space vectors become essential for describing quantum behavior and that's why those vectors are common throughout the scientific literature on quantum computing.

Scientists like to say that the physical system is "in a quantum state" as if the quantum state were a piece of clothing that the physical system puts on. That seems to imply that a quantum state is like a location: My electron is in a state in the northeast corner of the room. That metaphorical way of speaking is okay as long as you don't take the words literally. The quantum state is not a "thing" that the physical system has or does not have. We find it best to take the point of view that the quantum state is just a mathematical description of state describing the physical object.

There are lots of disagreements among physicists, chemists, mathematicians, and philosophers of science about the status of the concept "quantum state." Is it something real (like mass) or is it "just" a mathematical construct that helps us understand quantum behavior? We will stick with the idea that the quantum state is simply an abstract notion that can be represented by symbols, which if manipulated with certain rules allow us to understand and predict the behavior of quantum objects (qubits). In particular, we can predict the probabilities of various measurement outcomes.

4.4 Quantum States and State Space

ALICE: How do we describe the state of a qubit? One of the fundamental assumptions of quantum mechanics is that quantum states are represented by vectors in an abstract state space. As we mentioned in Chapter 3, these vectors, like ordinary force or position vectors, have both a magnitude (length) and a direction. However, unlike ordinary vector quantities, quantum state vectors live in an abstract state space. These kinds of vector spaces have been thoroughly studied by mathematicians, but for our purposes, we will not need to know most of those details.

The abstract space contains the state vectors and comes equipped with other mathematical objects, called operators (or, in QC language, quantum gates), that allow us to change one state vector into another. These state vectors are not themselves directly observable, but they allow us to predict, as mentioned previously, the probability of the outcomes of measurements made on the qubits.

BOB: To emphasize that the state space is an abstract space and not a physical space, let me quote a famous aphorism attributed to the quantum theorist Yakiv Aharonov in speaking

Fig. 4.2 Erwin Schrödinger's grave marker in Alpbach, Austria.
Source: R. Hilborn.

about Hilbert space—the abstract quantum state space: "In Hilbert space, no one can hear you scream."

ALICE: For a quantum system like those used in quantum computing devices, we write a general "state vector" as $|S\rangle$. Often, physicists like to use the notation $|\psi\rangle$ with the Greek letter ψ (psi, pronounced "sigh"). You can also remember the symbol as a version of Poseidon's trident (or Neptune's trident, if you prefer Roman to Greek gods): a three-pronged pitchfork.

BOB: ψ was a great choice of symbol because I always want to sigh when I am perplexed by the mysteries of quantum mechanics. In fact, there is a ψ sign on Erwin Schrödinger's grave in the village of Alpbach in Austria. Schrödinger (1887–1961) was one of the founders of quantum mechanics and is now most famous for his Schrödinger's cat paradox. Figure 4.2 is a photo I took several years ago when I was in Alpbach for a conference.

ALICE: That's on my list of places to visit, Bob, but in spite of Schrödinger's grave marker, we will use a variety of other labels to keep track of different kinds of quantum states. For example, for a trapped-ion qubit, the basis states are so-called energy eigenstates: states associated with a definite energy of the ion. Let's suppose that we focus on just two energy states: $|E_0\rangle$ and $|E_1\rangle$. Then we can write a general quantum state of the trapped ion as

$$|S\rangle = \underbrace{\beta_0|E_0\rangle}_{\text{part of the state associated with } E_0} + \underbrace{\beta_1|E_1\rangle}_{\text{part of the state associated with } E_1}. \qquad (4.1)$$

ALICE: Eq. (4.1) expresses the key concept of quantum mechanics: A general quantum state can be written as a combination (*superposition*) of other quantum states. β_0 and β_1 are called the *state coefficients* or *state amplitudes*, which are just numbers. The states $|E_0\rangle$ and $|E_1\rangle$ are called the *basis states* and the overall state $|S\rangle$ is said to be a superposition of $|E_0\rangle$ and $|E_1\rangle$. In Chapter 5, we will show you how the amplitudes are related to the probabilities of measurement outcomes.

Cardy, I urge you to write all the equations in this chapter by hand. That will help you "read" the equations and give you practice identifying state vectors and state coefficients. Cognitive researchers have found that you tend to understand things better when you write them down. The physical act of writing involves a lot of mental decisions which are helpful in reading the symbols and promoting understanding. In fact, it might be useful to write the vector "container" symbol $|\rangle$ in one color, the state vector label in a different color, and the amplitudes in a third color to increase your awareness of what you are writing.

CARDY: I'll do that. I found that to be an important exercise in my statistics classes. It really does help me "see" the structure and meaning of the equations.

BOB: We're going to use the computational basis states $|0\rangle$ and $|1\rangle$ instead of $|E_0\rangle$ and $|E_1\rangle$; so, we don't have to worry about the detailed physics of the qubits. Let me remind you that the symbols in the state vector containers are just names. They have no immediate connections to the *numbers* 0 and 1. In fact, if it were not for the connection to classical computing and classical bits, we should probably label those states as $|\text{Zero}\rangle$ and $|\text{One}\rangle$. However, since almost everyone in QC and QIS uses the 0 and 1 forms, we just need to get used to that notation. But, as we mentioned in Chapter 3, don't be tempted to write something like $5|1\rangle = |5\rangle$.

In many QC and QIS applications we will use the computational basis states $|0\rangle$ and $|1\rangle$. Any state vector $|S\rangle$ can be written as a superposition of those basis states:

$$|S\rangle = a_0 |0\rangle + a_1 |1\rangle, \tag{4.2}$$

where a_0 and a_1 are the state coefficients (amplitudes).

CARDY: How do I know what a_0 and a_1 are?

BOB: As we mentioned before, the state amplitudes a_0 and a_1 carry information about the results of measurements on the qubit whose state is $|S\rangle$. So, you can use think of the amplitudes as numbers determined by measurements or by theories that predict the results of measurements.

ALICE: Figure 4.3 shows a graphical representation of the basis state vectors $|0\rangle$ and $|1\rangle$ and a general state vector $|S\rangle$. Geometrically, a_0 is found by drawing a line from the tip of $|S\rangle$ to the line along the $|0\rangle$ direction. We call a_0 the perpendicular projection of $|S\rangle$. Similarly, for a_1. As shown in Figure 4.3 we could also write the state in terms of column vectors in three ways:

$$\begin{aligned} |S\rangle &\Rightarrow a_0 \begin{pmatrix} 1 \\ 0 \end{pmatrix} + a_1 \begin{pmatrix} 0 \\ 1 \end{pmatrix} \\ &= \begin{pmatrix} a_0 \\ 0 \end{pmatrix} + \begin{pmatrix} 0 \\ a_1 \end{pmatrix} \\ &= \begin{pmatrix} a_0 \\ a_1 \end{pmatrix}. \end{aligned} \tag{4.3}$$

As we mentioned in Chapter 3, we use the symbol $\Rightarrow$ to indicate mathematical equivalence between different representations.

ALICE: Let's pause to recall the different meanings of the 0s and 1s in these equations. In Eq. (4.2), the symbols inside the state vectors are merely names given to the two basis states.

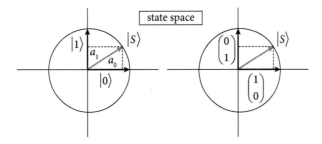

Fig. 4.3 A state space diagram showing the two computational basis states, $|0\rangle$ and $|1\rangle$ on the left, and a general state vector $|S\rangle$. a_0 and a_1 are the perpendicular projections of the vector $|S\rangle$ along the $|0\rangle$ and $|1\rangle$ directions, respectively. All three vectors have length equal to 1. Consequently, their tips lie on a circle with radius 1. On the right, the column vector form of the basis states is used.

In Eq. (4.3), the 1s and 0s are the coordinates of the tips of the basis state vectors in state space, assuming the tails are at (0,0).

In words, Eqs. (4.2) and (4.3) assert that the state $|S\rangle$ of the quantum object is a combination (superposition) of the computational basis states $|0\rangle$ and $|1\rangle$. Superposition is a critical characteristic of quantum states that makes them different from the classical computing states described in Chapter 3. Classical computer states are always one or the other of the individual basis states—never a superposition. In fact, superposition is the "secret sauce" of QIS and QC. We will see the power of superposition in almost all the quantum algorithms we will discuss. Remember that the state of a classical computer's wire (for example) is either $|0\rangle$ or $|1\rangle$. The wire's state can't be a combination of these two states; so, superposition makes no sense for a classical computer.

BOB: Another important quantum principle is that you have a choice of basis states. Describing the same quantum state in terms of different basis states is a powerful tool in QC. In Chapter 8, we will take up the details of using different basis states.

One notational issue is important: We will always list the coefficient associated with the computational basis state $|0\rangle$ first (the top entry in a column vector) and the other, associated with $|1\rangle$, second (the bottom entry in a column vector). This order is illustrated in Eq. (4.3) in the last line. We are then using what is called an "ordered basis set." Using an ordered basis set helps in the bookkeeping of state vectors and amplitudes. We do this routinely in everyday life: We almost always say "(x,y) coordinates" or, if you are a baseball fan, "the count is 3 and 2," with the understanding that the 3 refers to the number of balls and 2 is the number of strikes.

CARDY: But what does it mean to be in a combination of 0 and 1? Is it a number between 0 and 1?

ALICE: Cardy, that indeed is a crucial issue in quantum physics. It turns out to be a rather subtle notion and, in a way, it clashes with our everyday understanding of what it means for an object to "have" some property.

If we think back to the Boolean logic we used in Chapter 2, we see the conundrum: What would it mean for a proposition to be in a superposition of True and False? Is the proposition then partly True and partly False? Or for a wire in a classical computer to be partially V_{high} and partially V_{low}? It certainly does not mean that the wire has a voltage between those two values (as a steady state).

All of this seems rather abstract (and it is). So, let's look at a specific example: so-called *polarization states* of light. We choose this example for two reasons: (1) many QC and QIS experiments make use of light polarization, and (2) with only a few dollars' worth of equipment, you can carry out experiments at home that illustrate many of the basic features of light polarization. You can also do these experiments at the movies with the "3D" glasses before the show begins.

4.5   Polarized Light

BOB: For more than 150 years, we have known that a light beam (think of a laser-pointer beam) can be described as oscillating electric and magnetic fields that travel through space as illustrated in Figure 4.4. These are the same kinds of fields that are responsible for "static electricity," for sparks that jump from your fingers to a doorknob after you rub your shoes on a carpet, and for refrigerator magnets. We say light is polarized if the oscillation pattern of the electric and magnetic fields that constitute the light wave is systematically organized. We are interested in the case when the electric field in the light beam oscillates along a fixed direction. In that case, the light beam is said to be "linearly polarized." The *polarization direction* of light is the direction along which the light wave's electric field oscillates. For example, the electric field might oscillate along a horizontal axis or a vertical axis or along an axis tilted at 32.1° with respect to a horizontal axis, and so on. In Figure 4.4, the light is linearly polarized along the vertical axis, labeled by E.

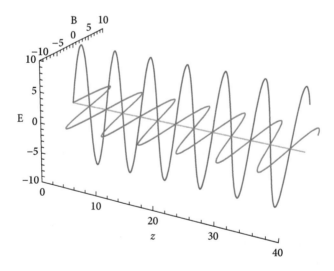

Fig. 4.4 A light beam, traveling to the right in the z direction, can be described by a combination of oscillating electric and magnetic fields. In this case the electric field (blue) E is oscillating in the vertical direction. The magnetic field (red) B is in the horizontal direction. The units of E and B have been chosen to make the amplitudes equal. The figure shows the field strengths as a function of position at a specific time.

ALICE: Linear polarization is relatively easy to detect using polarizing material read-ily available in "polarized" sunglasses and "3D" glasses used in movie theaters. Inexpensive sheets of polarizing material are also available from science equipment suppliers (see Further Reading).

The polarizing material in sunglasses is oriented to block light with electric field vectors oscillating in a horizontal direction while transmitting light polarized in the vertical direction. Why? You would like polarized sunglasses to block "glare" from horizontal surfaces: a road, the hood of a car, the surface of a lake or the ocean (see Figures 4.5 and 4.6). It turns out that when light reflects off a horizontal surface, the reflected light has most of its electric field oscillating in a horizontal direction, parallel to the horizontal surface from which it was reflected. We say that the light is horizontally polarized.

A polarizing sheet is a stretched polymer material that absorbs light polarized along the stretch direction and transmits (more or less) light polarized perpendicular to the stretch di-rection. By rotating the polarizing sheet (or your sunglasses), you can determine if a given light

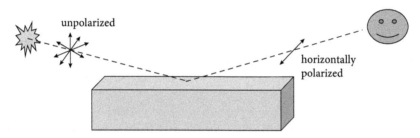

Fig. 4.5 A diagram showing how unpolarized light from the light source becomes horizontally polarized after reflecting off a horizontal surface such as a piece of dark polished granite.

Fig. 4.6 Photos of the surface of a shiny granite countertop. On the left, you can see light reflected from the top surface viewed through a polarizing filter with its transmission axis horizontal. On the right, the photo is taken through a polarizing filter whose transmission axis is vertical. The horizontally polarized glare light is absorbed by the filter and you can see the pattern in the granite more clearly, without the 'haze' from the glare light.
Source: R. Hilborn.

beam is "unpolarized." If the brightness of the light transmitted is unchanged when you rotate the filter, then the light is unpolarized (strictly speaking, it has no linear polarization). That usually means that there are equal amounts of light with all polarization directions.

You can use glare light reflected from a horizontal surface to find the transmission axis of your polarizing sheet or your sunglasses. Look at the glare light through your polarizing sheet and rotate the sheet so the glare light transmitted through the sheet is maximized. Then you know that the transmission axis of your sheet is horizontal. The difference between the maximum intensity and the minimum is most dramatic if the angle between the surface and the ray of light coming toward your eyes is about $30°$ (see Figure 4.6).

BOB: Once you know the polarization axis direction for transmission for your sunglasses (or your polarizing sheet), you can find a light beam's polarization direction by rotating the sunglasses until you find a maximum in the amount of light coming through. Here is how you do it: Rotate the sunglasses and find the orientation that leads to the maximum amount of light coming through the lenses. The orientation of the polarizing material's transmission axis is then the predominant electric field oscillation direction. If rotating the sunglasses $90°$ from the maximum intensity direction gives no light coming through, we say the light is linearly polarized in the axis direction that gives maximum intensity. If there is a small intensity of light at $90°$ from the maximum intensity direction, we say the light has at least partial linear polarization. Note that the two polarization axes (maximum and minimum transmission) are perpendicular to each other. In the next section, we will see this perpendicularity in the abstract state space.

ALICE: Figure 4.6 illustrates how a properly oriented polarizing sheet can remove glare light. But I want to raise one detail. Bob, what you said about unpolarized light is true, but there is another kind of light polarization called circular polarization, which might appear to be unpolarized by the criterion you posed.

BOB: You are correct, Alice. But for our current purposes, all we need is linearly polarized light.

Try It 4.1

Look at glare light reflected from a shiny floor through a polarizing sheet or polarizing sunglasses. Do Bob's statements describe what you see? Do you see a change in intensity of the light coming through the polarizing sheet when you rotate the sheet about the viewing direction? Now look at light from computer and tablet screens. Light from most computer and tablet screens is polarized. Light from smart phone screens is partially polarized. Is that what you find? What about your TV screen?

BOB: Polarizing sheets can be used both to analyze the polarization state of the light beam and also to produce polarized light with an electric field oscillation direction you want. (Actually, to completely characterize the polarization state, you need to use a few other devices in addition to the polarizing sheet, but the basic idea still holds.)

In other words, a polarizing sheet can be used both to prepare a light beam with its polarization direction along the sheet's transmission axis and to analyze ("measure") the polarization state, usually by rotating the polarization sheet about a direction perpendicular to the sheet and detecting how the power of the transmitted light varies with the orientation of the transmission axis.

4.6 Photons

ALICE: As we mentioned before, polarized light is often used in QIS and QC applications. The *quantum* nature of light manifests itself in the empirical observation that light is made up of small, discrete bundles of energy. Those bundles are called *photons*. The discovery of the photon nature of light was one of the most fundamental and important experiments leading to the development of quantum mechanics early in the 20th century. Ordinary light beams transmit billions and billions of photons per second, so in most everyday situations, we don't notice the individual photons.

CARDY: So, it's like rain? Rain is made up of individual raindrops. But if we are having a heavy rainstorm, all we hear is a whoosh of water hitting the roof of our house. However, during a light sprinkle, I can hear the individual raindrops hitting the roof.

ALICE: Exactly. Another analogy is the sound made by popping popcorn. When only a few corn kernels are popping, you hear the individual "pops," which seem to occur randomly. If lots of kernels pop at about the same time, you hear a continuous "roar."

The detailed quantum properties of photons become apparent only when we have the sensitivity to detect individual photons. So, when we use photon qubits for QIS and QC, we will want to think about manipulating light photon by photon. In particular, how do we describe the quantum polarization state of an individual photon?

BOB: Here is how that works: In our preferred language, a photon is a qubit: a quantum object for which there are just two polarization conditions. For linear polarization, we will choose two quantum basis states: one $|hlp\rangle$ for horizontal linearly polarized light and the other $|vlp\rangle$ for vertical linearly polarized light.[1]

BOB: Figure 4.7 shows a state space diagram for a linear polarization state (indicated by the red arrow) that is along neither the horizontal nor the vertical direction. A general quantum polarization state $|S\rangle$ for the photon can be written as a superposition of the two basis states. We use $|hlp\rangle$ and $|vlp\rangle$ in place of $|0\rangle$ and $|1\rangle$ in Eq. (4.2) and write

$$|S\rangle = a_h |hlp\rangle + a_v |vlp\rangle. \tag{4.4}$$

In words, the state is a combination (a superposition) of horizontal and vertical polarization.

Linearly polarized basis states constitute a special case in the sense that the directions associated with the basis states can be the same as the real (physical) space horizontal and vertical

---

[1] Note that many authors use $|H\rangle$ and $|V\rangle$ for the linear polarization basis state vectors, but H is used for several quantum gates and operators, as we shall see.

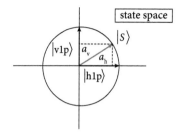

Fig. 4.7 A state space diagram for linearly polarized light (photons). The two basis state vectors are $|hlp\rangle$ for horizontal linear polarization and $|vlp\rangle$ for vertical linear polarization. The vectors all have unit length, so their tips fall on a circle of radius 1. a_h and a_v are the amplitudes of the state vector along the basis state directions.

directions. Later, we will meet up with other quantum basis states that have similar state space diagrams but very different physical space representations.

It might be tempting to call the state $|S\rangle$ a *mixture* of the two basis states, but that would be misleading. When I say "mixture" you might be thinking about what happens if I grab a handful of candies from a bowl. Some of the candies are red and some are blue. My handful will have a mixture of red and blue candies. A superposition state is more subtle. In the current context, the superposition state refers to just one qubit (just one piece of candy). Furthermore, it does not mean the piece of candy is purple, a color mixture of red and blue. As we shall see in the next chapter, the superposition state tells us that if we observe the candy's color, sometimes we will see red and sometimes we will see blue.

4.7 Ordinary Vectors

ALICE: Let's pause for a few minutes and remind ourselves that this way of writing a quantum state vector is like the way we describe other vectors in mathematics, physics, engineering, and other sciences. For example, we use vectors to describe velocity, position, acceleration, forces, and electromagnetic fields in physics.

Let's review the connection between an ordinary vector and its components relative to the traditional x and y axes. To be concrete, suppose we have a vector represented by the symbol $\vec{B}$ and, in keeping with our restriction to two-state systems, we assume that the vector "lives" in a two-dimensional space. The vector can be described as a sum of two vectors, $b_x\hat{x}$ in the x direction and $b_y\hat{y}$ in the y direction. $\hat{x}$ is a vector of unit length pointing in the x direction with an analogous definition for $\hat{y}$. The unit vectors $\hat{x}$ and $\hat{y}$ play the role of basis vectors. We could also use two column vectors or a column vector with two entries. So, $\vec{B}$ can be written in three equivalent ways:

$$\vec{B} = b_x\hat{x} + b_y\hat{y} \Rightarrow b_x \begin{pmatrix} 1 \\ 0 \end{pmatrix} + b_y \begin{pmatrix} 0 \\ 1 \end{pmatrix} = \begin{pmatrix} b_x \\ b_y \end{pmatrix}. \tag{4.5}$$

ALICE: Now comes a crucial question: How do we express the length of the vector $\vec{B}$ in terms of the coefficients b_x and b_y? The answer is given by the Pythagorean theorem: The square of

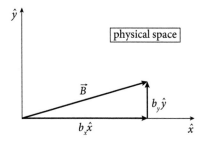

Fig. 4.8 A graphical representation of $\vec{B} = b_x\hat{x} + b_y\hat{y}$.

the hypotenuse of a right triangle (the square of the length of the vector $\vec{B}$ in Figure 4.8) is equal to the sum of the squares of the other two sides. For the right triangle in Figure 4.8, we have

$$\left\|\vec{B}\right\|^2 = B^2 = b_x^2 + b_y^2, \tag{4.6}$$

where the first two terms are different ways of writing the square of the length of the vector $\vec{B}$. The double vertical lines are the most common notation for the length of the vector—what mathematicians call the "norm" of a vector.

4.8 Back to Polarized Light

ALICE: Figure 4.7 is the analogous vector component diagram for polarized light. By comparing Figures 4.7 and 4.8, we see that we can express the square of the "length" of our quantum state vector in Eq. (4.4) as

$$\| \, |S\rangle \, \|^2 = a_h^2 + a_v^2. \tag{4.7}$$

I used quote marks around length for a state vector because that length is not something you can measure with a ruler. For almost all our work in quantum computing, we will use state vectors whose length is 1 (they are called "unit vectors" or "normalized vectors"). Following that convention, we will insist that

$$a_h^2 + a_v^2 = 1. \tag{4.8}$$

We will motivate that requirement in Chapter 5.

BOB: In addition to polarizing sheets and polarizing sunglasses, another way to measure the polarization state of a light beam makes use of devices called polarizing beam splitters (PBSs for short). Most PBSs are made from a transparent crystalline material that is sensitive to the polarization direction associated with the light traveling through the device. From a classical physics point of view, the light's electric field interacts with the PBS material in such a way as to have light with vlp travel in a direction different from light with hlp. When an unpolarized

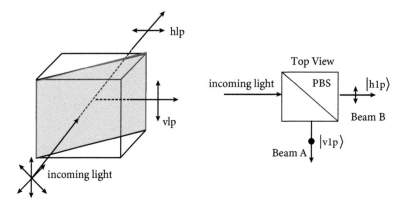

Fig. 4.9 A polarizing beam splitter (PBS). The diagonal plane marks the boundary between the two parts of the splitter. The incoming light beam is "unpolarized," an equal mix of all polarization directions. The beam that comes out of the PBS to the right has its electric field oscillating in the vertical direction; that is, it is linearly polarized in the vertical direction. The other beam has its electric field oscillating in the horizontal direction. On the right is a top view of the polarizing beam splitter. The black filled circle indicates that the electric field vector is oscillating perpendicular to the plane of the figure.

light beam interacts with a PBS, the beam is split into two parts. One of the parts is associated with vlp and the other with hlp. What is "vertical" and what is "horizontal" is determined by the orientation of the PBS.

Using photon language, we say that the PBS sorts a stream of incoming photons into two separated beams, one of which is described by $|hlp\rangle$ and the other by $|vlp\rangle$. Recall that a polarizing sheet transmits only one polarization state, whose direction is parallel to the sheet's transmission axis.

BOB: In the top view part of Figure 4.9, the beam going to the right is described by a linear polarization state $|hlp\rangle$ with the polarization direction parallel to a horizontal edge of the prism, while the other beam is described by a linear polarization state represented by $|vlp\rangle$ parallel to a vertical edge of the prism. The PBS "splits" the incoming beam into two beams: one with hlp and the other with vlp.

Try It 4.2

Write the state vector in Eq. (4.4) for each of the light beams coming out of the PBS in both state vector and column vector forms. Explain which coefficients (amplitudes) are zero and non-zero for each of the beams.

BOB: The PBS can be used two ways: First, the device can be used as what I like to call a quantum *state separator*. If you send in a light beam described by an arbitrary polarization state, the PBS splits up the light beam into sub-beams each of which is associated with one of the linear polarization basis states. The PBS has analyzed the light beam into its two linear polarization components. The intensity of the light in each beam (proportional to the number of photons in each beam) can be predicted from the quantum state sent into the PBS. The exact connection will be discussed in Chapter 5.

The PBS can also be used as a quantum *state preparation* device. For example, the light that emerges in the right-going beam shown in the top view part of Figure 4.9 has horizontal linear polarization no matter what the polarization state of the incoming beam. So, by using just that right-going beam, we know we are working with photons whose polarization quantum state is $|\text{hlp}\rangle$.

We could carry out a similar state preparation using a polarizing sheet whose transmission axis is aligned horizontally. The photons that get through the sheet can be described by the hlp state. As we said before, in the case of the polaroid sheet, the photons described by the vlp state are absorbed by the polaroid sheet.

4.9 Orthogonality of Two-State Vectors

ALICE: Now that we have developed a basic understanding of a quantum state vector, let's look at a few simple mathematical techniques that will be useful in understanding quantum computation. Many operations in QC involve multiplying two state vectors together to obtain a number (called a scalar), rather than another vector. A "scalar" is just a regular number with no direction associated with it. This kind of product is often called a *scalar product* or an *inner product*, or a *dot product* since the dot product notation $\vec{C} \cdot \vec{B}$ is often used in standard vector algebra. Figure 4.10 shows two vectors and their components in state space.

Let's see how the scalar product works with the notation we have introduced so far. Let's start with the column vector form for two state vectors $|B\rangle$ and $|C\rangle$:

$$|B\rangle \Rightarrow \begin{pmatrix} b_0 \\ b_1 \end{pmatrix}$$

$$|C\rangle \Rightarrow \begin{pmatrix} c_0 \\ c_1 \end{pmatrix}, \tag{4.9}$$

where the state amplitudes b_0, b_1, c_0, and c_1 are just regular numbers (scalars). To multiply $|B\rangle$ by $|C\rangle$ using the dot (inner) product $\vec{C} \cdot \vec{B}$, we first convert $|C\rangle$ to a *row vector* $\begin{pmatrix} c_0 & c_1 \end{pmatrix}$

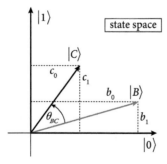

Fig. 4.10 A state space diagram showing two state space vectors $|B\rangle$ and $|C\rangle$ and the angle θ_{BC} between them. b_0 and b_1 are the components of $|B\rangle$ along the state space basis vectors. c_0 and c_1 are the corresponding components of $|C\rangle$. See Eq. (4.9).

keeping the order of the numbers the same (top to bottom in the column vector becomes left to right in the row vector).

CARDY: One of my physics major friends told me that we should be using c_1^* and c_0^* in the row vectors and said something about complex conjugation. That sounds slightly salacious to me. What did she mean?

ALICE: In the most general version of state vectors (as used in quantum mechanics), state coefficients are complex numbers. Complex numbers have two parts: historically called the "real" part and the other the "imaginary" part. In that situation, c_0 and c_1 are replaced by their complex conjugates c_1^* and c_0^* when writing the row vector. The complex conjugate of a complex number has the same real part as the original number, but its imaginary part changes sign.

For almost all situations in this book, the state coefficients will be "real" numbers. Then the complex conjugate is the same as the original number. We will look at the generalization to complex numbers in Chapter 15 and explain why they are needed in the most general form of quantum mechanics and quantum computation.

Now let's get back to the dot product. The row vector is represented by the "backwards" state vector symbol $\langle C|$. To help remember which is which, we call the original vector $|C\rangle$ a "right vector" because the angle bracket points to the right, while the "backwards" row vector $\langle C|$ is called a "left vector" because the angle bracket points to the left.

Aside

Some authors use another pair of somewhat strange names for these two types of state vectors. Those names were introduced by Paul Dirac, the inventor of Dirac notation. He had realized that expressions involving angular brackets, such as $\langle C|\ A\rangle$, are critical in understanding transformation rules in quantum mechanics (as we shall see in Chapters 7 and 8) and that those brackets could be "factored" into two parts, $\langle C|$ and $|A\rangle$; he apparently decided to split the word "bracket" to give names to those two parts. But instead of using "brac" and "ket," he used (whimsically? sophomorically?) "bra" and "ket," which, alas, continue to be widely used in QC and QIS. So, if you read or hear "ket" or "ket vector," think "right vector" or "column vector" and if you read or hear "bra" or "bra vector," think "left vector" or "row vector."

To form the scalar product of two state vectors $|B\rangle$ and $|C\rangle$, we use the column vector (right vector) $|B\rangle$ and the row vector (left vector) $\langle C|$ with the following rule:

$$\langle C|\,|B\rangle = \langle C|\ B\rangle \Rightarrow \begin{pmatrix} c_0 & c_1 \end{pmatrix} \begin{pmatrix} b_0 \\ b_1 \end{pmatrix} = c_0 b_0 + c_1 b_1. \tag{4.10}$$

The result, as promised, is just a number. In the first two terms in Eq. (4.10), you can see that the right vector $|B\rangle$ appears on the right of the angular bracket term $\langle C|\ B\rangle$ while the left vector $\langle C|$ appears on the left. The last two terms in Eq. (4.10) show how to multiply the row vector of state amplitudes times a column vector of state amplitudes to obtain the scalar product result.

Note also that the multiplication of the state vector symbols on the left in Eq. (4.10) involves two vertical line symbols, which are almost always replaced by a single vertical line as indicated in the second term in Eq. (4.10). The multiplication rule mimics the way regular vectors are multiplied using the scalar product (dot product) rule: $\vec{D} \cdot \vec{E} = d_x e_x + d_y e_y$.

CARDY: You brought in these left state vectors for this dot product multiplication. Why can't we just multiply two right vectors together?

ALICE: Another astute question! When we get to describing multi-qubit systems, we'll see that we can make use of the product of two (or more) right vectors to form what are called product states for the qubits.

Let's look at the scalar product $\langle C| \, B \rangle$ in more detail. If $\langle C| \, B \rangle = 0$ and if neither vector itself is a 0 vector—I'm beginning to sound like a mathematician—then we say that the two state vectors $|B\rangle$ and $|C\rangle$ are *orthogonal*. In that case, the Dirac bracket is

$$\langle C \, | B \rangle = c_0 b_0 + c_1 b_1 = 0. \tag{4.11}$$

In a state space diagram, two orthogonal state vectors are perpendicular to each other. The orthogonality is an expression of the mutual exclusivity of the quantum states. For example, if your basis states are $|vlp\rangle$ and $|hlp\rangle$, then if you are in state $|vlp\rangle$ with $a_v = 1$ you cannot also be in state $|hlp\rangle$; the states are mutually exclusive. Figure 4.7 illustrates the orthogonality of $|hlp\rangle$ and $|vlp\rangle$.

Try It 4.3

Check orthogonality with our standard basis vectors expressed in column vector format:

$$|0\rangle \Rightarrow \begin{pmatrix} 1 \\ 0 \end{pmatrix} \quad |1\rangle \Rightarrow \begin{pmatrix} 0 \\ 1 \end{pmatrix}. \tag{4.12}$$

Hint: Use the multiplication rule expressed in Eq. (4.10).

Try It 4.4

Create two state vectors that do *not* lie along the basis state directions by giving numerical values between 0 and 1 to the coefficients in Eq. (4.10) with $c_0 b_0 + c_1 b_1 = 0$. Are those state vectors orthogonal? Draw state space diagrams for the vectors and explain how the state space diagrams indicate orthogonality when $c_0 b_0 + c_1 b_1 = 0$.

ALICE: One more question: What is the scalar product (dot product) of a state vector with itself? Let's look at $|A\rangle = a_0 |A_0\rangle + a_1 |A_1\rangle$ and write the scalar product with itself as

$$\langle A | A\rangle \Rightarrow \begin{pmatrix} a_0 & a_1 \end{pmatrix} \begin{pmatrix} a_0 \\ a_1 \end{pmatrix} = a_0^2 + a_1^2. \tag{4.13}$$

We see that the scalar product of a vector with itself is the *square* of the length of the vector. For quantum state vectors, we will almost always use $a_0^2 + a_1^2 = 1$ as mentioned previously. This is just the Pythagorean theorem with the length of the hypotenuse equal to 1. If the square of the length is 1, then the length itself is 1, and we say that the vector is "normalized."

Similarly, if we have two state vectors, each with length 1 and if the two state vectors point in the same direction in state space, then the Dirac bracket will be

$$\langle C | B\rangle = c_0 b_0 + c_1 b_1 = 1. \tag{4.14}$$

Comparing Eqs. (4.11) and (4.14), we see that the Dirac bracket ranges from 0 (when the two normalized vectors are orthogonal) to 1 (when the two vectors are parallel). If the vectors lie along the same line, but point in opposite directions, then we get $\langle C | A\rangle = -1$. We shall see in Chapter 8 that the Dirac bracket is the same as the cosine of the angle between the two vectors in state space.

Try It 4.5

When the two state vectors are orthogonal, the angle between them in state space is 90°, and when they are parallel the angle between them is 0°. Look up $\cos 90^\circ$ and $\cos 0^\circ$ to check the previous paragraph's statement about the numerical range for the Dirac bracket for two normalized state vectors.

   CHAPTER SUMMARY

- Qubits are actual physical objects whose properties we can measure. For qubits, there are two possible results for the measurement outcomes.

- The associated quantum states are represented by state vectors in an abstract two-dimensional state space analogous to a classical vector in physical space with x and y spatial coordinates. A state vector can be represented by right vector symbols $|A\rangle$, by arrows in a state space diagram, and by column vectors.

- The quantum state carries all the possible information—the probabilities—about the outcomes when you measure that state.

- Quantum algorithms employ quantum states interacting with gates and measurement devices. The measurement devices make use of the qubit's physical properties.

- Unlike classical states, a general quantum state vector can be represented by a linear combination (superposition) of basis states. A state of superposition means the qubit is described by both basis states weighted with numerical coefficients (amplitudes).

- For the computational basis states, the superposition is written as

$$|S\rangle = a_0 |0\rangle + a_1 |1\rangle . \tag{4.15}$$

- Linearly polarized photons are examples of qubits. Their basis states are vertical linear polarization $|vlp\rangle$ and horizontal linear polarization $|hlp\rangle$.

- Polarizing sheets and polarizing beam splitters can be used both to prepare and to analyze linear polarization in a light beam. This dual use occurs often in quantum systems.

- Often a quantum algorithm will require state vectors to be multiplied together. The dot product (inner product or scalar product) of two state vectors is represented by the Dirac bracket $\langle A|\ B\rangle$ between a left vector and a right vector, which can be evaluated by finding the product of the row vector corresponding to $\langle A|$ and the column vector corresponding to $|B\rangle$. The result of this multiplication will be a single number, and if that number is zero, then the state vectors are orthogonal.

- The dot product of a state vector with itself, for example $\langle A|\ A\rangle$, is the square of the length of that state vector in state space. Normalized state vectors have a length equal to 1.

- The dot product of two normalized state vectors (each have length 1) is equal to the cosine of the state space angle between the two vectors in state space: $\langle C|\ A\rangle = \cos\theta_{AC}$.

 FURTHER READING

Michael G. Raymer, *Quantum Physics: What Everyone Needs to Know* (Oxford University Press, Oxford, 2017). This book has a nice section on linearly polarized light.

Linear polarizing sheets and 3D movie glasses are readily available online. Some suggested sources (the authors have no commercial connections to any of these sources) are Amazon (www.amazon.com), American Polarizers, Inc. (https://www.apioptics.com/product/ap38-006t/), Edmund Optics (https://www.edmundoptics.com), and Arbor Scientific (https://www.arborsci.com/products/polarizing-filters).

Schrödinger's grave marker has been updated since the photo in Figure 4.2 was taken. The circular plaque now has Schrödinger's equation $i\hbar\,\partial\psi/\partial t = H\psi$, which is a differential equation for the wave function ψ. The name plate also lists his wife, who, in fact is not buried there. To see the new marker, do a web search for "Schrödinger's grave."

5 Quantum Measurements

It is much easier to make measurements than to know exactly what you are measuring.

J. W. N. Sullivan (1928)

5.1 Quantum Measurements: What Are They?

ALICE: We are now ready to tackle quantum measurements, an integral part of quantum information science and quantum computing. A quantum measurement occurs when a qubit interacts with a measurement device designed to give you information about the quantum state of the qubit. Ideally, the measurement device is equipped with signals such as light bulbs or sound generators or electrical outputs that tell you which output channel has the outgoing qubit. For most purposes in QIS and QC, we don't need to worry about the details of the measurement devices, but for those who want to build quantum computers they are a big deal.

Let's start with a concrete example: linearly polarized light and a polarizing beam splitter (PBS) as shown in Figure 5.1. We have added photon detectors to the PBS setup. If a detector senses a photon, it will activate a light, indicating that the device has made a measurement. Just as in Chapter 4, we denote the linear polarization state associated with the incoming photon as the superposition of the two linear polarization basis states:

$$|S\rangle = a_{\mathrm{h}} |\mathrm{hlp}\rangle + a_{\mathrm{v}} |\mathrm{vlp}\rangle. \tag{5.1}$$

BOB: Let's unpack the measurement process illustrated in Figure 5.1. We read the diagram from left to right. A photon described by a quantum state $|S\rangle$ interacts with a linear polarization measurement device, the PBS. The PBS orientation angle is indicated by the reading of a display on the device. We say that the two linear polarization states (horizontal and vertical) are the "measurement basis states."

Suppose we have added the appropriate equipment to be able to detect single photons. If a single photon interacts with the measurement device, the output will indicate either red or blue, but never both at the same time. Those are fundamental principles of quantum mechanics. The qubit is not divisible, and vlp and hlp are mutually exclusive outcomes. The output of the measurement device for a single photon will be one or the other (but never both) of the two output states (the measurement device basis states), here the vertical (vlp) or horizontal (hlp) linear polarization states. For this measurement device, the blue square or the red square will light up to let us know which state was observed.

Quantum Computing: From Alice to Bob. Alice Flarend and Bob Hilborn, Oxford University Press.
© Alice Flarend and Robert C. Hilborn (2022). DOI: 10.1093/oso/9780192857972.003.0005

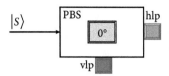

Fig. 5.1 A quantum measurement device for linearly polarized light. A photon qubit described by the state vector $|S\rangle$ interacts with a polarizing beam splitter (PBS), a linear polarization measurement device. The measurement device is set to sort out horizontal and vertical polarization relative to a "horizontal" axis at $0°$ indicated by the central display. If the qubit is detected in the vlp beam, the red light will be lit. If it is detected in the hlp beam, the blue light will be lit.

If we don't detect the photon but let it travel on from the PBS, the state that should be used to describe the outgoing photon is the device's basis state associated with the output channel the photon takes. This assertion is called the *measurement postulate* of quantum mechanics, though in some ways it is misnamed. We will explain that in a moment. According to the measurement postulate, the outgoing qubit state vector is one or the other of the basis states of the device. For linearly polarized light, after interacting with the PBS, the photon's state is either $|hlp\rangle$ or $|vlp\rangle$, but never both. As we mentioned in Chapter 4, the PBS has separated the stream of photons into two beams, but since we have not detected which beam a particular photon is in, we have not yet completed a quantum *measurement*. This may seem a bit fussy (and it is) but it highlights the importance of qubit detection in quantum measurements.

Measurement Postulate

Immediately, after a qubit interacts with a measurement device, the qubit's quantum state is the basis state associated with the device's output channel in which the qubit is found.

CARDY: But how do you know that is true?

ALICE: Great question! There are two parts to the answer. First, we need to understand the physics of the measurement device. For polarized light, in particular, we need to know how the device interacts with light and how that interaction depends on the polarization state of the photon. For our purposes, we will assume that work has been done and we know the physics. The second part of the answer is that you can check which state we have by letting the photon interact with a second linear polarization measurement device. (Of course, we need to be sure that the photon does not interact with anything that might change its state before it gets to the second device.) If the photon's quantum state after the first PBS is $|vlp\rangle$ for example, then the photon interacting with the second PBS will, assuming the two PBSs have the same settings, always end up in the second's vlp output beam and never in its hlp beam. If it *always* ends up in the vlp beam of the second PBS, then we know that its state before interacting with the second measurement device must have been $|vlp\rangle$.

BOB: That point raises another important issue mentioned in Chapter 4. If the qubit interacts with the measurement device and the measurement device indicates vlp, we really can't say much of anything about the qubit's state *before* it interacted with the measurement device. The state before might have been vlp or it could have been a superposition state $|S\rangle = a_h |hlp\rangle + a_v |vlp\rangle$. The only thing we know is that the state wasn't purely hlp, because if it had been we would never would have observed a vlp result. By way of contrast, the quantum "gates" we will meet up with in Chapter 6 are "reversible" in the sense that if we know the output state, we know what the input must have been. That's why we will make the distinction between "measurement devices" and "quantum gates."

ALICE: There are many subtleties associated with measurements in quantum mechanics. For now, we assume that we have a practical understanding of how a measurement device operates based on classical physics (which is all we need to describe how a polarizing beam splitter works). We will return to the subtleties later.

BOB: As an experimentalist, I want to point out that if we don't absorb the photon in a light detector after the PBS in Figure 5.1, we can let the photon interact with further quantum gates and measurement devices, which lie along one path or the other before we finally detect the photon. In essence, we have "tagged" the photon by the path that it follows and we can infer that if the photon is eventually detected in the vlp path, it had vlp when leaving the first PBS and, conversely, hlp for the other path.

Try It 5.1

Suppose that an unpolarized beam of photons interacts with a PBS with its "horizontal" direction set at $0°$. The hlp output beam from the PBS continues on and then interacts with a PBS whose "horizontal" direction is set at $90°$. What will be the output from the second PBS?

5.2 Light Intensity from a Two-Polarizer Setup

ALICE: To tease out the physical meaning of the state amplitudes in Eq. (5.1), let's use a polarizing sheet to produce a beam of linearly polarized photons. We then employ a PBS as a measurement device and measure how much of the linearly polarized light beam emerges in each of the PBS's output beams. We then ask a critical question: What happens if we rotate the PBS relative to the polarization transmission axis of the polarizing sheet?

Cardy, how do you think the two intensities (the number of photons per second) will vary as we change the angle between the polarizing sheet and the PBS direction as shown in Figure 5.2?

CARDY: Well, if $\theta = 0$, the transmission directions are aligned and that means that whatever comes through the first polarizing sheet will also come out in the hlp beam I_{2h} from the PBS. When the angle is $90°$, I expect nothing will be in the hlp beam from the PBS; so all the light must come out in the vlp beam I_{2v}. But I am not sure what happens in between.

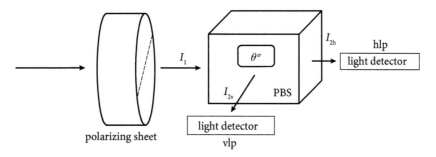

Fig. 5.2 A light beam on the left approaches a polarizing sheet with its transmission axis indicated by the dashed line. The beam with intensity I_1 then encounters a polarizing beam splitter (PBS) with its hlp transmission axis rotated by an angle θ relative to the transmission axis of the polarizing sheet. Light detectors produce signals proportional to the intensity (number of photons per second) of the light beams after they emerge from the PBS.

ALICE: That's a good starting analysis, Cardy. Just like a physicist, you went for the extremes of the angles where the analysis is more clear-cut. The middle is almost always a little messier and more complicated.

BOB: We find from actual experiments with light beams (which have lots of photons) that the intensity (power) of the light transmitted by the PBS in its hlp beam is given by

$$I_{2h} = I_1 \cos^2\theta \tag{5.2}$$

and in the other output beam

$$I_{2v} = I_1 \sin^2\theta, \tag{5.3}$$

where θ is the angle between the transmission axis of the sheet and the hlp direction of the PBS. I_1 is the intensity of the light beam before it interacts with PBS.

CARDY: I am a bit rusty at trig. Could you give me a graph or something like that?

ALICE: The graphs are shown in Figure 5.3. Let's check that the results given in the figure make sense for various cases. For $\theta = 0$, the two transmission axes are parallel and we would expect, as Cardy said, that everything that gets through the first sheet is also transmitted by the beam splitter in the horizontal polarization output beam (I_{2h}) and none in the vertical polarization output beam (I_{2v}). That is consistent with $\cos(0) = 1$ and $\sin(0) = 0$ in Eqs. (5.2) and (5.3). If we rotate the PBS by $\theta = 90°$, the beam splitter's hlp output gets no light since the hlp's polarization direction is now perpendicular to the polarizing sheet's transmission axis. We expect that none of the light will get through in that beam and all the light shows up in the vlp beam. For $\theta = 45°$, 50% of the light goes into one PBS output beam and 50% into the other, which is consistent with the square of $\cos 45°$.

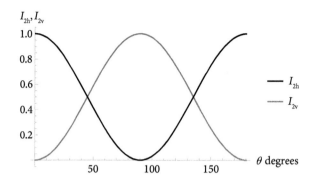

Fig. 5.3 A plot of the relative intensities I_{2h}/I_1 and I_{2v}/I_1 in the two output beams for the PBS as a function of the angle between the polarizing sheet transmission axis and the PBS axis. See Eqs (5.2) and (5.3) for the definitions of I_{2h} and I_{2v}.

Try It 5.2

Explain how those results are consistent with the measurement postulate of quantum mechanics. In particular, look at the cases $\theta = 0°$ and $\theta = 90°$.

5.3   State Amplitudes and Probabilities

BOB: Let's now assume we have the equipment (the detectors in Figure 5.2) needed to detect (and count) the number of photons in each beam (during a specified period of time). The equipment to detect individual photons is readily available in most college and university physics departments. Unfortunately, the equipment is too expensive to buy for yourself. (We give links to several videos about detecting individual photons in the Further Reading section.)

We will go through this process in some detail because it gives us a model we can use for getting information about the quantum state of other qubits. Using the setup illustrated in Figure 5.2, Alice prepares a beam of light in a linearly polarized state using the first polarizing sheet but she doesn't tell me the transmission axis direction. Our task is to find the polarization state of Alice's beam using the output of the detectors in Figure 5.2 relative to the PBS vertical and horizontal polarization directions. That is, we want to find the numerical values of the state amplitudes in Eq. (5.1).

Let's suppose a photon is detected in the vlp output beam of the polarization measurement device (the PBS). What does that result tell us about its state $|S\rangle$ before it encountered the PBS? We now know that its state $|S\rangle$ was not *purely* hlp. In terms of the amplitudes, we can say that a_h does not equal 1 and a_v does not equal 0. If the state $|S\rangle$ had been purely hlp, no photons would end up in the PBS vlp channel. But other than that, having detected just one photon, that is all we can say.

We can learn more about the state $|S\rangle$ by asking Alice to prepare more photons with that same state and send them towards our polarization measurement device. If we detect a photon in the vlp channel, we write down a "V." If it is detected in the hlp channel, we write down "H."

After a long time (depending on how rapidly Alice can fire photons at us), we end up with a long string of Vs and Hs: VHHVVVHVHHH ... with no apparent pattern. The sequence looks completely random. Let's count n_v, the number of Vs, and n_h, the number of Hs, in the list. The total number of photons detected is $n_v + n_h = N$.

The fraction of Hs, that is, n_h/N, is the probability that one of Alice's photons ends up in the hlp beam. Then we say that for Alice's photon state $|S\rangle$, the probability of detecting a horizontally polarized photon is

$$P_h \approx n_h/N. \tag{5.4}$$

Similarly, the probability for detecting a vertically polarized photon is

$$P_v \approx n_v/N. \tag{5.5}$$

We use approximate equality because with a finite number of photons received, the ratios are *estimates* of the probabilities (in the language statisticians like to use). For example, if we repeatedly flip a coin, we almost never exactly have the same number of heads and tails. But we expect to find $P_{tails} \approx 1/2$ and $P_{heads} \approx 1/2$.

Since there are only two possible outcomes for the photon polarization state (hlp and vlp), the two probabilities must add up to 1:

$$P_v + P_h = 1. \tag{5.6}$$

Eq. (5.6) is just like Chapter 4's normalization condition for state amplitudes $a_h^2 + a_v^2 = 1$. That suggests that we may interpret the square of each of the amplitudes individually as the probability for those particular outcomes and the square of the length of a state vector as the sum of the probabilities of the outcomes:

$$a_h^2 = P_h \qquad a_v^2 = P_v, \tag{5.7}$$

where P_h is the probability that the photon ends up in the hlp beam and P_v is the probability for its ending up in the vlp beam.

Try It 5.3

Supposing the two amplitudes are $a_h = 1/2$ and $a_v = \sqrt{3}/2$, what are the probabilities for measuring the photon with horizontal and vertical polarization? Answer the same question with $a_h = 1/4$ and $a_v = \sqrt{15}/4$.

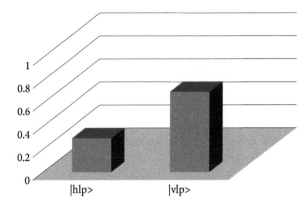

Fig. 5.4 A column chart for a single linearly polarized photon qubit. In this case, the probability of finding the photon in the hlp state is 0.3, and 0.7 for the vlp state.

BOB: Figure 5.4 shows a column chart, which is a common way of displaying the probabilities for a qubit quantum state. We deploy this kind of chart again in Chapter 10 when we work with multi-qubit states.

ALICE: Recognizing that the squares of the state amplitudes give us measurement probabilities is perhaps the most important result in quantum mechanics. That statement is called the *probability postulate of quantum mechanics*. This postulate is also known as the *Born postulate* (or *Born rule*). It was introduced by Max Born in 1926, early in the development of quantum mechanics.

Probability Postulate of Quantum Mechanics

The square of a basis state amplitude is the probability of the qubit's interaction with the measurement device yielding a result associated with that basis state.

There is a crucial conceptual issue here, which we alluded to before. Note that the probability interpretation does *not* say that the qubit had that value of the measured property before the interaction with the measurement device. As David Mermin has succinctly written in *Quantum Computer Science* (Mermin, 2007, p. 40) "[there is a] mistaken view that the quantum state encodes a property inherent in the Qbits [qubits]. The state encodes only the possibilities available for the extraction of information from those Qbits."

BOB: Did you know that Max Born's granddaughter is the singer Olivia Newton John? I find it ironic that she has "Newton" as part of her name given that her grandfather was a leader in the revolution that showed the limits of Newtonian physics.

ALICE: I don't know if people of Cardy's age even have a clue who that singer is unless they have seen the movie *Grease*! Let's get back to quantum measurements. We have argued that making a long series of observations (the Hs and Vs) allows us to determine the *square* of the quantum state amplitudes. Can we find the "un-squared" amplitudes a_h and a_v? The state amplitudes are sometimes called "probability *amplitudes*" to distinguish them from the probabilities themselves. It seems easy enough. All we have to do is take the square root. But the difficulty is that there are two possible square roots: $\sqrt{a_h^2} = \pm a_h$. How do we know which is correct?

In other words, to get the complete quantum state (that is, to "measure" the quantum state), we also need to find out if the coefficients are positive or negative. In later chapters, we will see how to do that. Finding those signs (+ or −) is called finding the "relative phase" of the coefficients. Those relative phases turn out to be important in QC and QIS. In Chapter 15, we will show you what happens when the amplitudes are complex numbers, something that is necessary in full-blown quantum mechanics. Until then, all we will need are amplitudes that are positive or negative real numbers or 0.

Let's pause for a moment to review what we have learned. We "measure" a quantum state by determining the numerical values of the state vector coefficients. This almost always requires repetitions of (1) preparing the qubit in a specified quantum state (though we may not know what the state is) and (2) observing the result of having the qubit interact with a measurement device and recording the channel into which the qubit goes (e.g. horizontal polarization versus vertical polarization for a photon, or spin-up versus spin-down for a spin system). The collection of individual observations then constitutes (at least a part of) a state measurement. Those repetitions give information that can be used to find the state vector coefficients.

BOB: Here is one of the situations in which the terminology Alice used can help avoid confusion. I like to say that for each interaction of a qubit with the measurement device we are making an *observation*. Recognizing the limits on what we have learned from a single observation, we know we need to repeat the observations many times with identically prepared qubits to be able to say anything about the probabilities for getting the various outcomes from the measurements.

An analogy might be helpful. If you flip an unknown coin and it comes up with a "head," you can't say much about the coin except that the probability of observing a head is not zero. You don't know if both sides of the coin are "heads," or if the coin is weighted in such a way to make observing a head much more likely than observing a tail, or if the probabilities for heads and tails are the same. To *measure* the probability of getting heads for that coin, we need many *observations* of individual coin flips.

CARDY: Photons! Heads or tails! I am beginning to feel light-headed. Ha ha!

ALICE: I like that one, Cardy. From a single observation of a qubit, as Bob pointed out, you learn only a little about what the qubit's quantum state was before the observation. That doesn't seem too bad, but it has an implication that is extremely important for QIS and QC. The implication is that we cannot clone a quantum state because, after one observation, we have only limited information about the amplitude associated with the basis state for that observation. We have no information about the amplitude associated with the other basis state. The result is often called the *no-cloning theorem*.

The no-cloning theorem does not prevent us from preparing several qubits with the same quantum state. What it does prohibit is having someone else learn all the details of a qubit's unknown quantum state from one observation on that qubit alone. In other words, what they do learn from an observation is insufficient to allow then to prepare another qubit in that same state or to restore the observed qubit to its state before the observation. By "unknown," we mean that we do not know the state amplitudes and relative signs (or phases). If we knew them, of course, we could prepare another qubit in that state by the preparation procedures we will learn about later, or we could restore the observed qubit's original state.

In Chapter 8, we will see how to use the no-cloning theorem to detect if someone is eavesdropping on a quantum information channel. In Chapter 10, we will show how it is intimately tied to the fundamental concepts of quantum states.

As an example, let's go back to photons. Suppose we have prepared the qubit state to be one with linear polarization at $45°$ with respect to the axis of a PBS. It is important to remember that for any single photon observation, we find the photon in either the vlp beam or the hlp beam of the PBS. As far as we know, according to the rules of quantum mechanics, there is no way to predict which beam a particular photon ends up in (except for the special case in which the photon happens to have been prepared with a state corresponding to vlp or hlp). All that quantum mechanics gives us is the probability for the photon's ending up in one or the other of the two measurement output states. In the $45°$ polarization state case, the probabilities are 0.5 for vlp and 0.5 for hlp.

BOB: Some authors write about "measuring" the quantum state of a photon by simply putting a PBS in the light beam path (or analogous devices for other types of qubits). That may be necessary, but it is not sufficient. We haven't measured anything until we know which path a particular photon takes and that requires that we detect the presence of that photon. Usually this means detecting the photon by absorbing it with a photodetector.

If we let the photon exit the measurement device without detection, all we can legitimately say is that *if* the photon is in the vlp beam, the appropriate quantum state vector is $|vlp\rangle$ and *if* it is in the hlp beam, the appropriate quantum state vector is $|hlp\rangle$; but without further detection, we don't know which has actually occurred.

Unfortunately, the term "measure" is abused in many descriptions of quantum computing. You will often read that "a measurement collapses the quantum system's state vector," or that "the measurement device causes the qubit to *jump* from the state $|S\rangle$ to the state $|vlp\rangle$" (for example). At best, these statements try to give you a mental picture of what happens in a quantum measurement. At worse, they muddle thinking about quantum mechanics. Alice and I prefer to say that after the qubit interacts with a measurement device, we have new information about its state. Once the qubit leaves the device, its state is now one or the other of the device's basis states and so it is appropriate to use that information to describe the qubit's state. We are simply updating our knowledge about the qubit's state.

From this point of view, thinking about measurements reenforces the notion that state vectors are not things. For those who insist that state vectors are real things, wouldn't all those collapsing state vectors make a terrible noise? And what a mess to clean up!

Many QIS and QC people avoid the physics of what happens in measurement devices and say simply "We measure a qubit's state and get 0 or 1," without worrying about the many technical difficulties of carrying out measurements on quantum systems and how those measurements are reflected in our formal description of quantum states.

ALICE: There is another way of thinking about quantum measurements. Instead of repeatedly preparing a state and recording the outcomes of the observations, we could think of preparing identically at one time the states of a large number of qubits and then observing them all at once. The qubits are identically prepared in the sense that the quantum states of all the qubits are the same. If the qubits do not interact, then we may assume that the results of observing the system of many qubits with measurement devices are the same as those obtained from a sequence of qubits measured one at a time. Either way, from those observations we can calculate ("estimate") the probabilities of observing the qubit in each of the measurement device's basis states.

No matter what terminology you use, measurements are key to all applications of quantum mechanics, including QIS and QC. The theoretical physicist Julian Schwinger, who won the Nobel prize for his development of relativistic quantum electrodynamics, characterized quantum mechanics as "the symbolism of atomic measurements" (Schwinger, 2001). By that he meant that quantum state vectors carry information about quantum measurements, not about inherent properties of the quantum objects.

CARDY: Can't we do better than just probabilities? Can't we know the "actual state" of the object?

ALICE: As far as we know, those probabilities are a manifestation of how nature works at the quantum level. Probabilities show up in almost all quantum situations. For example, if I shine a brief pulse of light from a laser on a collection of atoms, the atoms will make transitions from a lower energy state to a higher energy state as they absorb the light. Later, each atom will emit a photon as it makes a transition from the higher energy state to a lower energy state. But some of the atoms will emit their photons shortly after the laser pulse and others will do so later. We can use quantum mechanics to calculate the probability per unit time that an atom will emit a photon, but there is no way, as far as we know, to predict when any particular atom will emit its photon.

Try It 5.4

You can observe this spread-out-in-time emission of light directly. If you have a TV set turned on in a darkened room, hit the on/off button (or the equivalent remote control button) and watch the screen. It will continue to "glow" for a short period of time after the set turns off. The light-emitting pixels in the TV screen take some time to "decay" back to their "off" states. Some do that more quickly than others. This is an example of quantum randomness. Another example is the clicking of a Geiger counter when it is exposed to radioactive material. Some of the atomic nuclei in the radioactive material emit subatomic particles, which are registered by the Geiger counter. But the decays occur randomly, hence the erratic clicking of the counter.

Try It 5.5

Suppose Prof. Blah claims that her measurements indicate that $\cos\theta$ and $\sin\theta$ should replace $\cos^2\theta$ and $\sin^2\theta$ in Eqs (5.2) and (5.3). Draw graphs of both the original results and Prof. Blah's results. How might you distinguish between them? See Figure 5. 5. Are there theoretical reasons that might rule out Prof. Blah's results? Hint: what happens to $\cos\theta$ for angles between $90°$ and $180°$?

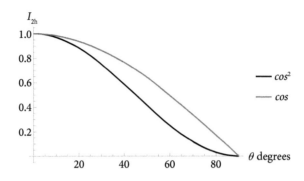

Fig. 5.5 A plot of $\cos^2\theta$ and $\cos\theta$.

BOB: We have seen that a quantum state vector encodes information about the results of measurements made on the qubit (the physical device). But we need to remind ourselves that in the quantum world, we can't legitimately say that the measurements tell us anything definite about the dynamical properties (properties that can change) associated with the qubit other than the probabilities of various measurement outcomes. For example, as we discussed before, if we prepare a beam of photons with linear polarization at $45°$ with respect to a PBS's axes, the photon state can be expressed as a superposition of the hlp and vlp states associated with the PBS directions. Then half the photons come out in the vlp beam and half in the hlp beam. We are tempted to say that before they met the measurement device the photons that come out in the vlp beam must have had some property different from those that come out in the hlp beam. That seems like common sense. But as far as we know, there are no such "hidden properties." The photons are randomly distributed between the two output beams. Pretty scary, huh?

CARDY: But if randomness rules at the quantum level, doesn't that mean that the world is not deterministic, effect does not follow cause, and we don't know for sure if the sun will rise tomorrow? Doesn't that show that the answers produced by a QC will be random? How do we know what the actual answer is?

ALICE: You have hit exactly on some of the most puzzling aspects of quantum mechanics. In the subject's early days, a few scientists tried to "explain" this randomness by saying that there are "hidden variables"—properties of the quantum system that are hidden from us. Atoms that emit photons at different times have different values of those hidden variables. That certainly sounds plausible. However, since the 1980s, experiments have shown that almost all plausible hidden variable theories are ruled out. Quantum randomness reigns supreme!

5.4 State Preparation

CARDY: Before we end up for the day, I have another question: how do we prepare qubits in the quantum states we want them to have?

BOB: You packed a lot of important and somewhat controversial issues into that question! Let's see if we can untangle them. First, there is the fundamental issue: Are qubits "in" quantum states, do they "have" quantum states, or are they simply "described by" quantum states?

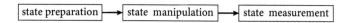

Fig. 5.6 A schematic representation of many QIS and QC processes.

As we will soon discover, there are conceptual traps if we claim that the qubit "has" or is "in" a quantum state. So, we will stick with the notion of "description."

Now, to state preparation itself. There are two parts to the process. We will see how this plays out in QIS and QC throughout the book. Almost always, we prepare a qubit so its corresponding quantum state is one of the basis states for the appropriate physical property (or collection of properties) we are concerned with. For example, for photon polarization, we might start with $|vlp\rangle$ by having the photon pass through a PBS and keeping only the photons that are in the vlp outgoing beam.

In many QIS and QC situations, we then manipulate that basis state to convert it into a desired superposition state. (We will see how to do this in Chapter 6.) The important point is that preparing an initial state is in some cases just like a measurement process. The physics of the state preparation device is often the same as the physics of the measurement device except that we are producing photons, say from a laser or an LED or a flame, and using a polarizing device to keep only those that ended up in the desired polarization beam. So, in quantum physics, state preparation and state measurement are two sides of the same coin. After a measurement, we know almost nothing about the state before the measurement. Similarly, preparing a qubit in a state wipes out information about the state before the preparation operation. Figure 5.6 illustrates the overall process:

ALICE: Bob is right, of course, but this kind of conceptual analysis is often taken for granted in QIS and QC, where we assume we know how to do state preparation and state measurement and we get on with the real work of building QCs and quantum communication devices and inventing algorithms to carry out quantum information tasks. But, keep in mind there are many subtleties in both state preparation and state measurement. We will return to these subtleties in Chapter 14.

CHAPTER SUMMARY

- Quantum measurement is different from classical measurement because it does not generally yield a deterministic result. If you measure the velocity of identically prepared classical (non-quantum) objects, you will get the same value each time, within experimental uncertainty. If you measure identically prepared quantum objects described by a superposition state, the outcomes vary randomly from one observation to the next. You will not get the same answer each time, regardless of experimental uncertainty.

- When a quantum measurement is made, the outcome will be the value of the physical property associated with one of the measurement device's basis states. If the qubit is not absorbed during the observation, the quantum state of the qubit after the interaction is the basis state of the device associated with the observed outcome. All other information about its prior state is lost.

- The numerical coefficients in a superposition quantum state are measurement outcome probability amplitudes—not probabilities.

- The probabilities of observation outcomes are given by the squares of the coefficients of the basis states in the superposition state. This result is is known as the Born rule or the probability interpretation of quantum mechanics. For the state $|S\rangle = a_0 |0\rangle + a_1 |1\rangle$, the probability of getting the state $|0\rangle$ as a result of a measurement is $P_0 = a_0^2$ and $P_1 = a_1^2$ for the $|1\rangle$ state.

- Classical computers are devices for which the amplitudes of the basis states are either 0 or 1. A wire in a classical computer is, with certainty, in one of the states, but not the other.

- Because information about the original state is lost upon measurement, the observer cannot make a clone of the original quantum state with only one observation. This is the no-cloning theorem.

FURTHER READING

Michael G. Raymer *Quantum Mechanics: What Everyone Needs to Know* (Oxford University Press, Oxford, 2017). This book has clear and thoughtful discussions of quantum measurements, along with excellent introductions to other aspects of quantum physics—all done with almost no mathematics.

N. David Mermin, *Quantum Computer Science: An Introduction* (Cambridge University Press, Cambridge, 2007).

Julian Schwinger and B.-G. Englert (editor), *Quantum Mechanics: Symbolism of Atomic Measurements* (Springer, Berlin, Heidelberg, and New York, 2001). The title indicates that Schwinger emphasizes that quantum mechanics is a theory of quantum measurements and their correlations. This text is written at the upper-level undergraduate or graduate-student level.

ONLINE VIDEOS

These videos show experimental evidence that light is composed of individual quanta called photons:

https://www.youtube.com/watch?v=GzbKb59my3U.

https://www.youtube.com/watch?v=I9Ab8BLW3kA.

6 Quantum Gates and Quantum Circuits

Probable impossibilities are to be preferred to improbable possibilities.

Aristotle, *Poetics*

6.1 What is a Quantum Gate?

ALICE: To do quantum computing, we need devices that will operate on and change the quantum state associated with a qubit (the physical object). In quantum mechanics, those devices are called *operators*. However, in analogy to the language used to describe the manipulation of bits in classical computing, we will call the corresponding quantum devices quantum *gates*. For example, we might think about a system with two qubits as input to some device, which then produces some output. A generic quantum gate is shown in Figure 6.1.

We read Figure 6.1 from left to right. We say that $|S_A\rangle$ is an input qubit state, prepared by Alice for example, while $|S_B\rangle$ is an input qubit state prepared by Bob. The gate operates on those states to produce output states $|O_1\rangle$ and $|O_2\rangle$. A combination of such gates is a quantum circuit.

CARDY: Cool. I've always wanted to prepare a quantum state. But how do I actually do that?

BOB: Remember, we talked about that issue in Chapter 5, where we saw that state analyzer devices such as PBSs can also be used as state preparation devices. Although the details depend on the physical nature of the qubits being used, the state analyzers result in states that are the basis states associated with the device. In this chapter we will first see how to use quantum gates (operators) to manipulate those states to produce other states that eventually will be used in QC and QIS. We will also return to the distinction between operators and state analyzer or measurement devices.

The horizontal lines in Figure 6.1 are called "quantum wires" in analogy with the regular wires used to bring voltage signals to gates and integrated circuits in traditional computers. In most QC cases, they are just a visual indication of how a qubit interacts with a series of gates, not physical wires.

You should read the quantum circuit as a timeline, proceeding from left (earlier in time) to right (later in time). The quantum wires "carry" the quantum state vector information. The gate acts on those quantum states and produces output states shown on the quantum wires on the right side of the diagram. Those wires (that is, the associated quantum states) may then go on to other gates for further processing.

Quantum Computing: From Alice to Bob. Alice Flarend and Bob Hilborn, Oxford University Press.
© Alice Flarend and Robert C. Hilborn (2022). DOI: 10.1093/oso/9780192857972.003.0006

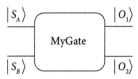

Fig. 6.1 A generic two-qubit quantum gate. There are two input qubits whose quantum states are shown on the left and two output quantum states (on the right).

CARDY: How does the quantum gate know what those states are and how the input states are to be related to the output states? How do we carry out computational tasks in this scheme?

BOB: Those are in fact the fundamental questions of QC; we will address them step by step as we go along.

6.2 Single-Qubit Quantum Gates

ALICE: Let's start with some quantum gates that manipulate single-qubit quantum state vectors. The gate shown in Figure 6.1 is called a two-qubit gate, with two inputs and two outputs. We will tackle those kinds of gates in Chapter 9. In this chapter, we will focus on how gates manipulate the computational basis states $|0\rangle$ and $|1\rangle$. From there we can figure how they process any other single-qubit state.

CARDY: Oh, yeah. I remember that any quantum state can be represented as a superposition of the computational basis states.

ALICE: Perhaps the simplest quantum gate is the NOT gate. Recall that in binary logic, the NOT gate changes a 0 to a 1 and vice versa. A quantum NOT gate changes the basis state vector $|0\rangle$ into the other basis state vector $|1\rangle$ and vice versa. We represent that process graphically as a quantum circuit (Figure 6.2):

ALICE: We can write these results in a "logic (truth) table" just the way we did for a classical computing NOT gate (Table 6.1).

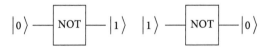

Fig. 6.2 Quantum NOT gate.

Table 6.1 The input/output table for a quantum NOT gate.

NOT gate input	NOT gate output
$|0\rangle$	$|1\rangle$
$|1\rangle$	$|0\rangle$

It turns out to be useful to represent the quantum gate by a matrix, which multiplies the column vectors representing the basis states. For the NOT gate, the matrix is the following:

$$\text{NOT} \Rightarrow \begin{pmatrix} 0 & 1 \\ 1 & 0 \end{pmatrix} \tag{6.1}$$

and the action on the basis states is

$$\text{NOT} \,|0\rangle \Rightarrow \begin{pmatrix} 0 & 1 \\ 1 & 0 \end{pmatrix} \begin{pmatrix} 1 \\ 0 \end{pmatrix} = \begin{pmatrix} 0 \\ 1 \end{pmatrix}$$

$$\text{NOT} \,|1\rangle \Rightarrow \begin{pmatrix} 0 & 1 \\ 1 & 0 \end{pmatrix} \begin{pmatrix} 0 \\ 1 \end{pmatrix} = \begin{pmatrix} 1 \\ 0 \end{pmatrix}. \tag{6.2}$$

Reading Eq. (6.2) from left to right, we see that the 2×2 NOT matrix acts on the column vector for a basis state vector to produce the other basis state vector. In QC, the NOT matrix is usually called the X matrix:

$$X = \begin{pmatrix} 0 & 1 \\ 1 & 0 \end{pmatrix}. \tag{6.3}$$

CARDY: I need a reminder about a matrix multiplying a column vector.

BOB: We introduced that idea in Chapter 3. A matrix multiplying a column vector gives a new column vector. Here is the multiplication rule:

$$\begin{pmatrix} a & b \\ c & d \end{pmatrix} \begin{pmatrix} e \\ f \end{pmatrix} = \begin{pmatrix} ae + bf \\ ce + df \end{pmatrix}. \tag{6.4}$$

As a mnemonic, you can think of the matrix on the left of Eq. (6.4) as a stack of two row vectors $(a\ b)$ and $(c\ d)$. The *top* entry in the column vector on the right after the equal sign of Eq. (6.4) is just the product of the matrix's *top* row vector multiplying the column vector on the left of Eq. (6.4). The *bottom* entry on the right is the product of the *bottom* row vector on the left times the column vector on the left. Neat?

CARDY: Yes, it does help me remember the way to multiply a column vector by a matrix.

Try It 6.1

Check that multiplying a generic qubit quantum state vector $|S\rangle = \alpha_0 |0\rangle + \alpha_1 |1\rangle$ by the NOT Matrix gives a state vector with the amplitudes swapped between the two basis states.

ALICE: In addition to NOT, another simple but surprisingly useful matrix is called the identity matrix. It is the matrix equivalent of multiplying a mathematical object by the number one: You just get the same mathematical object as output. The identity matrix, often indicated by I, is the following:

$$I = \begin{pmatrix} 1 & 0 \\ 0 & 1 \end{pmatrix}. \tag{6.5}$$

Try It 6.2

Check that multiplying the computational basis states (in column vector form) by the identity matrix gives the states back again.

Try It 6.3

Check that multiplying a generic qubit quantum state vector $|S\rangle = \alpha_0 |0\rangle + \alpha_1 |1\rangle$ by the identity matrix gives the same state vector back again.

The X matrix and the identity matrix are part of a set of four matrices, called the Pauli matrices, commonly used in quantum mechanics. They are named after Wolfgang Pauli (1900–58), one of the pioneers in quantum mechanics. The other two are called the Y and the Z matrices:

$$Y = \begin{pmatrix} 0 & -1 \\ 1 & 0 \end{pmatrix} \tag{6.6}$$

and

$$Z = \begin{pmatrix} 1 & 0 \\ 0 & -1 \end{pmatrix}. \tag{6.7}$$

CARDY: I vaguely remember seeing the Pauli matrices in a quantum computing talk and I think there was an i in the Y matrix. But I don't see it here.

ALICE: You're right, Cardy. In the most general form of the Pauli matrices, we need complex numbers, which involve $i = \sqrt{-1}$, but for our purposes, until we get to Chapter 15, the Y in Eq. (6.6) will be perfectly fine. In Chapter 15, we will show you when and why complex numbers are important for quantum mechanics.

This is also a good point to note that physicists like to denote the Pauli matrices by the lower-case Greek σ (sigma):

$$\sigma_x = X \quad \sigma_y = Y \quad \sigma_z = Z. \tag{6.8}$$

BOB: Let's see what happens when Z operates on the computational basis states:

$$Z|1\rangle \Rightarrow \begin{pmatrix} 1 & 0 \\ 0 & -1 \end{pmatrix} \begin{pmatrix} 0 \\ 1 \end{pmatrix} = \begin{pmatrix} 0 \\ -1 \end{pmatrix} = -\begin{pmatrix} 0 \\ 1 \end{pmatrix} = -|1\rangle \tag{6.9}$$

$$Z\,|0\rangle \Rightarrow \begin{pmatrix} 1 & 0 \\ 0 & -1 \end{pmatrix} \begin{pmatrix} 1 \\ 0 \end{pmatrix} = \begin{pmatrix} 1 \\ 0 \end{pmatrix} = |0\rangle. \tag{6.10}$$

CARDY: What does minus a quantum state vector mean in Eq. (6.9)?

ALICE: Let me give you a partial answer. In your high school math or physics classes, you probably have seen that in standard vector algebra, the negative of a regular vector is another vector with the same length but pointing in the opposite direction. For example, if you are moving eastward, we described your velocity with a vector pointing east. The negative of that vector would represent your velocity if you were traveling west. The same is true for quantum state vectors, but the directions are in abstract state space. We will address the physical implications of the minus sign later.

BOB: While we are introducing single-qubit gates, we should show you one more, the Hadamard gate, which turns out to be very useful in QIS and QC. The Hadamard gate H has the matrix representation

$$H \Rightarrow \frac{1}{\sqrt{2}} \begin{pmatrix} 1 & 1 \\ 1 & -1 \end{pmatrix}. \tag{6.11}$$

Suppose H acts on $|0\rangle$, then we have

$$H\,|0\rangle = \frac{1}{\sqrt{2}} \{|0\rangle + |1\rangle\} = |S_+\rangle = |+\rangle, \tag{6.12}$$

and for the other computational basis state, we get

$$H\,|1\rangle = \frac{1}{\sqrt{2}} \{|0\rangle - |1\rangle\} = |S_-\rangle = |-\rangle. \tag{6.13}$$

We have introduced special names $|S_+\rangle$ or $|+\rangle$ and $|S_-\rangle$ or $|-\rangle$ for the resulting states because they occur frequently in QIS and QC. Eqs. (6.12) and (6.13) tell us that if we start with a quantum state that is just one of the basis states, then when a Hadamard gate acts on the state, the result is a superposition of the basis states. This superposition state will be very useful in the quantum algorithms in the next chapters. Note that the question of a minus sign in front of a state vector comes up here as well as with the Z operator.

CARDY: Why did you include the $1/\sqrt{2}$ in the definition of the Hadamard gate?

Try It 6.4

Answer Cardy's question. Hint: What is the length of the superposition states produced by the Hadamard gate in Eqs. (6.12) and (6.13).

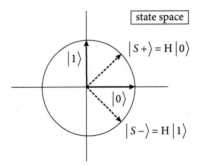

Fig. 6.3 A state space diagram illustrating the state vectors that result from having a Hadamard gate act on the computational basis states.

CARDY: Those states in Fig. 6.3 look just like states for polarized light tilted at $45°$ which we can think of as a superposition of hlp and vlp states!

ALICE: You nailed it, Cardy! The important point of the Hadamard gate is that it acts on the basis states individually to produce a quantum state that is a superposition of the basis states. That superposition is then very useful, as we shall see, in quantum information and quantum computing.

What happens if H acts on a general quantum superposition state represented by the column vector $\begin{pmatrix} \alpha_0 \\ \alpha_1 \end{pmatrix}$? In that case, we get

$$
\begin{aligned}
H \left|\psi\right\rangle \Rightarrow \frac{1}{\sqrt{2}} \begin{pmatrix} 1 & 1 \\ 1 & -1 \end{pmatrix} \begin{pmatrix} \alpha_0 \\ \alpha_1 \end{pmatrix} &= \frac{1}{\sqrt{2}} \begin{pmatrix} \alpha_0 + \alpha_1 \\ \alpha_0 - \alpha_1 \end{pmatrix} \Rightarrow \frac{1}{\sqrt{2}} (\alpha_0 + \alpha_1) \left|0\right\rangle \\
&\quad + \frac{1}{\sqrt{2}} (\alpha_0 - \alpha_1) \left|1\right\rangle \\
&= \frac{1}{\sqrt{2}} \alpha_0 \{\left|0\right\rangle + \left|1\right\rangle\} + \frac{1}{\sqrt{2}} \alpha_1 \{\left|0\right\rangle - \left|1\right\rangle\} \\
&= \alpha_0 \left|S_+\right\rangle + \alpha_1 \left|S_-\right\rangle.
\end{aligned}
\tag{6.14}
$$

Try It 6.5

Write out the details of all the matrix multiplications in Eq. (6.14) and check the results.

Try It 6.6

Suppose $\alpha_0 = -\alpha_1$ in Eq. (6.14). What is the result for $H \left|\psi\right\rangle$?

BOB: As an experimentalist, I ought to point out that the equivalent of the Hadamard gate was developed back in the 1930s and '40s in the field that became known as nuclear magnetic resonance. The quantum systems (what we would now call qubits) were atomic nuclei with spin angular momentum and magnetic dipole moments (magnetic north pole–south pole combinations). Putting those quantum systems into superposition states is a key component of magnetic resonance imaging, which, as you probably know, is a big deal in medical diagnostics. Analogous superposition states can be produced in atoms and molecules using bursts of laser light. So, the technology needed to produce superposition states has long been part of every quantum mechanic's toolbox.

Try It 6.7

Can the Hadamard gate H be represented as a combination of Z and X gates H $= (1/\sqrt{2})\,(Z+X)$? Hint: Remember that to add matrices, you add the elements that are in the corresponding positions, that is $\begin{pmatrix} a & b \\ c & d \end{pmatrix} + \begin{pmatrix} e & f \\ g & h \end{pmatrix} = \begin{pmatrix} a+e & b+f \\ c+g & d+h \end{pmatrix}$.

Try It 6.8

Draw quantum circuits that show how each of the X, Y, Z, and H gates processes the computational basis states $|0\rangle$ and $|1\rangle$. See Figure 6.2.

Try It 6.9

Show how each of the X, Y, and Z gates processes a general single-qubit quantum state $|S\rangle = \alpha_0\,|0\rangle + \alpha_1\,|1\rangle$ as we did with Eq. (6.14).

6.3 Successive Quantum Gates

BOB: A single quantum gate is neat, but to do something useful in QIS and QC, we need more than one gate. We can combine quantum gates in succession to do some interesting things. As a first look at such a situation, consider the quantum circuits shown in Figure 6.4.

In the left circuit of Figure 6.4, an input quantum state $|\psi\rangle$ is first acted on by a Hadamard gate and then the output from the Hadamard gate is acted upon by an X gate. In the circuit on the right, the order in which the gates act is reversed. In abstract form we write

$$\begin{aligned}
|\psi_{\mathrm{HX}}\rangle &= \mathrm{X}\,(\mathrm{H}|\psi\rangle) = \mathrm{X}\,\mathrm{H}|\psi\rangle \\
|\psi_{\mathrm{XH}}\rangle &= \mathrm{H}\,(\mathrm{X}|\psi\rangle) = \mathrm{H}\,\mathrm{X}|\psi\rangle.
\end{aligned} \tag{6.15}$$

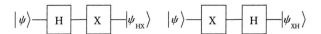

Fig. 6.4 Two quantum circuits. On the left, the Hadamard gate H acts on the state $|\psi\rangle$ first and then the NOT (X) gate acts to produce the state $|\psi_{HX}\rangle$. On the right, the order of operations is reversed.

In the first equality in Eq. (6.15), we group the operations within parentheses to remind ourselves that H acts first and then X. As shown on the far right of Eq. (6.15), the operations are usually written without the parentheses, which is cleaner typographically, but we need to remember the convention about the order of operations.

Try It 6.10

Draw the quantum circuit diagram for $XHXH\,|\psi\rangle = |\Phi\rangle$.

ALICE: Please note that by convention, in the *equations*, we read the operations from right to left, the opposite of how we read the *quantum circuit diagram*. That seems to be a bother, but that is what almost all QIS and QC people do; so, we need to get used to that. It is probably worth making a note of that because we tend to read text from left to right.

Let's see if it actually matters what direction we read the equations in. The matrix representation of those operations is just the product of the two matrices:

$$XH \Rightarrow \frac{1}{\sqrt{2}} \begin{pmatrix} 0 & 1 \\ 1 & 0 \end{pmatrix} \begin{pmatrix} 1 & 1 \\ 1 & -1 \end{pmatrix} = \frac{1}{\sqrt{2}} \begin{pmatrix} 1 & -1 \\ 1 & 1 \end{pmatrix} \tag{6.16}$$

$$HX \Rightarrow \frac{1}{\sqrt{2}} \begin{pmatrix} 1 & 1 \\ 1 & -1 \end{pmatrix} \begin{pmatrix} 0 & 1 \\ 1 & 0 \end{pmatrix} = \frac{1}{\sqrt{2}} \begin{pmatrix} 1 & 1 \\ -1 & 1 \end{pmatrix}. \tag{6.17}$$

Surprise! The two products are not the same. With quantum gates, the order matters! We say that the two operators (matrices) X and H do not *commute*.

Try It 6.11

Apply XH to each of the standard computational basis states. Then do the same with HX. Do you get the same result?

6.4 **Mathematical Interlude**

ALICE: It turns out that we can't use any combination of numbers in the 2×2 matrices representing quantum gates. To see why, we need to look at some of the math associated with matrices. If you don't remember the details or you haven't seen these manipulations in a math course, don't worry. We'll show you what you need to know.

The matrices allowed to represent quantum gates have to satisfy the following requirements:

1. The determinant of the matrix must be equal to $+1$ or -1. This guarantees that the output column vector is a unit vector if the input vector is a unit vector. Remember that we use vectors with length 1 because we want our probabilities to sum to 1.

2. The product of the matrix and its transpose must equal I, the identity matrix.

3. The columns in the matrix form a set of orthogonal unit vectors. So do the rows.

Let's check the Hadamard gate matrix to see if it satisfies those conditions and, along the way, we will explain what determinant and transpose mean. For almost all of this book, we will restrict ourselves to gates that satisfy these conditions, and you won't need to worry about the mathematical details. However, it is good to know some of the mathematical terminology in case you come across these terms in other books and articles.

The determinant of a 2×2 matrix is the product of upper-left and lower-right elements minus the product of the lower-left and the upper-right elements. In particular

$$\mathrm{Det}\begin{pmatrix} a & b \\ c & d \end{pmatrix} = ad - bc. \tag{6.18}$$

Here is an example with numerical entries in the matrix:

$$\mathrm{Det}\begin{pmatrix} -\frac{1}{2} & 0 \\ 0 & \frac{4}{2} \end{pmatrix} = -\frac{4}{4} - 0 = -1. \tag{6.19}$$

Try It 6.12

Evaluate the determinant of the Hadamard gate matrix.

ALICE: The transpose of a matrix is a matrix of the same size but with the elements swapped across the primary diagonal (upper-left to lower-right). In other words, in the transpose the old rows become the new columns. The transpose will be designated by a superscript T attached to the symbol for the matrix:

$$A = \begin{pmatrix} a_{11} & a_{12} \\ a_{21} & a_{22} \end{pmatrix}$$

$$A^{\mathrm{T}} = \begin{pmatrix} a_{11} & a_{21} \\ a_{12} & a_{22} \end{pmatrix}. \tag{6.20}$$

Here is a numerical example for which the transpose is not equal to the original matrix:

$$\begin{pmatrix} 0.5 & -0.2 \\ 0.1 & 0.4 \end{pmatrix}^{\mathrm{T}} = \begin{pmatrix} 0.5 & 0.1 \\ -0.2 & 0.4 \end{pmatrix}. \tag{6.21}$$

Aside

In Chapter 15, we will find that if the elements of the matrix are complex numbers, we will need to consider the transpose combined with the complex conjugation of all the elements.

Let's try this out with the Hadamard gate's matrix representation and calculate the product of the matrix and its transpose.

$$\mathrm{H}^{\mathrm{T}}\mathrm{H} = \frac{1}{2}\begin{pmatrix} 1 & 1 \\ 1 & -1 \end{pmatrix}\begin{pmatrix} 1 & 1 \\ 1 & -1 \end{pmatrix} = \frac{1}{2}\begin{pmatrix} 2 & 0 \\ 0 & 2 \end{pmatrix} = \begin{pmatrix} 1 & 0 \\ 0 & 1 \end{pmatrix} = \mathrm{I}. \tag{6.22}$$

Matrices that satisfy Eq. (6.22) are called *orthogonal* if the elements are real and *unitary* if the elements are complex numbers. We see that the Hadamard gate matrix satisfies condition 2. Now consider condition 3. In analogy with our definition of orthogonal state vectors, the orthogonality condition for the Hadamard gate's two column vectors is:

$$\begin{pmatrix} 1 & 1 \end{pmatrix}\begin{pmatrix} 1 \\ -1 \end{pmatrix} = 1 - 1 = 0. \tag{6.23}$$

Try It 6.13

Show that these three conditions are met by the other Pauli matrices X, Y, and Z introduced in the previous section.

Later, we will find that the matrices representing quantum gates can be built out of combinations of the basis state vectors in their right vector and left vector forms. So, there is a close connection between these matrices and the column and row representations of the right and left state vectors.

6.5 Successive Application of the Same Quantum Gate

BOB: Most quantum gates have another interesting property. If you apply the gate to a quantum state and then apply it again to the result of the first application, you end up with the quantum state you started with. These gates are in a sense "reversible": You can get back to your starting state if you apply the gate twice (assuming you haven't done anything else to the state produced by the first application).

Here is another piece of terminology that occurs in QIS and QC, though we won't be making much use of it. The inverse of a matrix A, written symbolically as A^{-1}, is defined to be that matrix which when multiplied by the original matrix gives the identity matrix I:

$$A^{-1}A = I = AA^{-1}. \tag{6.24}$$

If the transpose of a real number matrix is equal to its inverse, we say that the matrix is orthogonal and if the elements are complex numbers, we call it *unitary*. Unitary matrices are important in quantum mechanics because when they operate on a quantum state vector, they produce a new state vector with the same length as the original state vector. The name "unitary" is used because we almost always use state vectors of unit length—their length $= 1$.

CARDY: That seems like rather strange terminology. Could we try that out with some of the gates?

ALICE: Excellent strategy, Cardy. It is always a good idea to try out new ideas with specific examples. Let's start with the Hadamard gate interacting with a superposition of the computational basis states $|0\rangle$ and $|1\rangle$ as shown in Eq. (6.12). The operation inside the square brackets on the left produces a superposition state; then the second H acts on that superposition state:

$$H\,[H\,|0\rangle] \Rightarrow \frac{1}{\sqrt{2}}\begin{pmatrix} 1 & 1 \\ 1 & -1 \end{pmatrix}\left[\frac{1}{\sqrt{2}}\left\{\begin{pmatrix} 1 \\ 0 \end{pmatrix}+\begin{pmatrix} 0 \\ 1 \end{pmatrix}\right\}\right]$$

$$= \frac{1}{2}\left\{\begin{pmatrix} 1 & 1 \\ 1 & -1 \end{pmatrix}\begin{pmatrix} 1 \\ 0 \end{pmatrix}+\begin{pmatrix} 1 & 1 \\ 1 & -1 \end{pmatrix}\begin{pmatrix} 0 \\ 1 \end{pmatrix}\right\}$$

$$= \frac{1}{2}\begin{pmatrix} 1 \\ 1 \end{pmatrix}+\frac{1}{2}\begin{pmatrix} 1 \\ -1 \end{pmatrix} = \frac{1}{2}\begin{pmatrix} 2 \\ 0 \end{pmatrix} = \begin{pmatrix} 1 \\ 0 \end{pmatrix} \Rightarrow |0\rangle. \tag{6.25}$$

In the first line of Eq. (6.25), we applied H to $|0\rangle$ and got the superposition state expressed in Eq. (6.12). In the second line, we applied the second H to each of the components of the superposition state. We then grouped terms and found that we end up with the quantum state we started from.

Try It 6.14

Work through Eq. (6.25) step by step to make sure you understand the various matrix and column vector operations. Is the H matrix equal to its transpose? Is it equal to its inverse? Is it an orthogonal matrix?

Try It 6.15

Repeat the previous Try It but use an X gate rather than a Hadamard gate. Explain what you find.

Try It 6.16

Go through the same steps with a general superposition state as the starting state:
$|S\rangle = a_0\,|0\rangle + a_1\,|1\rangle \Rightarrow \begin{pmatrix} a_0 \\ a_1 \end{pmatrix}$. Do you get $\mathrm{H}\,[\mathrm{H}\,|S\rangle] = |S\rangle$?

6.6   Measurement Devices are not Reversible

ALICE: In Chapter 5 we introduced measurement devices and noted that we will avoid calling them gates because they are not (in almost all circumstances) represented by matrices that satisfy the conditions laid out in the previous two sections. In particular, they are not reversible. For example, after a photon interacts with a polarization state analyzer, its state might be $|\mathrm{vlp}\rangle$. As we have emphasized before, from that information, we know very little about the state describing the photon before it arrived at the device. So, there is no way to apply the measurement operation to get back to the original state.

State analyzers can, however, be represented by "projection operators" constructed as products of right vectors and left vectors. For example, the operation of a state analyzer that puts the qubit in the device's $|0\rangle$ state is represented by

$$M_0 = |0\rangle\,\langle 0|\,. \tag{6.26}$$

If a state $|S\rangle = a_0\,|0\rangle + a_1\,|1\rangle$ interacts with (is acted upon by) the device, we write

$$M_0 \underbrace{|S\rangle}_{\text{input state}} = \underbrace{|0\rangle}_{\text{output state}} \overbrace{\langle 0|S\rangle}^{\text{projection operator}} \tag{6.27}$$

The output state (consistent with the measurement postulate) is the basis state $|0\rangle$. The probability of getting that result is the square of $\langle 0|\,S\rangle$, which we recognize as the amplitude a_0 of $|S\rangle$ associated with the basis state $|0\rangle$. If the qubit is subsequently detected, we have completed a quantum observation.

As we discussed in Chapter 5, the probability is found experimentally by repeating the state preparation–measurement cycle many times. But from a single observation, which is the important point here, we don't know anything about the amplitude except that it is not 0.

What happens if we apply M_0 to the state on the right side of Eq. (6.27)?

$$M_0 \; \underbrace{|0\rangle}_{\text{state after 1st } M_0} \langle 0|S\rangle = |0\rangle \underbrace{\langle 0|0\rangle}_{=1}\langle 0|S\rangle = |0\rangle\langle 0|S\rangle. \qquad (6.28)$$

For the last equality in Eq. (6.28), we used $\langle 0\,|0\rangle = 1$ since the basis state is a unit vector. We see that we don't get back to the original state $|S\rangle$. As we mentioned in Chapter 5, some authors like to say the measurement "collapses the state" and that collapse is irreversible. I prefer to say that the state analyzer device projects the state of the incoming qubit along one of the device's basis states. Since there are many (in fact infinitely many) different incoming states that get projected to the same output state, the measurement device projection is not reversible.

BOB: It is important to note that the projection operator simply gives a description of the results of interaction with a state analyzer device; it does not tell us the physical mechanism by which the projection occurs. As we mentioned in Chapter 5, even though quantum mechanics is nearly 100 years old, there is still no unique formulation of how a state analyzer measurement device works.

We certainly know how to carry out the measurements, but translating their operation into the formal theory of quantum mechanics remains an open question. I should point out that similar interactions occur for qubits even outside the organized measurement devices found in laboratories and quantum computers around the world. In "real life" qubits are always interacting with billions and billions of other qubits, and those interactions are much like the interactions with our measurement devices. That means we don't have a widely accepted way of describing those kinds of interactions either. The projection operator and its extensions do give us a workable formal description, but that is different from having a detailed mechanistic theory.

 CHAPTER SUMMARY

- Quantum computers make use of quantum gates. The resulting gate arrays can be drawn as circuits just like classical computers. Gates can be represented by formal operators or by matrices. Here the matrix representations of the X (or NOT), the Y, and the Z gates are

$$X \Rightarrow \begin{pmatrix} 0 & 1 \\ 1 & 0 \end{pmatrix} \quad Y \Rightarrow \begin{pmatrix} 0 & 1 \\ -1 & 0 \end{pmatrix} \quad Z \Rightarrow \begin{pmatrix} 1 & 0 \\ 0 & -1 \end{pmatrix}.$$

- An important gate in QC is the Hadamard gate which puts an input basis state into a state of superposition. The matrix representation of this gate is

$$H \Rightarrow \frac{1}{\sqrt{2}} \begin{pmatrix} 1 & 1 \\ 1 & -1 \end{pmatrix}.$$

- Just like in classical computing, multiple gates are employed for most computational tasks. Learning to read these quantum circuits is like learning a different language. In English, the adjectives come before the noun while in Spanish the noun comes first. The order of the words is important in these languages and the order of the gates is important in QC. For better or worse, the order in which the gates operate is left to right in a circuit diagram. Read the left side of the following diagram as "$|\psi\rangle$ meets up first with H and after that with X."

$$|\psi_{\text{HX}}\rangle = \underset{\text{X acts second}}{X} \left(\underset{\text{H acts first}}{H|\psi\rangle} \right) = X\,H|\psi\rangle$$

$$|\psi_{\text{XH}}\rangle = \underset{\text{H acts second}}{H} \left(\underset{\text{X acts first}}{X|\psi\rangle} \right) = H\,X|\psi\rangle.$$

In the state vector representation, we read from right to left:

- Not all possible matrices represent quantum gates. The matrix's columns must represent orthogonal unit vectors. The matrix determinate must equal ± 1 to ensure that a unit vector results when the matrix acts on a unit vector. Unit vectors are needed so that the probabilities will sum to 1.

- One of the most important requirements is that the gate must be reversible, meaning that if the gate is applied to a state vector twice, the result is the original state vector.

- Measurement devices are not reversible. Measuring a quantum state only tells us that the state's amplitude for a specific basis vector direction is not zero. Another way of stating this is that the measurement device projects the state vector onto a particular device basis vector. All other information is lost.

FURTHER READING

Chris Bernhardt, *Quantum Computing for Everyone* (MIT Press, Cambridge, MA, 2019) has a nice introduction to quantum gates and their matrix representations.

7 Putting a Spin on Spin

If you do not change direction, you may end up where you are heading.

LaoTzu

7.1 Quantum Spin Systems

CARDY: Hi, Alice and Bob. I just got back from a physics colloquium on quantum computing. I need to thank you guys because I understood a lot more about the talk than I expected. But I was puzzled by the speaker's emphasis on spin. She was talking about spin-up and spin-down and spin-one-half and Bloch spheres and lots more. My head was spinning! Could you please fill me in about this spin stuff?

ALICE: Sure, Cardy. A quantum "spin" system is just another example of a qubit, another physical system. And in fact, the types of quantum spin systems usually talked about in quantum computing are two-state systems, whose formal description is just like photon polarization and other two-state systems we mentioned before. Physicists love spin systems because they are relatively simple theoretically but nevertheless have many practical applications. Physicists have studied such systems in detail from the early days of quantum mechanics in the 1920s and continue on today. Most of the representations of the quantum states of those systems are exactly like those we used before for photon polarization and for our abstract computational basis states, with only one or two minor modifications.

Here is a quick introduction to the basics of spin systems. Many quantum particles (such as electrons, protons, neutrons, and quarks) and systems built from them (such as nuclei, atoms, and molecules) behave as if they were spinning tops. Exactly how that comes about, particularly for "elementary particles" like electrons and quarks, is not entirely understood, but it is a well-established empirical fact.

We can associate a direction with spinning motion: For example, the axis about which the disk in Figure 7.1 is spinning. The usual convention is to use a right-hand rule to choose a direction along the axis: Wrap the fingers of your right hand around the top in the direction in which the edge of the disk is moving. The thumb of your right hand will then point along the spin axis in the direction associated with the angular momentum vector $\vec{L}$. When the angular momentum is due to the usual rotation motion like a rotating top or a planet orbiting the Sun, we use the symbol $\vec{L}$ for its angular momentum. When the angular momentum is associated with "spin" in the quantum mechanics sense, we use the symbol $\vec{S}$.

Quantum mechanics tells us that if we measure the orientation of the spin vector of an electron along some specified direction (say north), we get only two possible results:

Quantum Computing: From Alice to Bob. Alice Flarend and Bob Hilborn, Oxford University Press.
© Alice Flarend and Robert C. Hilborn (2022). DOI: 10.1093/oso/9780192857972.003.0007

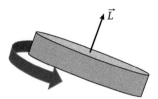

Fig. 7.1 A rotating disk has angular momentum represented by the vector $\vec{L}$. The direction of $\vec{L}$ is along the rotation axis and its orientation (up or down) is given by the right-hand-rule.

The spin vector points along that direction (we call that "spin-up") or it points opposite to that direction (we call that "spin-down").

For other quantum objects, there may be more than two possible results. The number of possible orientation values is specified by what historically is called a spin "quantum number" s. The number of possible orientation states is $2s + 1$. For electrons, protons, neutrons, and quarks, measurements show that $s = \frac{1}{2}$, so the number of orientation states is $2 \times (1/2) + 1 = 2$. For nuclei and atoms, for example, built up from protons, neutrons, and electrons, quantum mechanics tells us that the possible values for s are 0, 1/2, 1, 3/2, 2, and so on.

In quantum computing with spin systems, we almost always assume that we are using qubits with spin quantum number $s = \frac{1}{2}$. So, these are called "spin-1/2" systems (pronounced "spin-one-half"). As we shall see, having qubits with just two possible states is sufficient to do all sorts of interesting quantum information and quantum computing operations. Hence, we stick with spin-1/2 systems.

7.2   Representations of Spin-1/2 Systems

ALICE: To describe the states of spin-1/2 qubits, we can use the three representations introduced for other quantum states: abstract quantum state vectors (right vectors), column vectors, and graphical representations in state space. The state vectors are often written as $|\uparrow\rangle$ for spin-up and $|\downarrow\rangle$ for spin-down. Those are usually chosen to be the basis states for the system. A spin-1/2 state space diagram is shown in Figure 7.2.

The two basis states can be represented by column vectors in the usual way:

$$|\uparrow\rangle \Rightarrow \begin{pmatrix} 1 \\ 0 \end{pmatrix} \qquad |\downarrow\rangle \Rightarrow \begin{pmatrix} 0 \\ 1 \end{pmatrix}. \tag{7.1}$$

CARDY: I am confused. The spin-up and spin-down directions are opposite each other, but the basis state vectors are perpendicular—oops, I should say orthogonal—to each other.

ALICE: You are right, but you have to get used to the basis state vectors living in a quantum state space and not in physical space. In the quantum state space, they are always chosen to be orthogonal to each other. Recall that the orthogonality is a way of enforcing the fact that the properties associated with the basis states are mutually exclusive: If you observe the spin orientation of the qubit you get either spin-up or spin-down, but never both.

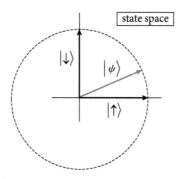

Fig. 7.2 The state space representation of the quantum state of a spin-1/2 qubit. $|\uparrow\rangle$ (spin-up) and $|\downarrow\rangle$ (spin-down) are the two basis states. $|\psi\rangle$ is a general state vector. The dashed circle has a radius $= 1$.

However, we often use the physical-space orientation of the spin to label the states. Experiments show that for a spin-1/2 system, the spin vector orientation observations give us the two results—either along the observation direction or opposite the observation direction, or more succinctly, up or down.

Try It 7.1

Before looking at the following figures, draw an arrow representing a spin-up state and another arrow representing a spin-down state in a state space diagram like Figure 7.2.

ALICE: As Eq. (7.1) shows, the spin-up state $|\uparrow\rangle$ of a spin-1/2 qubit is customarily associated with the $|0\rangle$ computational basis state (which lies along the "horizontal" state-space axis as shown in Figs. 7.2 and 7.3) and $|\downarrow\rangle$ with the $|1\rangle$ computational basis state (which lies along the "vertical" state-space axis), although there is no fundamental reason you couldn't use the opposite association if you wanted to. The set of correspondences is

$$|\uparrow\rangle = |0\rangle \Rightarrow \begin{pmatrix} 1 \\ 0 \end{pmatrix} \qquad |\downarrow\rangle = |1\rangle \Rightarrow \begin{pmatrix} 0 \\ 1 \end{pmatrix}. \tag{7.2}$$

Figure 7.2 represents a state $|\psi\rangle$ that is a superposition of spin-up and spin-down.

Cardy, did you notice that in Figure 7.3, the dark state vector points in the direction opposite to the label in the state vector for spin-down?

CARDY: Yes, I saw that. Why does that work out that way?

ALICE: This happens because of the collision of two conventions: (1) associating spin-down with the $|1\rangle$ computational basis state, and (2) associating the direction of $|1\rangle$ in state space with the vertical axis. That is just a reminder that the directions in state space need not be related to directions in physical space and that the labels in the state vectors are not necessarily related to directions in state space. You might want to keep a drawing of the spin-1/2 state-space diagram handy to consult because that representation is so counterintuitive.

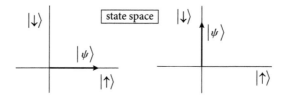

Fig. 7.3 On the left, the dark arrow represents a spin-up state for a spin-1/2 system. On the right, the dark arrow represents a spin-down state.

CARDY: How do you measure the spin orientation?

BOB: To measure the spin orientation, we make use of the fact that common elementary particles such as electrons, protons, and neutrons behave as if they are tiny bar magnets with a north magnetic pole (N) and a south magnetic pole (S). The critical point is that the direction associated with the spin lies along the particle's north–south direction. So, if we can carry out measurements to determine if the north pole is up or down, we have determined whether the spin orientation is up or down. Those elementary particles produce magnetic fields that are much like the magnetic fields produced by small bar magnets. Their N and S poles respond to external magnetic fields almost the way a compass needle responds to Earth's magnetic field.

ALICE: Cardy, note the cautious "as if" language Bob is using. In some aspects the quantum particle's north pole–south pole behavior is like a toy bar magnet. In other respects, it is quite different. We say "as if" because simple mechanical models of rotating spheres of electrical charge or other circulating charges cannot explain the magnetic properties of elementary particles. In fact, current science does not have completely satisfactory explanations of why elementary particles have spin. We can describe spin and we can do calculations involving spin, but we can't totally explain *why* particles have spin. Regardless of where spin comes from, we can measure those north pole–south pole orientations using the magnetic fields associated with them. For a proton, the N pole is in the same direction as the spin angular momentum vector. For an electron (negative charge), they are in opposite directions.

CARDY: Oh, yeah! I see that if the spin-1/2 state goes from up to down, we can think of the bar magnetic rotating $180°$ in our three-dimensional world, but in state space, the state vector rotates only $90°$ from one orthogonal state to the other. That is cool, but pretty weird. But that didn't happen for our photon polarization states. Why not?

ALICE: Wow, you got it, Cardy! That observation is a critical one in quantum mechanics and its applications in atomic, molecular, optical, nuclear, and particle physics. The important observation is that if you use a PBS to measure the polarization state of a photon, the two mutually exclusive outcomes are vertical linear polarization and horizontal linear polarization. In that case the physical space directions associated with those outcomes are orthogonal and, of course, so are the state space basis vectors. One of the reasons we decided to start our study of qubits using photon polarization states was to avoid just that added bit of complexity with spin-1/2 states.

CARDY: But I would like to know *why* it works out that way for photons.

BOB: That's a great *why* question! The standard textbook explanation is that photons have spin-1 states. How that is tied to horizontal and vertical linear polarization is a bit complicated;

so, we won't go into the details, but the Further Reading section points you to books where you can read about those details.

CARDY: I guess I am supposed to ask *why* photons have spin-1 states.

BOB: Spot on. The textbook explanation is that photons are the quantum particles associated with electric and magnetic fields. Those fields can be described as vectors and it turns out that the quantum theory of vector fields requires spin-1 particles. That may sound a bit circular, but it turns out the spin of quantum particles is associated with the scalar, vector, or tensor nature of the underlying fields. However, for our QC work, we don't need those details.

7.3 Quantum Gates and Spin-1/2 Systems

BOB: Let's take a quick look out how the quantum gates introduced in Chapters 5 and 6 apply to spin-1/2 systems. Cardy, what do you expect to happen if a NOT (X) gate acts on a spin-up basis state?

CARDY: Hmm. The NOT is supposed to flip the state to its orthogonal partner. So, I would expect that we get a spin-down state.

BOB: Just right! In more formal terms, we write

$$\text{NOT}\,|\!\uparrow\rangle = |\!\downarrow\rangle \qquad \text{NOT}\,|\!\downarrow\rangle = |\!\uparrow\rangle . \tag{7.3}$$

What about the Z gate? The Z gate action is easily found using the matrix multiplication form

$$\text{Z}\,|\!\uparrow\rangle = \text{Z}\,|0\rangle \Rightarrow \begin{pmatrix} 1 & 0 \\ 0 & -1 \end{pmatrix} \begin{pmatrix} 1 \\ 0 \end{pmatrix} = \begin{pmatrix} 1 \\ 0 \end{pmatrix} = |0\rangle = |\!\uparrow\rangle \tag{7.4}$$

$$\text{Z}\,|\!\downarrow\rangle = \text{Z}\,|1\rangle \Rightarrow \begin{pmatrix} 1 & 0 \\ 0 & -1 \end{pmatrix} \begin{pmatrix} 0 \\ 1 \end{pmatrix} = \begin{pmatrix} 0 \\ -1 \end{pmatrix} = -\begin{pmatrix} 0 \\ 1 \end{pmatrix} = -|1\rangle = -|\!\downarrow\rangle . \tag{7.5}$$

ALICE: One piece of quantum jargon you should know: If an operator acts on a state vector and returns the same state vector multiplied by a number, then that state vector is said to be an *eigenvector* of that operator and the corresponding number is called an *eigenvalue*. The adjective eigen is German for "characteristic." That jargon is used a lot in quantum mechanics. From Eqs. (7.4) and (7.5), we see that the spin-up and spin-down basis states $|\!\uparrow\rangle$ and $|\!\downarrow\rangle$ are eigenvectors of the operator Z and the eigenvalues are $+1$ for $|\!\uparrow\rangle$ and -1 for $|\!\downarrow\rangle$.

Try It 7.2

Are either of the spin-1/2 basis states eigenvectors of the NOT operator?

Try It 7.3

Figure out the results of the Y operator acting on the spin-up and spin-down states. Draw the results in a state space diagram. Are the spin-1/2 basis states eigenvectors of the Y operator?

Try It 7.4

Figure out the results of the Hadamard operator (Chapter 6) acting on the spin-up and spin-down states. Draw the results in a state space diagram.

ALICE: I want to point out another peculiar feature of spin systems. It turns out that a "spin-up" state does not have the spin pointing exactly along the "up" direction, as illustrated in Figure 7.4. Although the z component of the spin is certain for the spin-up state, its x and y components are unknown. This result is due to what is called the uncertainty principle in quantum mechanics. If the spin pointed exactly along the z axis (for example), then its x and y components would both be exactly zero. The quantum theory of spins tells us that we can't prepare a state with both the x and y components of spin known precisely. That rules out having the spin pointing exactly along the z axis. This peculiar feature is expressed by the mathematical fact that the operators (gates) X, Y, and Z do not commute with each other. That is, XY $\neq$ YX, XZ $\neq$ ZX, and ZY $\neq$ YZ.

I hope we have you given the basic concepts of quantum spin systems. We will return to them later, but for now the crucial take-home message is that spin-1/2 qubits can be treated formally using quantum state vectors exactly the way we treated any other two-state quantum system.

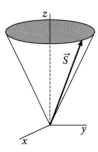

Fig. 7.4 A sketch of a quantum spin vector $\vec{S}$ with a definite component along the z direction. The vector, with its tail anchored at the origin, should be thought of as smeared around the outside surface of the cone, so that its x and y components average to 0. The angle between $\vec{S}$ and the z axis stays fixed for this state.

CHAPTER SUMMARY

- Spin states are a paradigmatic example of qubits with the two basis states being spin-up and spin-down. They offer a testbed to practice and review the fundamentals we have covered so far.

- Spin-1/2 states are an example of the difference between physical states and quantum space states. A particle flipping from a spin-up to a spin-down has, in a sense, changed its physical spin orientation by 180° (but see the caveat about spin-up and spin-down at the end of section 7.3). In the quantum state space, however, the change is only 90° because the abstract state-space basis vectors are orthogonal.

- Spin states can be operated on by quantum gates. For example, the NOT gate changes the spin state to its orthogonal partner. The Hadamard gate acting on a basis spin state creates a state of superposition of both spin states.

- There are many more candidates for qubits other than spin and polarization, but they all follow the same principles that we have laid out in Chapters 4–7.

 FURTHER READING

For more information on spin in quantum mechanics, we recommend Michael G. Raymer, *Quantum Physics: What Everyone Needs to Know* (Oxford University Press, Oxford, 2017). This book also has a fine discussion of polarization states of photons.

Daniel F. Styler, *The Strange World of Quantum Mechanics* (Cambridge University Press, Cambridge, 2000).

8 Your Basis, My Basis

Mathematics is the art of giving the same name to different things. Jules Henri Poincaré,
Science et Méthod (1908)

8.1 Taking Stock

ALICE: Cardy, I hope our previous discussion has stopped your head from spinning so much.

CARDY: The spinning has subsided. I think I have a reasonable idea of what is going on.

ALICE: Just keep in mind that most of what we are going to be talking about with quantum computing and quantum information can all be formulated in terms of the computational basis states, which, in some sense, are independent of the physical nature of the qubits. They could be spin-1/2 particles like electrons, protons, or appropriate atoms; they could be photons; or they could be circulating superconducting currents or more. From now on, we will stick pretty much to the basis state language and only bring in the physical details like spin-1/2 particles or photons when we need explicit examples.

This is a good time for a quick review. In Chapter 4, we showed you how to represent an abstract qubit state vector in terms of a superposition of two basis vectors $|A_o\rangle$ and $|A_1\rangle$:

$$|S\rangle = a_0 |A_0\rangle + a_1 |A_1\rangle. \tag{8.1}$$

This state can also be expressed in column vector form

$$\begin{pmatrix} a_0 \\ a_1 \end{pmatrix}_A = a_0 \begin{pmatrix} 1 \\ 0 \end{pmatrix}_A + a_1 \begin{pmatrix} 0 \\ 1 \end{pmatrix}_A. \tag{8.2}$$

Note that we have put a subscript A on the column vectors to remind us of the basis we are using. A is the general basis name, while A_0 and A_1 label the two basis states. The state coefficients a_0 and a_1 are just numbers, and we saw that a_0^2 is the probability of getting the state $|A_0\rangle$ and a_1^2 is the probability of getting the state $|A_1\rangle$ when we make a measurement associated with the A basis.

We also noted that we can choose different basis vectors (states) for a given type of qubit. The different basis vectors correspond to different configurations of preparation and measurement devices—for example, the orientation of a polarizing beam splitter for linear polarized light.

Quantum Computing: From Alice to Bob. Alice Flarend and Bob Hilborn, Oxford University Press.
© Alice Flarend and Robert C. Hilborn (2022). DOI: 10.1093/oso/9780192857972.003.0008

Using different basis states to analyze qubits turns out to be a critical component in quantum information processing and quantum computing.

In this chapter, we will show you how to set up a relatively simple way of translating from the representation of the qubit's state in one set of basis states to the representation in another. We will slow down the pace and provide more detail in this chapter than we have done previously because the results are so important to QC and QIS.

We will work through several examples that we will be used in later chapters. And we will see that using different basis sets to observe qubits prepared in the same quantum state will allow us to learn more about the quantum state. For QIS, we will also see that using two (or more) basis sets gives us a method to find out if Cardy is eavesdropping on my communications with Bob.

CARDY: Wait a minute! I would never eavesdrop.

BOB: We didn't mean you personally, Cardy. But we know that people do eavesdrop on communications. One of the promising uses of QIS is developing more secure communications processes.

8.2 Changing Basis States

ALICE: Before we get too far off in the QIS direction, let's tackle the fundamental issue of representing quantum states using different sets of basis vectors. In fact, a superposition state in one basis set will be a single basis state in some other appropriately chosen basis set. More importantly, as we have mentioned, we get more information from a quantum state if we make measurements on it relative to several basis sets. For example, we might start with the description of the polarization state for photons using horizontal and vertical polarization basis states but want to make measurements in terms of $+45°$ and $-45°$ polarization orientations. So, how do we express a qubit state in terms of these new basis states?

Let's warm up by considering ordinary two-dimensional vectors, like force vectors in physics. The actual physical interpretation is not important, but it may be helpful to have a concrete example in front of us. If you are familiar with dot products and changing unit vectors for describing ordinary vectors, you should feel free to jump ahead to section 8.3.

Let's look at Figure 8.1, where we can see the x and y components of the vector $\vec{F}$.

CARDY: I remember something like that from my high school physics class.

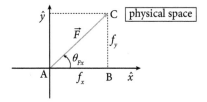

Fig. 8.1 A vector $\vec{F}$ expressed with components f_x and f_y along the $\hat{x}$ and $\hat{y}$ directions, respectively, where we used the common '∧' (caret) notation to indicate a vector of unit length. ABC is a right triangle.

ALICE: We also used some of this in Chapter 4, but a review won't hurt. First, the vector $\vec{F}$ can be thought of as a weighted superposition of unit vectors pointing along the x and y axes:

$$\vec{F} = f_x\,\hat{x} + f_y\,\hat{y}. \tag{8.3}$$

Those coefficients are the weights associated with how much of the vector $\vec{F}$ is along each unit vector direction. How do you find those coefficients? You could project the vector $\vec{F}$ along each unit vector direction by drawing a line perpendicular to the unit vector from the tip of $\vec{F}$ to the appropriate axis. You then use a ruler to measure f_x and f_y. Of course, the accuracy of your results depends on how carefully you construct and measure those lines. In many cases, we would like to use a formula to find f_x and f_y. Not only does that allow us to postpone putting numbers into the equations, it also allows us to use math to derive interesting relationships among the results.

Thinking back to the definitions of cosines and sines and looking at the right triangle ABC in Figure 8.1, you may recall that the coefficients f_x and f_y can be written as

$$f_x = \left\|\vec{F}\right\|\cos\theta_{Fx} \qquad f_y = \left\|\vec{F}\right\|\sin\theta_{Fx}.$$

Note if $\left\|\vec{F}\right\| = 1$, which is the choice used in QC and QIS as discussed in Chapter 4, then the x component of the vector is just $\cos\theta_{Fx}$ and the y component is $\sin\theta_{Fx}$.

By the Pythagorean theorem, we have

$$f_x^2 + f_y^2 = \left\|\vec{F}\right\|^2. \tag{8.4}$$

CARDY: Okay, that all looks fairly familiar.

ALICE: Now let's figure out how this works out if we use a different set of unit vectors $\hat{q}$ and $\hat{r}$ as shown in Figure 8.2. Focus on the new set of axes, $\hat{q}$ and $\hat{r}$ and ignore the original $\hat{x}$ and $\hat{y}$ directions. We find the representation of our vector $\vec{F}$ and the coefficients in the qr basis in the same way we found x and y components of $\vec{F}$, but now $\vec{F}$ lies at an angle θ_{Fq} from the q axis:

$$\vec{F} = f_q\hat{q} + f_r\hat{r}, \tag{8.5}$$

with

$$f_q = \left\|\vec{F}\right\|\cos\theta_{Fq} \qquad f_r = \left\|\vec{F}\right\|\sin\theta_{Fq}. \tag{8.6}$$

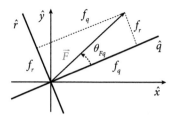

Fig. 8.2 The vector $\vec{F}$ and its components relative to the qr set of axes.

CARDY: I remember seeing something like that in my high school physics course when we were dealing with things sliding down inclined planes. My big question is whether $\vec{F}$ in Eq. (8.5) is the same as the $\vec{F}$ in Eq. (8.3).

ALICE: This is where math notation becomes a bit slippery; the right sides of the equations look different, but they are just two different ways of representing the same vector $\vec{F}$. In going from one set of axes to the other, $\vec{F}$ doesn't change—only the components along the axes change. To drive home that point, I can write

$$\vec{F} = f_q\hat{q} + f_r\hat{r} = f_x\hat{x} + f_y\hat{y}. \tag{8.7}$$

CARDY: That helps, but I will need to keep reminding myself that we have just a single vector $\vec{F}$.

ALICE: That is important and we will remind you of that as we go along. Now, let's look at how the unit vectors in the two bases are related to each other, ignoring the vector $\vec{F}$. Once we have that information, it is relatively easy to find the relationships among $\vec{F}$'s components.

In Figure 8.3, we show the two sets of unit basis vectors and the angle θ between them. The dashed circle has a radius $= 1$ and reminds us that each of the basis vectors has a length 1. Note that the qr set of axes is rotated relative to the xy axes by the angle θ. We will use the convention preferred by physicists: the angle θ is taken to be positive if the rotation is counterclockwise. Figure 8.3 shows the geometry that allows us to write $\hat{q}$ and $\hat{r}$ in terms of $\hat{x}$ and $\hat{y}$:

$$\hat{q} = \cos\theta\,\hat{x} + \sin\theta\,\hat{y}$$
$$\hat{r} = -\sin\theta\,\hat{x} + \cos\theta\,\hat{y} \tag{8.8}$$

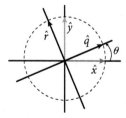

Fig. 8.3 This plot shows the two sets of unit vectors: (1) $\hat{x}$ and $\hat{y}$ (2) $\hat{q}$ and $\hat{r}$. The dashed circle has a radius $= 1$. θ is the angle between $\hat{x}$ and $\hat{q}$.

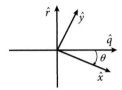

Fig. 8.4 The *xy* axes have been rotated clockwise by the angle θ relative to the *qr* axes.

or the other way around

$$\hat{x} = \cos\theta\,\hat{q} - \sin\theta\,\hat{r}$$
$$\hat{y} = \sin\theta\,\hat{q} + \cos\theta\,\hat{r}. \tag{8.9}$$

Try It 8.1

Check that Eqs. (8.8) and (8.9) make sense if $\theta = 0°$ and $\theta = 90°$ and check that the minus signs are in the correct positions. Remember that a $0°$ rotation is no rotation at all and that a $90°$ rotation means that the horizontal and vertical axes are interchanged.

BOB: Eq. (8.9) can be derived from Eq. (8.8) with a bit of algebra and trigonometry. Alternatively, we can be clever and think of the *xy* axes as rotated *clockwise* by the angle θ relative to the *qr* axes as shown in Figure 8.4. Given our convention, the clockwise angle is counted as negative. Remember that $\sin(-\theta) = -\sin\theta$ and that $\cos(-\theta) = \cos\theta$ is all we need to go from Eq. (8.8) to Eq. (8.9) without doing any algebra!

ALICE: Cardy, please be aware that some authors use the opposite convention (clockwise angles are positive), which means that they have minus signs in different places than we do.

CARDY: Thanks for the warning! I have also seen unit vectors $\hat{i}$ and $\hat{j}$ used in place of $\hat{x}$ and $\hat{y}$.

ALICE: Yes, there are lots of variations in notation throughout science and math. You just need to get used to that and feel comfortable adjusting the notation to match what you like the best. It's a little like spelling the same word different ways, such as grey versus gray. Those two words look different but mean the same thing.

8.3   Changing Quantum Basis States

BOB: Now that we have ordinary vectors under control, we're ready to apply the same ideas to quantum basis state vectors in state space. To make the connection obvious, let's use similar symbols to get started. Suppose that $|F\rangle$ is a quantum state vector and the unit vectors $|x\rangle$ and $|y\rangle$ are basis state vectors in a state space (see Figure 8.5).

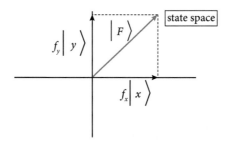

Fig. 8.5 A state space diagram with basis states $|x\rangle$ and $|y\rangle$ and a state vector $|F\rangle$.

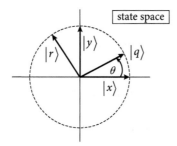

Fig. 8.6 A state space diagram with two sets of basis vectors. The dashed circle has a radius $= 1$.

BOB: We are restricting ourselves to basis state changes that can be expressed in term of a single angle θ. In Chapter 15, we will show you what to do in more general cases.

Suppose the qr axes are rotated relative to the xy axes as shown in Figure 8.6. What is the state space equivalent of Eq. (8.8) which relates the two sets of basis vectors? It's just what you might expect by comparing Figures 8.3 and 8.6.

$$|q\rangle = \cos\theta\,|x\rangle + \sin\theta\,|y\rangle$$
$$|r\rangle = -\sin\theta\,|x\rangle + \cos\theta\,|y\rangle\,, \tag{8.10}$$

where the angle θ is the state space angle between the corresponding state space basis vectors. Looking at this from a quantum perspective, we are writing $|q\rangle$ and $|r\rangle$ as superpositions of the states $|x\rangle$ and $|y\rangle$.

CARDY: That is cool. The math is the same; all you have done is change the symbols from regular vectors to state vectors.

ALICE: You have hit upon what I whimsically call the fundamental theorem of mathematics: If a result holds for symbols x, y, and z, then changing the symbols to a, b, and c still gives a valid result. Of course, the symbols must represent the same type of mathematical object: If x is a vector, then so must be a. But as long as you respect that rule, you have a very powerful theorem!

We can also use our computational basis state notation for Eq. (8.10) with $|0\rangle_{qr} = |q\rangle$, $|1\rangle_{qr} = |r\rangle$ and so on:

$$|0\rangle_{qr} = \cos\theta|0\rangle_{xy} + \sin\theta|1\rangle_{xy}$$
$$|1\rangle_{qr} = -\sin\theta|0\rangle_{xy} + \cos\theta|1\rangle_{xy}, \tag{8.11}$$

where we used subscripts on the basis state vectors to remind us of the basis we are using. The "reverse" relationship is then

$$|0\rangle_{xy} = \cos\theta|0\rangle_{qr} - \sin\theta|1\rangle_{qr}$$
$$|1\rangle_{xy} = \sin\theta|0\rangle_{qr} + \cos\theta|1\rangle_{qr}. \tag{8.12}$$

Note that Eqs. (8.11) and (8.12) can be used to express any change of basis for a two-state qubit. They are all we will ever need!

CARDY: But why do we need all of this stuff? I though you said we could use any set of basis states to describe a qubit's state. One is as good as any other.

ALICE: At some level, you are right, Cardy. All basis sets are equally fine. However, as we shall see, making measurements based on projections along different state space basis sets will give us important information about a quantum state that we can't get from just one basis set. Almost all interesting QIS and QC methods make use of multiple basis sets. An example of this is changing the orientation of the polarizer for the light experiments in Chapter 5. We are almost ready to look into this problem specifically, after a bit more preparation. Take it away, Bob.

BOB: We now tie all this to qubit quantum states and show explicitly what happens when we have a change of basis states. The basic procedure we want to describe is preparing a qubit in a state expressed in one set of basis states and then carrying out a measurement on that state with a measurement device whose set of basis states is different from the basis set used in preparing the state. The overall process is the following:

1. Express the qubit state as a superposition of the basis states used to prepare the state.
2. Transform the state description from the preparation basis states to the measurement basis states using the equivalent of Eqs. (8.11) and (8.12).
3. Use the squares of the measurement state coefficients to calculate the probabilities for the measurement outcomes of each of the measurement basis states. These probabilities will be used in almost every QIS and QC application.

Here is a specific example. Suppose that the qubit state $|\psi\rangle$ is prepared as a superposition state in the xy basis set:

$$|\psi\rangle = a_0|0\rangle_{xy} + a_1|1\rangle_{xy}. \tag{8.13}$$

We decide we want to measure the state using the qr basis; so we use Eq. (8.12) to express the xy basis vectors in terms of the qr basis vectors:

$$|\psi\rangle = a_0 \underbrace{\left\{ \cos\theta |0\rangle_{qr} - \sin\theta |1\rangle_{qr} \right\}}_{|0\rangle_{xy}} + a_1 \underbrace{\left\{ \sin\theta |0\rangle_{qr} + \cos\theta |1\rangle_{qr} \right\}}_{|1\rangle_{xy}}$$

$$= |0\rangle_{qr} \left\{ a_0 \cos\theta + a_1 \sin\theta \right\} + |1\rangle_{qr} \left\{ -a_0 \sin\theta + a_1 \cos\theta \right\}. \tag{8.14}$$

Note that in the second line of Eq. (8.14), we grouped the terms for $|0\rangle_{qr}$ and $|1\rangle_{qr}$ so that we can find the state coefficients—the terms inside the curly braces.

CARDY: I kinda see that, but I need to write it out for myself to make sure I know how the various pieces got moved around.

Try It 8.2

Follow Cardy's advice and write out the details of Eq. (8.14) starting from Eq. (8.13) and using the basis conversions in Eq. (8.11) or Eq. (8.12) (you decide which to use) in Eq. (8.14).

BOB: Our final step is to use the Born rule to calculate the probabilities that the measurement outcome states will be $|0\rangle_{qr}$ or $|1\rangle_{qr}$.

CARDY: Oh, yeah! I remember. We just calculate the squares of the various coefficients.

BOB: Exactly! We use $P\left(|0\rangle_{qr}\right)$ for the probability that the measurement outcome state will be $|0\rangle_{qr}$. So, the two probabilities are

$$P(|0\rangle_{qr}) = \left\{ a_0 \cos\theta + a_1 \sin\theta \right\}^2$$

$$P(|1\rangle_{qr}) = \left\{ -a_0 \sin\theta + a_1 \cos\theta \right\}^2. \tag{8.15}$$

A specific numerical example might be helpful. Let's use $\theta = 60°$ and $a_0 = 1/\sqrt{2} = a_1$.

Try It 8.3

Show that with Bob's numbers, we get $P(|0\rangle_{qr}) = 1/2 + \sqrt{3}/4$ and $P(|1\rangle_{qr}) = 1/2 - \sqrt{3}/4$. Also, show that the two probabilities add up to 1 as they ought to.

8.4 Three Basis Sets: $0°$, $60°$, and $−60°$

ALICE: After all of that abstract math, let's apply what we have learned to the change of basis for linearly polarized photons, which we will need in Chapter 14. And in fact, the change of basis states for photons is an operation commonly used in many quantum information processes.

We will consider three state-space sets of basis vectors, one with the $|hlp\rangle$ along the direction labeled $0°$ and the other two with the corresponding basis vectors rotated by $60°$ and $−60°$ from the initial direction $0°$. We will label the basis vectors as follows.

A. Basis vectors $|hlp\rangle_{0°}$ and $|vlp\rangle_{0°}$.

B. Basis vectors $|hlp\rangle_{60°}$ and $|vlp\rangle_{60°}$.

C. Basis vectors: $|hlp\rangle_{−60°}$ and $|vlp\rangle_{−60°}$.

We can use Eq. (8.10) to relate the various basis vectors. For example,

$$|hlp\rangle_{60°} = \cos 60° |hlp\rangle_{0°} + \sin 60° |vlp\rangle_{0°} = \frac{1}{2}|hlp\rangle_{0°} + \frac{\sqrt{3}}{2}|vlp\rangle_{0°}$$

$$|vlp\rangle_{60°} = -\sin 60° |hlp\rangle_{0°} + \cos 60° |vlp\rangle_{0°} = -\frac{\sqrt{3}}{2}|hlp\rangle_{0°} + \frac{1}{2}|vlp\rangle_{0°}. \tag{8.16}$$

Try It 8.4

Check Eq. (8.16) by using the state space diagram and the column vector representations. Then use Eq. (8.10) to write the basis vectors in terms of each other. Verify that the transformed basis states have unit length.

ALICE: We will also need the equations linking the $−60°$ basis states to the $0°$ basis states. As we mentioned before, you should be able to write those down immediately by replacing $60°$ in Eq. (8.16) with $−60°$.

Try It 8.5

Carry out the replacement as Alice has suggested. Keep that result handy for later work.

BOB: Let's see what happens to the measurement probabilities with these basis vector transformations. If I prepare a photon in the $|vlp\rangle_{60°}$ state, for example, and then have it interact with a linear polarization measurement device operating with $0°$ basis states, then the measurement device will give hlp in the $0°$ basis with probability $\left(-\sqrt{3}/2\right)^2 = 3/4$ and vlp for the $0°$ basis with probability $(1/2)^2 = 1/4$.

Try It 8.6

Check Bob's numbers and verify that they are correct.

In Chapter 14, we will see that the results developed in this section can be used to show that quantum systems have probability correlations that cannot be correctly described by classical physics even though the classical predictions are based on what seem to be obvious and intuitive ideas about the nature of objects and their properties. In other words, these quantum correlations show that our assumptions about the nature of reality are wrong!

8.5 Dot and Inner Products of State Vectors

BOB: It often turns out to be useful to express the trigonometric functions in the change of basis equations such as Eq. (8.10) in more abstract ways involving the basis state vectors themselves. We met up with this idea in Chapter 4, where we introduced the dot product of two state vectors. Recall that the dot product is also called the "scalar" or "inner" product.

Let's start with a quick review of the dot product for ordinary vectors and then switch to quantum state vectors.

As we mentioned in Chapter 4, the "dot product" is called "dot" because it is represented symbolically by a dot between two vectors. It gives us a number:

$$\vec{B} \cdot \vec{C} \equiv \left\| \vec{B} \right\| \left\| \vec{C} \right\| \cos\theta_{BC}, \tag{8.17}$$

where θ_{BC} is the angle between the two vectors. (See Figure 8.7). As before, the double vertical bars mean the length of the vector. Eq. (8.17) can be read as telling us how much of vector $\vec{B}$ lies along the direction of vector $\vec{C}$ multiplied by the length of vector $\vec{C}$ or vice versa; it treats $\vec{B}$ and $\vec{C}$ symmetrically.

If $\vec{C}$ is a unit vector, that is, if $\left\| \vec{C} \right\| = 1$, then

$$\vec{B} \cdot \vec{C} = \left\| \vec{B} \right\| \cos\theta_{BC} = B_C, \tag{8.18}$$

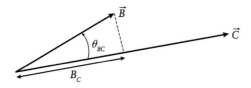

Fig. 8.7 Two vectors $\vec{B}$ and $\vec{C}$ and the angle θ_{BC} between them. B_C is the projection of the vector $\vec{B}$ along vector $\vec{C}$'s direction. B_C is called the perpendicular projection of $\vec{B}$ along $\vec{C}$.

and the dot product gives us directly the component of $\vec{B}$ that lies along the $\vec{C}$ direction.

If both the vectors have unit length, Eq. (8.17) becomes

$$\vec{B} \cdot \vec{C} = \cos \theta_{BC}. \tag{8.19}$$

In words: The dot product of two unit vectors is just the cosine of the angle between them. Using Eq. (8.19), we can write down some important results for the unit vectors in Figure 8.3. Here are a few representative ones:

$$\hat{x} \cdot \hat{x} = 1 \quad \hat{x} \cdot \hat{y} = 0 \quad \hat{q} \cdot \hat{x} = \cos \theta \quad \hat{q} \cdot \hat{y} = \sin \theta. \tag{8.20}$$

CARDY: Wait a minute! Your dot products in Eqs. (8.17) and (8.18) always had cosines, but you dropped a sine on me in Eq. (8.20). What's going on?

BOB: Yes, I was a bit sneaky there. Remember that the angle θ is the angle between $\hat{q}$ and $\hat{x}$. Then the angle between $\hat{q}$ and $\hat{y}$ is $90° - \theta$ and $\cos(90° - \theta) = \sin \theta$. That's how I got the $\sin \theta$ in the last term of Eq. (8.20).

CARDY: Now I see it. Thanks. My trigonometry is a bit rusty.

BOB: Don't worry about that. We have in front of us all the trig we are going to need. And most of the time, we will need only the numerical values of the cosines and sines. Those you can get from the calculator on your phone.

Try It 8.7

Explain each of the results in Eq. (8.20) by thinking about the length of the vectors and the angle between them.

CARDY: Whew! I haven't done so much trigonometry since high school. But I think I got the basic idea.

ALICE: If you are still a bit shaky, I suggest coming back to this section in a day or two and working through the details with paper and pencil. Fortunately, once we finish the following section, you will see that most of the trigonometry can be subsumed into one relatively simple expression.

BOB: Now let's use scalar products of quantum state basis vectors. In particular, we want to translate Eq. (8.20) to our notation for quantum state vectors. Here, we replace the scalar products with the Dirac brackets:

$$\hat{x} \cdot \hat{q} \Rightarrow \langle x | q \rangle = \cos \theta_{xq} \quad \hat{y} \cdot \hat{q} \Rightarrow \langle y | q \rangle = \cos \theta_{yq} = \sin \theta_{xq}. \tag{8.21}$$

> **Try It 8.8**
>
> Use the same ideas to translate $\hat{x} \cdot \hat{r}$ and $\hat{y} \cdot \hat{r}$ into Dirac bracket form.

We can now rewrite Eq. (8.10) as follows:

$$|q\rangle = \cos\theta|x\rangle + \sin\theta|y\rangle = |x\rangle \underbrace{\langle x|q\rangle}_{\text{projects } q \text{ along } x} + |y\rangle \underbrace{\langle y|q\rangle}_{\text{projects } q \text{ along } y}$$

$$|r\rangle = -\sin\theta|x\rangle + \cos\theta|y\rangle = |x\rangle \underbrace{\langle x|r\rangle}_{\text{projects } r \text{ along } x} + |y\rangle \underbrace{\langle y|r\rangle}_{\text{projects } r \text{ along } y} .$$

(8.22)

CARDY: How did the bracket terms get moved to the right of the basis state vectors?

BOB: I did that to show that there is an easy way to remember Eq. (8.22), which is the key expression for changing basis states. For example, for the first line, start with $|q\rangle$ on the far right and read to the left: Form the Dirac bracket with $\langle y|$ and then multiply the bracket with the basis state $|y\rangle$. Repeat for q and x. The second line does the same thing for r.

CARDY: Ah! Very nice. I just need to remember which of the brackets is a cosine and which a sine.

BOB: That's where a diagram like Figure 8.6 will help out.

> **Try It 8.9**
>
> Change the symbols in Eq. (8.22) to solve for $|x\rangle$ and $|y\rangle$ in terms of $|q\rangle$ and $|r\rangle$.

ALICE: Eq. (8.22) is all you will need to carry out basis state changes for qubits.

BOB: Let's look at Figure 8.8. In our quantum state vector formulation, the bracket $\langle q|\,F\rangle$ gives us the amplitude (the projection) of the state vector $|F\rangle$ along the $|q\rangle$ basis vector direction and similarly for $\langle r\,|F\rangle$. So, we now have an abstract, but highly useful, way of writing a general state vector

$$|F\rangle = |q\rangle\,\langle q|\,F\rangle + |r\rangle\,\langle r|\,F\rangle\,,$$

(8.23)

or in terms of basis vectors $|x\rangle$ and $|y\rangle$ we have

$$|F\rangle = |x\rangle\,\langle x|\,F\rangle + |y\rangle\,\langle y|\,F\rangle\,.$$

(8.24)

If you prefer using the versions with trig functions, simply use $\langle k\,|F\rangle = \cos\theta_{kF}$, where k is any of the basis state labels (q, r, x, or y).

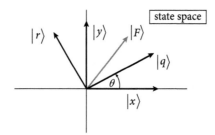

Fig. 8.8 A state vector $|F\rangle$ and two sets of orthogonal basis state vectors.

Mathematical aside

In our restricted quantum environment, $\langle F|\,w\rangle = \langle w|\,F\rangle$. In more general situations, one of the brackets is the complex conjugate of the other. See Chapter 15 for details.

8.6 **Spin-1/2 States**

ALICE: Now let's apply these ideas to a spin-1/2 system with two different measurement bases rotated by 90° in real space (45° in state space). See Figure 8.9. This is a combination we will see frequently in QIS and QC.

CARDY: If I am remembering correctly, we saw the difference between the state space and the real space angles in Chapter 7.

ALICE: That's right, Cardy. That angle relation is one of the peculiar features of spin-1/2 systems in quantum mechanics. In quantum state language, for a spin-1/2 system, a common measurement basis set will be the spin-up and spin-down states associated with the physical-space z-direction (0° basis states). With $2\theta = 90^\circ$, the measurement direction is now the physical space horizontal direction, and we indicate those basis states as $|\rightarrow\rangle$ and $|\leftarrow\rangle$ instead

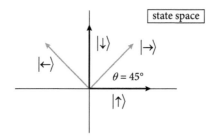

Fig. 8.9 A state space representation with spin-up and spin-down 0° basis states $|\uparrow\rangle$ and $|\downarrow\rangle$ and with 90° (physical space) basis states $|\rightarrow\rangle$ and $|\leftarrow\rangle$.

of the up and down arrows. Note that the arrows indicate the physical space direction, not the state space direction of the basis states. See Figure 8.9.

The basis states $|\rightarrow\rangle$ and $\langle\leftarrow|$ can of course be written as a superposition of the spin-up and spin-down states using the state space angle θ:

$$|\rightarrow\rangle = \cos\theta\,|\uparrow\rangle + \sin\theta\,|\downarrow\rangle$$
$$|\leftarrow\rangle = -\sin\theta\,|\uparrow\rangle + \cos\theta\,|\downarrow\rangle. \tag{8.25}$$

ALICE: Let's look at a specific example: Suppose Bob decides to use a measurement basis for the spin direction that is at angle 90° (counterclockwise) in physical space from the standard z-direction (0°). In state space, the rotation angle is then just 45° (half of 90°) as shown in Figure 8.9.

Try It 8.10

Use Eq. (8.25) to show that for $2\theta = 90°$, we get

$$|\rightarrow\rangle = \frac{1}{\sqrt{2}}|\uparrow\rangle + \frac{1}{\sqrt{2}}|\downarrow\rangle$$
$$|\leftarrow\rangle = -\frac{1}{\sqrt{2}}|\uparrow\rangle + \frac{1}{\sqrt{2}}|\downarrow\rangle. \tag{8.26}$$

Try It 8.11

Show using Eqs. (8.11) and (8.12) that following give the "reverse" connections:

$$|\uparrow\rangle = \frac{1}{\sqrt{2}}|\rightarrow\rangle - \frac{1}{\sqrt{2}}|\leftarrow\rangle$$
$$|\leftarrow\rangle = \frac{1}{\sqrt{2}}|\rightarrow\rangle + \frac{1}{\sqrt{2}}|\leftarrow\rangle. \tag{8.27}$$

ALICE: To get some practice with changing basis states, let's rewrite some of the previous results using the column vector form of the basis state vectors. To keep things straight, we will add subscripts to the column vectors to indicate the basis system being used. Eq. (8.26) is then written as

$$\begin{pmatrix} 1 \\ 0 \end{pmatrix}_{90°} = \frac{1}{\sqrt{2}} \begin{pmatrix} 1 \\ 0 \end{pmatrix}_{0°} + \frac{1}{\sqrt{2}} \begin{pmatrix} 0 \\ 1 \end{pmatrix}_{0°} = \begin{pmatrix} \frac{1}{\sqrt{2}} \\ \frac{1}{\sqrt{2}} \end{pmatrix}_{0°}$$

$$\begin{pmatrix} 0 \\ 1 \end{pmatrix}_{90°} = -\frac{1}{\sqrt{2}} \begin{pmatrix} 1 \\ 0 \end{pmatrix}_{0°} + \frac{1}{\sqrt{2}} \begin{pmatrix} 0 \\ 1 \end{pmatrix}_{0°} = \begin{pmatrix} -\frac{1}{\sqrt{2}} \\ \frac{1}{\sqrt{2}} \end{pmatrix}_{0°}. \tag{8.28}$$

The final equality in each line of Eq. (8.28) reminds us of the simple way of picturing the basis state: The entries in the column vectors are the "coordinates" of the tip of the basis state vectors.

Try It 8.12

Repeat the basis transformation used in section 8.4, but now for three spin-1/2 basis sets, where the angles in state space are $0°$, $120°$, and $-120°$. What are the angles in physical space?

8.7 General Basis State Transformations

BOB: So far, we have used rather specific basis states. Let's now write out a more general relationship. Suppose that Alice wants to use a basis labeled A, and I want to use one labeled B. The basis states are $|A_0\rangle$ and $|A_1\rangle$ for basis A and $|B_0\rangle$ and $|B_1\rangle$ for basis B.

CARDY: Before we go on, what is the meaning of 0 and 1 in the subscripts of those state vectors?

ALICE: Remember those symbols are just names; we could have used 1 and 2 or red and blue or Ted and Carol, but we include 0 and 1 to remind us of the convention (ordered basis states) that connects those states to bits 0 and 1 and to the computational basis states $|0\rangle$ and $|1\rangle$.

If we express my basis states in terms of Bob's basis states, we get the equivalent of Eq. (8.11):

$$|A_0\rangle = \cos\theta_{AB} |B_0\rangle + \sin\theta_{AB} |B_1\rangle$$
$$|A_1\rangle = -\sin\theta_{AB} |B_0\rangle + \cos\theta_{AB} |B_1\rangle. \tag{8.29}$$

In Eq. (8.29), θ_{AB} is the *state space* angle between the $|A_0\rangle$ basis direction and the $|B_0\rangle$ basis direction with $|A_0\rangle$ located counterclockwise from $|B_0\rangle$. The equation will be used over and over again as we discuss quantum algorithms

It may take a while to unpack these equations, but the effort will be worthwhile because variations on these equations show up in many quantum information processing situations. You should be congratulated once you feel reasonably comfortable with them. This is a major accomplishment!

Try It 8.13

Draw the state space diagram showing Alice's and Bob's basis vectors with the relative orientation implied by Eq. (8.29).

BOB: Suppose we want to go the other way around and switch from Alice's basis to mine. That is easy to do by just reversing the algebraic sign of the rotation angle θ. In other words, to go from Alice's $|A_0\rangle$ to my $|B_0\rangle$ requires a clockwise rotation in state space by $-\theta$. So, all we need to do is to switch A and B state labels and replace θ in Eq. (8.29) with $-\theta$ while remembering that $\cos(-\theta) = \cos\theta$ and that $\sin(-\theta) = -\sin\theta$. So, we end up with

$$|B_0\rangle = \cos\theta_{AB}|A_0\rangle - \sin\theta_{AB}|A_1\rangle$$
$$|B_1\rangle = \sin\theta_{AB}|A_0\rangle + \cos\theta_{AB}|A_1\rangle. \tag{8.30}$$

Try It 8.14

Rewrite Eqs. (8.29) and (8.30) using the symbolic bracket form of Eq. (8.22).

ALICE: How do these results allow us to express a general state vector $|\psi\rangle$ in terms of various basis states? Let's write $|\psi\rangle$ explicitly in terms of the basis states $|A_{0,1}\rangle$ and $|B_{0,1}\rangle$ that we have used in this section. To be specific, let's look at the case when Bob has prepared the state $|\psi\rangle$ in the $|B_{0,1}\rangle$ basis

$$|\psi\rangle = b_0|B_0\rangle + b_1|B_1\rangle, \tag{8.31}$$

but I decide to make measurements in terms of my $|A_{0,1}\rangle$ basis. So, I use Eq. (8.30) to replace the B basis states with the A basis states:

$$|\psi\rangle = b_0\underbrace{\{\cos\theta_{AB}|A_0\rangle - \sin\theta_{AB}|A_1\rangle\}}_{|B_0\rangle} + b_1\underbrace{\{\sin\theta_{AB}|A_0\rangle + \cos\theta_{AB}|A_1\rangle\}}_{|B_1\rangle}$$
$$= \{b_0\cos\theta_{AB} + b_1\sin\theta_{AB}\}|A_0\rangle + \{-b_0\sin\theta_{AB} + b_1\cos\theta_{AB}\}|A_1\rangle. \tag{8.32}$$

In the second line of Eq. (8.32), I grouped the terms associated with $|A_0\rangle$ separately from those associated with $|A_1\rangle$.

Using the Born rule, we find the following probabilities for the two measurement outcomes:

$$P(|A_0\rangle) = \{b_0\cos\theta_{AB} + b_1\sin\theta_{AB}\}^2$$
$$P(|A_1\rangle) = \{-b_0\sin\theta_{AB} + b_1\cos\theta_{AB}\}^2. \tag{8.33}$$

These results are just a straightforward generalization of Eq. (8.15).

BOB: We now have all we need to know about changing sets of basis states. In the remainder of the chapter, we will show you an important application of polarized light states, and then take up an important issue of plus and minus signs in quantum states. We will wrap up with an application of these ideas to quantum encryption.

8.8   The Three-Polarizing-Sheets Experiment

BOB: I want to show you a simple experiment you can do at home. I believe this experiment should be done by everyone who is interested in QC and QIS. We will use changes in basis states to illustrate what I think is the fundamental feature of quantum mechanics: superposition states. All you need are three sheets of linear polarizing material and a source of light. Your eyes will serve as the measurement detector. As mentioned before, for the polarizing sheets you could use polarizing sunglasses. Relatively inexpensive linear polarizing sheets are available from many suppliers online. Some suggestions are provided in the Further Reading section of Chapter 4.

First, you need to find the transmission axis directions for the polarizing sheets. This is easy to do using the setup shown in Figure 4.5, where you view light reflected from a horizontal surface such as a shiny floor or the surface of water in a lake or pond. That light will have predominantly horizontal linear polarization. View the light reflected from that surface with the polarizing sheet placed between the horizontal surface and your eyes, with the sheet perpendicular to the line from the surface to your eyes. Then rotate the polarizing sheet around that line until you see the maximum amount of light coming through the polarizing sheet. The transmission axis of the polarizing sheet is now parallel to that horizontal surface. Usually (but not always) the edges of the polarizing sheet are cut parallel or perpendicular to that axis. Use a marker or a small piece of tape to mark that axis. Do that for all three polarizing sheets. In what follows, we will call that the hlp direction for each sheet.

The next step is to place sheet A, with its transmission axis horizontal, in front of an illuminated surface. (You could put sheet A in front of a lamp or flashlight, but the illuminated surface provides a more uniformly distributed light intensity.) Then arrange sheet B, placed between A and your eyes, with its transmission axis perpendicular to that of sheet A (Figure 8.10 center image). You should find that almost no light from the illuminated surface comes through that combination.

We can describe that situation using our linear polarization basis states. The light from the illuminated surface first hits sheet A, whose transmission axis is horizontal. That means that the light exiting sheet A is described by the state vector $|\text{hlp}\rangle_A$. Some people would say that sheet A has collapsed the light's state function from whatever it was before entering A to $|\text{hlp}\rangle_A$ for the light that gets through the sheet. I prefer to say that we use $|\text{hlp}\rangle_A$ to describe the polarization state of the light after it has passed through sheet A; so, I don't have to worry about the mechanism by which sheet A changed the light's polarization state.

Sheet B has its transmission axis vertical. That means that it will not transmit any light with the state vector $|\text{hlp}\rangle_A$ since the basis states $|\text{hlp}\rangle_A$ and $|\text{hlp}\rangle_B$ (equivalent to $|\text{vlp}\rangle_A$) are orthogonal. So far, so good. Just what we expect for "crossed" linear polarizers.

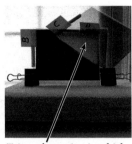

Triangular region in which
C overlaps with both A and B

Fig. 8.10 The three polarizing sheets experiment. On the left is polarizing sheet A with its transmission axis horizontal. In the center, polarizing sheet B, with its transmission axis vertical, is placed between A and the camera. On the right, sheet C, with its transmission axis at about 45° is placed *between* sheets A and B. The arrow points to the region in which sheet C overlaps sheets A and B. The gray vertical and horizontal bars in the background are mullions in the window behind the sheets.

Now comes the exciting part. Place the third polarizing sheet C *between* sheets A and B (Figure 8.10 right-side image). You should see something truly amazing. (It is amazing if you think about it in the right way, of course!) You should see (Figure 8.10) that light now comes through the region where sheet C is inserted between A and B. I find that amazing because adding sheet C has somehow allowed light to get through, even though sheets A and B have their transmission axes perpendicular and block all the other light. How can that happen?

Let's think about it this way: In order for light coming from sheet A to get through sheet B, sheet C must change the linear polarization of the light so that at least some of the light has linear polarization along the transmission axis of sheet B. To keep the explanation relatively simple, let's assume that you have oriented sheet C's transmission axis so that it is at 45° with respect to A's and B's transmission axes. Of course, light that comes through C has its linear polarization direction along C's transmission axis and the appropriate polarization state is represented by $|\text{hlp}\rangle_C$.

Now let's use polarization state vectors to describe the situation more formally. After the light leaves sheet A, its linear polarization state is $|\text{hlp}\rangle_A$. From Eq. (8.29) we can write that state as a superposition of the sheet C linear polarization states

$$|\text{hlp}\rangle_A = \frac{1}{\sqrt{2}} \left(|\text{hlp}\rangle_C + |\text{vlp}\rangle_C \right). \qquad (8.34)$$

We see that there is a linear polarization component of that light along the C transmission axis $(\text{hlp})_C$. The amplitude squared $= \frac{1}{2}$; so, we predict that 50% of the light from A gets through C.

Try It 8.15

Translate Eq. (8.29) into Eq. (8.34). Hint: Evaluate the appropriate cosines and sines and change the names of the states.

Next the light will enter sheet B. We write the state $|hlp\rangle_C$ as a superposition of sheet B's basis states

$$|hlp\rangle_C = \frac{1}{\sqrt{2}} \left[|hlp\rangle_B + |vlp\rangle_B \right]. \tag{8.35}$$

So, we see that when the light encounters sheet B, it has a component along the transmission direction (hlp)$_B$ of sheet B. We predict that 50% of the light from C will emerge from sheet B. So, overall with sheet C's transmission axis at $45°$ to those of A and B, we predict that $0.5 \times 0.5 = 0.25$ (25%) of the light that gets through A will emerge from C. That is what you should see on the far right part of Figure 8.10.

ALICE: Let me summarize what Bob showed: The light entering sheet C from A is in a superposition state of the C basis states; so, there is a probability of light getting through. The same is then true for the light entering sheet B; it is in a state that can be written as a superposition of B's linear polarization basis states. If there is no sheet C, the light transmitted through A is described by the state $|hlp\rangle_A = \cos\theta_{AB}|hlp\rangle_B + \sin\theta_{AB}|vlp\rangle_B$. But since $\theta_{AB} = 90°$, there is no projection of $|hlp\rangle_A$ along the transmission state $|hlp\rangle_B$ of polarizer B. Hence, none of the light gets through the "crossed" polarizer combination of A and B.

Now let's generalize the result to any angle between the A and C transmission axes. Rotate sheet C to see what happens as the angle between C's transmission axis and those of A and B changes. You should see a change in the amount of light that gets through. For the situation in which the A and B transmission axes are perpendicular, the more general case can be expressed in terms of the angle θ_{AC} between the A sheet and C sheet transmission axes:

$$I_B = \frac{1}{4} I_A \sin^2 2\theta_{AC}. \tag{8.36}$$

I have used I_A and I_B to indicate the intensity of light that emerges from sheet A and from sheet B, respectively, in the three-sheet setup.

Try It 8.16

Rotate sheet C relative to A and qualitatively check that Eq. (8.36) describes what is going on. Hints: $\sin 0° = 0$, $\sin 90° = 1$, $\sin 180° = 0$.

CARDY: That seems to work because it matched what I see when I rotate sheet C, but I don't understand how you got Eq. (8.36).

ALICE: It's not too hard to figure that out. Recall that in Chapter 5 we showed that when we rotated just two polarizing sheets, the state amplitudes got multiplied by the cosine of the angle of rotation in state space and the transmitted light intensities change as the square of the cosine. For linearly polarized light, that angle is the same angle by which we rotate the polarizing sheets relative to each other in physical space. In fact, we saw that the result for the new amplitude is just the old amplitude multiplied by the cosine of the angle by which the state vectors were rotated. Using a_A for the amplitude of the light coming through sheet A, with similar terms for sheets B and C, we have

$$a_C = a_A \cos \theta_{AC} \text{ and } a_B = a_C \cos \theta_{BC}$$
$$a_B = a_A \cos \theta_{BC} \cos \theta_{AC}. \tag{8.37}$$

From the geometry we know that $\theta_{AC} + \theta_{BC} = 90°$. From trigonometry, we remember that $\cos \theta = \sin (90° - \theta)$ and $\sin 2\theta = 2 \sin \theta \cos \theta$ (the "double angle" formula). Using those results, we find that the second line of Eq. (8.37) becomes

$$a_B = a_A \sin \theta_{AC} \cos \theta_{AC} = \frac{a_A}{2} \sin 2\theta_{AC}. \tag{8.38}$$

CARDY: I kinda remember what you did from trig, but I will need to check that out myself.

ALICE: Always good to do that. The final step is to recall that the probability of a photon being detected is given by the square of the amplitude. For light sources with lots of photons, the intensity of the light is proportional to that probability. Assembling all those details gives us the result stated in Eq. (8.36).

What you have seen is that linearly polarized light seems to be well described by superposition states. These effects were known long before the quantum nature of light (photons) was discovered. The pre-quantum explanations are based on the connection between the linear polarization direction and the direction of the electric field (a vector) according to the wave model of light.

BOB: The demonstration by itself does not show that the same "superposition effects" would work for individual photons: the particle model of light. However, experiments have been done in which photons are detected individually, and the results show that the superposition quantum state model is correct. If you had the equipment to observe the light photon by photon for this experiment, you would have direct evidence that the superposition state vector and probability interpretation of the squares of the amplitudes are correct descriptions of how nature behaves. The classical electric field model is fine when you are dealing with large numbers of photons in your light beam, but to account for the behavior of individual photons and the associated probabilities, we need quantum mechanics.

ALICE: We should also point out that this experiment demonstrates that polarizing sheets can be used both as part of measurement devices and as state preparation devices. In particular, we can say that sheet A prepares the photons it transmits in the $|hlp\rangle_A$ state associated with its transmission axis. Similarly, for sheets B and C. We could also say that sheet B can be used to measure the probability that photons coming from sheet C will have a component of their linear polarization along B's transmission axis.

CARDY: That's a cool experiment. I'm going to get myself some polarizing material and show this to my friends.

8.9   A "Minus 1" Mystery

ALICE: There remains another quantum state detail that we should discuss. In manipulating quantum state vectors, we will often find an expression like $-\left|\downarrow\right\rangle$, for example, in Eq. (8.26) in the first term of the lower line. What does the negative of a quantum state vector mean? For ordinary vectors $-\vec{F}$ is a vector that has the same length (magnitude) as the vector $\vec{F}$ but points in the opposite direction. The "obvious" generalization to quantum states is that in state space $-\left|\downarrow\right\rangle$ points in the direction opposite to that of $\left|\downarrow\right\rangle$. Note that $-\left|\downarrow\right\rangle \neq \left|\uparrow\right\rangle$. The direction is reversed in state space, not in physical space. In physical space, spin-up is in the direction opposite to spin-down, but not in state space.

CARDY: So, it seems that a minus sign just reverses the direction of the vector in state space. Is that right?

BOB: That's it! (See Figure 8.11.) Having the minus sign does not change the length of the state vector. It does change the sign of the components of the state vector along the basis vector directions, but remember that it is the square of those components that gives the probabilities that measurements will give results along the basis state directions. So, those are unchanged by an overall minus sign. Let's elaborate on that important result.

The distinction between $-\left|\downarrow\right\rangle$ and $\left|\uparrow\right\rangle$ is even more obvious if we write things out using column vectors:

$$-\left|\downarrow\right\rangle \Rightarrow -\begin{pmatrix} 0 \\ 1 \end{pmatrix} = \begin{pmatrix} 0 \\ -1 \end{pmatrix} \neq \begin{pmatrix} 1 \\ 0 \end{pmatrix}. \tag{8.39}$$

ALICE: Since quantum state vectors are meant to describe the quantum states of a qubit and those states are supposed to carry all the information about possible measurements, we need to look at how the probabilities associated with the quantum state change if the state vector is replaced by its negative. Bob stated the results, but let's look at the finer details. Remember that the probabilities are the only properties of a quantum state we can measure. Let's express the spin-up state vector in terms of Bob's basis state vectors:

$$\left|\uparrow\right\rangle = b_0 \left|B_0\right\rangle + b_1 \left|B_1\right\rangle. \tag{8.40}$$

By the Born rule, the probability of observing the spin along the direction associated with B_0 is $(b_0)^2$, while the probability of observing the spin along B_1 is $(b_1)^2$. Now let's look at the

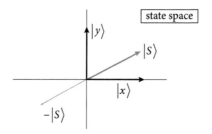

Fig. 8.11   A state space diagram showing a state vector $\left|S\right\rangle$ and its negative $-\left|S\right\rangle$.

negative of the state in Eq. (8.40):

$$-\left|\uparrow\right\rangle = -b_0\left|B_0\right\rangle - b_1\left|B_1\right\rangle. \tag{8.41}$$

Again, applying the Born rule, we find that the probability of observing the spin along B_0 is $(-b_0)^2 = (b_0)^2$, the same result we found before. The same holds true for spin along B_1. We conclude that the two states $\left|\uparrow\right\rangle$ and $-\left|\uparrow\right\rangle$ are *equivalent* in the sense of giving the same probabilities for observing the spin along the two directions associated with the basis set. In other words, measurements of the spin direction do not provide us with a way of distinguishing between the two states.

Let's look at a related situation that shows where a minus sign does in fact lead to a distinguishable result. This distinction turns up quite frequently in QC applications. To illustrate this point, we use the spin-1/2 states

$$\left|\Psi\right\rangle = \frac{1}{\sqrt{2}}\left\{\left|\rightarrow\right\rangle + \left|\leftarrow\right\rangle\right\}$$

$$\left|\Phi\right\rangle = \frac{1}{\sqrt{2}}\left\{\left|\rightarrow\right\rangle - \left|\leftarrow\right\rangle\right\}. \tag{8.42}$$

We might be tempted to say, based on our previous examples, that the two states in Eq. (8.42) are equivalent (in the sense we specified) because we showed that $\left|\leftarrow\right\rangle$ and $-\left|\leftarrow\right\rangle$ are equivalent. But, in fact, the two states in Eq. (8.42) are not equivalent: As we shall see, if we observe the spin orientation (up or down) along the $0°$ axis, we get spin-down with probability 1 and spin-up with probability 0 for $\left|\Psi\right\rangle$ in Eq. (8.42). For $\left|\Phi\right\rangle$, we get spin-up with probability 1 and spin-down with probability 0. In other words, by making measurements using the $0°$ basis states, we can distinguish between the two.

CARDY: Ah! I see the difference. For the equivalent states, you are changing the sign of both terms in the state expression while in Eq. (8.42) only the relative signs change.

ALICE: That is absolutely right! Quantum mechanicians say that there is a relative "phase difference" between the two terms and that relative phase has consequences for measurements. That turns out to be a big deal in quantum mechanics and in QC. In other words, two superposition states with different relative signs between the states in the superposition are not equivalent states.

BOB: Now let's work out what you get if these states in Eq. (8.42) interact with a measurement device with its measurement direction in the z-direction ($0°$). The simplest way to figure this out is to express the states in Eq. (8.42) in terms of the basis states in the measurement orientations $\left|\uparrow\right\rangle$ and $\left|\downarrow\right\rangle$. Eq. (8.26) has the results we need. Making the appropriate substitutions, we find that

$$\left|\Psi\right\rangle = \frac{1}{\sqrt{2}}\left\{\frac{1}{\sqrt{2}}\left|\uparrow\right\rangle + \frac{1}{\sqrt{2}}\left|\downarrow\right\rangle - \frac{1}{\sqrt{2}}\left|\uparrow\right\rangle + \frac{1}{\sqrt{2}}\left|\downarrow\right\rangle\right\}$$

$$= \left|\downarrow\right\rangle. \tag{8.43}$$

Try It 8.17

Carry out the same calculation for the state $|\Phi\rangle$ and show that spin measurements along the $0°$ direction give spin-up with probability 1.

Eq. (8.43) tells us that if we prepare the qubit in the state $|\Psi\rangle$ and then observe the spin along $0°$, we get spin-down every time (that is, with probability 1).

ALICE: The take-home message is that states with different relative signs can lead to different probabilities for measurement outcomes, or conversely, we can measure relative signs by making measurements with different basis states. Thus, those states are not equivalent. As we shall see in later chapters, many quantum algorithms make use of relative phases.

8.10 Bennett–Brassard-84 Encryption Protocol

BOB: Now that we understand changes of basis states, we want to reward you by looking at an actual application of these ideas in QIS. The method is called the BB84 encryption protocol after its inventors Charles Bennett and Gilles Brassard, who developed it—surprise—in 1984. The basic idea is that Alice and I, located some distance apart, want to establish a secret key (a string of 0s and 1s) to encrypt messages. Alice needs to send the key to me, but we are worried that someone, perhaps you, Cardy, is eavesdropping on our communications. The process we will use is an example of quantum key distribution (QKD).

CARDY: I have heard about encrypted messages, but I am not sure how a key is used in the encryption.

BOB: Encryption is a big business in the digital age and there are many, many different encryption schemes. Let me give you an example of one that I made up. Suppose the message we want to send is a plain English text. We might encrypt the message by replacing each letter in the message by the letter located a certain number of places to the right in the alphabet. For an example, we might replace an "a" in the message with a "d" three places to the right of "a" in the alphabet. This process is called a substitution encryption. Unfortunately, it is relatively easy to crack. But you can make the encryption more secure by varying how many places to the right you go to find the replacement letter. That is where the key comes in. Again, there are many ways you might do this, but here is a simple one. As I mentioned before, a key is often just a string of 1s and 0s. Let's suppose we arrange the key's 1s and 0s into groups, each with three bits. So, our key might look like

$$110 \quad 001 \quad 101 \quad 100 \quad 111 \quad 010 \ldots . \tag{8.44}$$

We then carry out the replacement of the first letter in the original message using the number represented by the three bits. For a three-bit binary number, these go from 0 to 7. The key in Eq. (8.44) tells us to replace the first letter of the message with the letter six ($= 110$ in binary) places to the right, the second letter one place to the right, the third letter five places to the right, and so on. Then when the encrypted message is received the key can be used to replace the letters with the letter located the appropriate number of places to the left. That process allows you to reconstruct the original message. If you choose the key bits randomly, you will have an encryption method that is difficult to crack.

ALICE: Of course, the key must be kept from everyone except those who are worthy or deemed worthy to have access to the secret message, namely Bob and me. The BB84 protocol is a way to use the properties of quantum states to allow Bob and me to generate a shared key while checking that no one else has gained access to it.

Try It 8.18

Suppose that Alice and Bob decide to use a classical communications channel with ordinary 0s and 1s; that is, the only states are 0 and 1. Describe what happens if Cardy is eavesdropping on that channel by intercepting a bit, recording its value, and then sending the same bit value to Bob. Can Alice and Bob detect that eavesdropping?

ALICE: The BB84 quantum protocol is based on the idea of sending qubits whose states are described by superpositions in one basis set but not in another. If both Bob and I measure the qubit state using the set of basis states in which there is no superposition, our measurement results are guaranteed to be the same. If, however, Bob measures the qubit state using a basis set different from the one I used, the mathematics that we just learned will come into play and Bob's measurement result will not always match my result. Let's see how this works.

I start with a long list of randomly chosen classical 0s and 1s. I send Bob a linearly polarized photon qubit for each bit in the list. I will prepare the qubit state in either the $0°$ or the $45°$ linear polarization basis, chosen at random and with equal probability. If the bit I want to send is a 0, I prepare the qubit in the $|0\rangle$ (hlp) state in that chosen basis and launch it to Bob. If the bit I want to send is a 1, I prepare the qubit in the $|1\rangle$ (vlp) state in my chosen basis and send that qubit off. I continue until I have sent the entire sequence of bits, choosing the preparation basis at random for each bit. I record both the basis used and the bit that is sent. An example of this is in Table 8.1.

BOB: When I receive one of Alice's photons, I choose at random one of the two basis sets to observe the linear polarization state of the qubit. That is, I am choosing a measurement basis with no communication with Alice. I won't always match her choice of basis states because I don't know in advance what basis set she used to prepare the qubit state.

Table 8.1 Alice's record. The top row indicates the basis set that Alice chose and the second row is the classical bit she wants to send in her message.

Basis	$0°$	$45°$	$0°$	$45°$	$45°$	$0°$	$0°$	$0°$	$45°$	$45°$
Bit	1	0	1	1	0	1	1	1	0	0

Table 8.2 Bob's record. The top row indicates the basis set Bob chose and the second row is the result of his observation.

Basis	$45°$	$0°$	$0°$	$0°$	$45°$	$45°$	$0°$	$45°$	$0°$	$45°$
Bit	1	1	1	0	0	0	1	1	0	0

I observe the polarization state when the photon arrives and write down my choice of measurement basis, a 0 if the observation is hlp, and a 1 if it is vlp. To be concrete, think of using a polarizing beam splitter as the measurement device. Table 8.2 shows a sample of the information I record.

BOB: If I am using a measurement basis that is the same one Alice used to prepare the qubit, my result will agree with what Alice sent. However, for the $0°$ and $45°$ basis sets, if I chose the other basis set, 50% of the time I will get 0 and 50% of the time I will get 1, no matter what Alice sent. Look at the first four entries in Tables 8.1 and 8.2. Our measurement bases are the same only once and, of course, I observed the same bit that Alice sent. We did not choose the same measurement basis for three of those entries, but for one we still had matching results. Overall, my results will agree with Alice's only 75% of the time. However, I don't yet know which runs have matching basis states and which don't.

CARDY: Whoa! Where did 75% come from?

BOB: There are two parts to the answer. First, given the way we choose basis sets, Alice and I choose the same basis set only 50% of the time. When we do, we get the same bit results. We agree 100% of the time.

CARDY: Okay, I see that. But how do we get to 75%?

BOB: To see how we get to 75%, we need to look at the other 50% of the runs where we don't choose the same basis set. In that situation, our measurement results agree 50% of the time and disagree 50% of the time. Now let's put this together to find the overall probability of the runs that lead to our getting the same results. Let's write this as the product of probabilities. For example, the probability that Alice and I get the same results given that we use the same basis states is 1 and the probability that we use the same basis states is 0.5.

$$P(A \text{ \& } B \text{ same results}) = P(A \text{ \& } B \text{ same results given the same basis states})$$
$$\times P(\text{same basis states})$$
$$+ P(A \text{ \& } B \text{ same results given different basis states})$$
$$\times P(\text{different basis states})$$
$$= 1 \times 0.5 + 0.5 \times 0.5 = 0.75.$$

Overall, we should agree 75% of the time and disagree 25% of the time.

ALICE: To show where the different basis state probabilities come from, we may use the math we learned about superpositions with different basis states. For example, suppose Bob is going to use the $45°$ basis for his measurements but I used the $0°$ basis to prepare the qubit state. I decide to send a vlp photon in that basis. Then we can use Eq. (8.29) to relate the state for a vlp photon in the 0° basis to hlp and vlp states in the 45° basis:

$$|\text{vlp}\rangle_{0^\circ} = \cos 45^\circ |\text{hlp}\rangle_{45^\circ} + \sin 45^\circ |\text{vlp}\rangle_{45^\circ} = \frac{1}{\sqrt{2}}|\text{hlp}\rangle_{45^\circ} + \frac{1}{\sqrt{2}}|\text{vlp}\rangle_{45^\circ}. \qquad (8.45)$$

Using the Born rule, we find that the probability that Bob gets hlp is ½; that is, 50% of the time for those conditions, Bob will observe a horizontally polarized photon in the 45° basis.

Try It 8.19

Examine the results in Tables 8.1 and 8.2 and see if they agree with the claims in the previous paragraphs.

CARDY: I did Try It 8.19 and the results agree with the claims.

BOB: That's good, but remember the tables give only the results of a limited set of trials. Good statistics needs many trials.

Now, let's suppose that Alice and I want to have a secret key with N bits. The following procedure will do two things: (1) determine if an eavesdropper has been listening in to our communications, and (2) determine the key to be used for subsequent communications. Alice transmits $4N$ qubits using the procedures described above. Alice and I then send each other the sequence of basis sets chosen, keeping the bits sent and received (the lower rows in Tables 8.1 and 8.2) secret. We go through the list of basis sets and keep only the results for which we both selected the same basis set. This will leave us with about $2N$ results.

CARDY: Why did Alice send $4N$ bits if you need only N?

BOB: We will explain this in just a minute. The quick answer is that we are going to throw away ¾ of the results and keep only ¼. You will soon see why after we address eavesdropping.

Since we sent the list of basis state choices over an insecure communications channel, it is possible that an eavesdropper might have listened in or tapped the channel. However, the eavesdropper only gets information about the bases used to prepare and read the bits and nothing about the actual bits in the key. Knowing that Alice and I used the 45° basis to generate a particular bit in the key is as useful as knowing two spies are using Dostoevsky's *Crime and Punishment* to generate the key. You still do not know how they are using the hundreds of pages in the book to generate and read the cypher, so the information is useless.

But there is an advantage to this quantum protocol: We can use the results to check if someone is eavesdropping. It works like this: From the $2N$ results where we used the same basis, we select N entries and share that list of results (key bits) using an unencrypted, insecure communications channel (e.g. our smart phones). We expect that if no one is eavesdropping or otherwise messing with our qubit communications channel (the one we used to send the photons), if Alice sent a 1, I should have received a 1 and if she sent 0, I should have received a 0.

ALICE: What happens if someone is indeed eavesdropping on the qubit channel? Let's suppose it is Cardy, even though we know Cardy would never do such a thing. Cardy wants to determine the sequence of 0s and 1s that I am sending to find our secret key. So Cardy lets the qubit I sent interact with a measurement device using either the $0°$ or $45°$ basis set. On average, half the time Cardy will choose the basis states I used and will successfully read the 0 or 1 which I sent. Cardy can then launch the qubit on to Bob without being detected. Bob observes the bit he would have seen if Cardy had not been eavesdropping.

However, half the time Cardy chooses the "wrong" basis set (the one I didn't use) and has only a 50% chance of getting the right 0 or 1 because the qubit will be in a state of superposition with respect to Cardy's basis. In this scenario, when Cardy sends a qubit on to Bob, his measurement result will be what I sent only 50% of the time.

In summary, we would expect to have 100% agreement between my bits and Bob's bits in the N selected runs if there is no eavesdropping. If we find that result, we will then keep the remaining N bits secret and use them as the encryption key. However, if Cardy has been eavesdropping, we will get only about 75% agreement instead of having 100% agreement, assuming that Cardy has been intercepting every photon. In other words, about ¼ of the time, we will get different results. It is ¼ because the probability of Cardy's choosing the wrong basis states is ½, and given the wrong basis, the probability of sending the wrong bit to Bob is ½.

CARDY: But why couldn't I, if I were eavesdropping, simply copy the qubit state that Alice sent? I measure one of the copies and send the other to Bob, who will get what he would have received with no eavesdropping?

ALICE: Here is where quantum mechanics enters. You recall the no-cloning theorem we mentioned in Chapter 5. There we showed that there is no way to copy a quantum state. So, your scheme won't work. If you are eavesdropping on our qubit channel, we can detect you and then choose not to use that sequence of measurements to define our key. We will just repeat the process until we get results that indicate no significant eavesdropping. The bottom line is that if Cardy or anyone else is eavesdropping on the qubit channel, we can use our results to detect them.

BOB: Cardy might try to be subtle and not measure every qubit that goes by. Cardy, perhaps you measure only every other qubit. Do you think that will go undetected? Do you get enough information about the bits being sent to figure out the key?

CARDY: Hmm. That sounds like a moderately complicated question in probability theory, but it seems like I will still be up against the superposition aspect with 50% chance of sending the wrong photon if I choose the wrong basis.

BOB: You're right. We will likely be able to detect your "part-time" eavesdropping because the agreement between what I observe and what Alice sent will drop below 100%. If we detect eavesdropping, we can always drop that sequence of bits and try another to generate a key. But deciding how much eavesdropping we can tolerate depends on exactly how we are using the key. That is a much more complex problem.

ALICE: Let's look at the encryption procedure in various ways to check the conclusions we have drawn. For example, suppose I chose the $0°$ basis and sent a 1, but Cardy chose the $45°$ basis, as shown in Figure 8.12. My 1 is described by a superposition of Cardy's states giving Cardy a 50% chance of reading a 1 and a 50% chance of reading a 0. Suppose Cardy reads a 1 in the $45°$ basis and sends that on to Bob. If Bob has chosen the $45°$ basis, he will get a 1, which matches what I sent but with the "wrong" basis (it doesn't match mine). We will throw away that result when we compare our basis lists. If Bob chooses the $0°$ basis (the one I chose), he

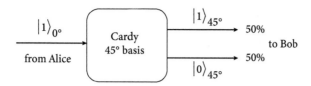

Fig. 8.12 A schematic diagram of what happens when Alice chooses to send 1 in the $0°$ basis but Cardy uses a $45°$ basis for measurements.

Table 8.3 The eavesdropping results when Alice, Cardy, and Bob have chosen the same basis set (either $0°$ or $45°$). Alice's bit is the one she wants to send to Bob. Bob's bit is the result of his measurement.

Alice's basis	Alice's bit	Cardy's basis	Cardy's bit	Bob's basis	Bob's bit
$0°$	0	$0°$	0	$0°$	0
$0°$	1	$0°$	1	$0°$	1
$45°$	0	$45°$	0	$45°$	0
$45°$	1	$45°$	1	$45°$	1

has a 50% chance of getting a 1 and a 50% chance of getting a 0. If he gets a 1, we won't know that eavesdropping has occurred, but if he reads a 0, then we know there was eavesdropping.

CARDY: Will I know I am measuring a superposition state or a single basis state?

BOB: No, you will just read your measurement device, and by the measurement postulate the qubit will end up being described by the measurement basis state corresponding to your observation. You cannot know about the state of the qubit before you measure it. You then send that qubit to me where I measure it with my chosen basis. If I choose the same basis as you, Cardy, my measurement will match your measurement 100% of the time. If my basis is the other one, then the qubit is in a superposition state of my basis states and I have a 50% chance of matching what you sent.

ALICE: Table 8.3 summarizes the eavesdropping situation when Cardy, Bob, and I have used the same basis states: either $0°$ and $45°$. Note that Bob's measurements always yield the bit I sent.

In Table 8.4, I have listed the four cases in which Cardy was unlucky and chose a basis different from mine and I sent either a 0 or 1. Bob has a 50% chance of getting a result that matches what I sent. Recall that we have selected just those results in which Bob used the same basis I used.

To summarize, when Bob and I look at just the cases in which we have chosen the same basis states, Bob and I should always agree on the bit sent and received if there is no eavesdropping. The cases listed in Table 8.4 show that when the eavesdropper and I have chosen different basis sets, Bob will have only a 50% chance of getting the bit I sent because his basis set will not match Cardy's.

Table 8.4 The eavesdropping results when Alice and Cardy have chosen different basis sets.

Alice's basis	Alice's bit	Cardy's basis	Cardy's bit	Bob's basis	Bob's bit
$0°$	0	$45°$	0, 1 (50:50)	$0°$	0, 1 (50:50)
$0°$	1	$45°$	0, 1 (50:50)	$0°$	0, 1 (50:50)
$45°$	0	$0°$	0, 1 (50:50)	$45°$	0, 1 (50:50)
$45°$	1	$0°$	0, 1 (50:50)	$45°$	0, 1 (50:50)

Try It 8.20

Bob's basis was given along with Alice's and Cardy's in Table 8.4. Suppose instead that Bob chose the other basis. Would the result be any different? If so, how?

Try It 8.21

There are, of course, other basis sets beyond those given in Tables 8.3 and 8.4. For example, there is the basis in which the polarizers are turned by $60°$. If you replace the $45°$ basis with a $60°$ basis, will the results of Table 8.3 change? If so, how? What about Table 8.4? Hint: Recall that $\cos(45°) = \sin(45°) = 1/\sqrt{2}$ while $\cos(60°) = 1/2$ and $\sin(60°) = \sqrt{3}/2$.

ALICE: To summarize, if there is eavesdropping, Bob will measure what I sent only about 75% of the time if Cardy is intercepting and observing each qubit, compared to 100% if there is no eavesdropping. If the bits match 100% of the time when Bob and I use the same basis set, we can be almost certain that no one has eavesdropped on our communications channel and the key is secure. This is a nice example of how the behavior of quantum states can be used to determine if a communications channel is being spied upon.

If we determine that Cardy (or someone more nefarious) is eavesdropping on our communications channel, we can try to send the message again. Of course, if the eavesdropper is persistent, we will realize that the communications channel is corrupted, and we will have to send the message some other way.

CARDY: Why aren't you completely sure?

BOB: There is always some probability that I will get the same result I would have gotten without eavesdropping, but the probability that all N cases will work out that way gets smaller as N gets larger. Let's suppose Cardy is eavesdropping on every qubit sent from Alice to me and that Alice and I compare N cases in which we had chosen the same basis set. For $N = 7$, there is less than a 1% chance that all of Alice's and my results agree. But the probability is not zero and so we could get complete agreement even with eavesdropping about 1% of the time if the whole process were repeated many times. That means that 99% of the time we would have at least some results that disagree. If $N = 20$, the chance for getting all matches with eavesdropping going on is about 10^{-4}% and we will almost certainly detect that eavesdropping is happening.

CARDY: How did you get those results so quickly?

BOB: I thought about the question in a different way and that leads to relatively easy calculations. Here is how I figured it out. The situation is like a coin toss for the eavesdropper: What is the chance of tossing a coin and getting a long run of heads? If we think of Cardy's choosing the same basis set that Alice had chosen as a "head," we need the probability of getting N heads in a row. The calculation is pretty easy because we know that the probability of getting N heads is just the probability of getting one head (0.5) multiplied by the probability of getting another head (0.5), and so on. So, the probability of getting N heads is just $1/2^N$. After dealing with binary numbers for a long time, I remember $1/2^7 = 1/128$; that probability is just a bit less than 1%. I also remember that $1/2^{10} = 1/1024 \approx 10^{-3} = 0.1\%$ and finally $1/2^{20} = \left(1/2^{10}\right)^2 \approx 10^{-6} = 0.0001\,\%$. So, using the coin analogy and simple numerical results, I got the answer.

To get a more precise estimate of the probability that the eavesdropper is not detected, we should also include the probability of choosing the "wrong" basis states but nonetheless having my measurement results match what Alice sent. We know that Cardy chooses the basis set that does not match Alice's about half the time, and in those cases Bob's measurements would give the right bit result about half the time. So, the probably of each of those "correct" bits being measured is 0.25. That leads to an interesting probability calculation: The two possibilities (Cardy chooses the matching basis, and Cardy chooses the other basis but I nevertheless get the right bit from my measurement) could occur in any order. Working out the details would take us rather far afield. In practical terms, if the number of bits sent is reasonably large (typically in the 100s), the possibility of *not* detecting the eavesdropper is so small that we need not worry about it.

 CHAPTER SUMMARY

- This chapter developed the mathematics needed to express one set of basis vectors in terms of another. This is important in QIS because measuring a quantum state in terms of a basis that is different from the one in which it was prepared can yield additional information about the state.

  The following equations express a quantum state originally in basis A in terms of basis B. The two basis vector sets differ, in state space, by an angle θ_{AB},
  $$|A_0\rangle = \cos\theta_{AB}\,|B_0\rangle + \sin\theta_{AB}\,|B_1\rangle$$
  $$|A_1\rangle = -\sin\theta_{AB}\,|B_0\rangle + \cos\theta_{AB}\,|B_1\rangle\,.$$

- Changing the basis from A to B expresses the quantum basis states $|A_0\rangle$ and $|A_1\rangle$ as a superposition state in the other basis B. If a qubit is prepared in one of the A basis states, the coefficients of the B basis states give, according to the Born rule, the probabilities of obtaining measurement outcomes of that state with a measurement device designed to observe B basis states. These changes in the probabilities can be leveraged for applications.

- The power of measuring in a basis different from the preparation basis was illustrated using three linear polarizers. The light passes through a horizontal polarizer first, putting it into a completely horizontal polarization state. If that light were to encounter a vertical polarizer, none of the light would get through because there is no projection of the completely horizontal basis state vector onto the vertical state vector. If, however, a 45° polarizer is inserted between the horizontal and vertical filters, light does indeed pass through the vertical filter. Seen from the 45° state space, a vlp state is an equal superposition of 45° basis states. This tells us that light will pass through with some probability given by the coefficients of that superposition. Next, according to the third polarizing sheet's state space, the 45° basis states are superpositions of the horizontal/vertical basis states. Therefore, there is a coefficient in the vertical component of that state vector and some light will pass through the third filter. This probability explanation is necessary for single-photon measurements because the classical wave explanation involving an electric field vector is not applicable to a single photon.

- Quantum cryptography also utilizes the superposition that results from changing basis states. Alice and Bob both randomly choose from two basis sets to prepare (Alice) and read (Bob) the photons they are using to communicate. If they both choose the same set, they will both agree 100% on the quantum state of the photon and they can use their measurements to prepare a key with which to encrypt their message. They can openly communicate which basis they used because knowing the measurement basis alone does not tell us the actual measurement result.

- The danger in sending information is that it might be intercepted by an eavesdropper, named Eve in quantum information parlance. To eavesdrop undetected, Eve needs to intercept Alice's photon, read its state, and then send a photon in the same state on to Bob. If Eve reads and prepares the photon using a basis different from Alice, she will not have a 100% probability of sending Bob a photon in the exact state that she received it in, because she cannot know the exact state. Since Eve is receiving a superposition state in terms of her basis states, her measurement cannot tell her the amplitudes of the components of the superposition. Hence, what she sends on to Bob will lead in general to a change in outcome probabilities for Bob's measurement. To check for eavesdropping, Alice and Bob compare not only their basis sets but also the preparation and measurement results. If eavesdropping has occurred, they will not get the same results 100% of the time even when they choose the same basis states.

 FURTHER READING

Aayam Bista, Baibhav Sharma, and Enrique J. Galvez, "A demonstration of quantum key distribution with entangled photons for the undergraduate laboratory," *American Journal of Physics* **89**, 111 (2021).

9 Multi-Qubit Systems, Entanglement, and Quantum Weirdness

They [atoms] move in the void and catching each other up jostle together, and some recoil in any direction that may chance, and others become entangled with one another in various degrees according to the symmetry of their shapes and sizes and positions and order, and they remain together and thus the coming into being of composite things is effected.

Simplicius, *De Caelo*

9.1 Two-Qubit Quantum States

ALICE: Hi, Cardy. In today's lesson we are going to talk about systems of two qubits. For useful QCs we will need systems with at least two qubits. How do we describe the states of those systems? What happens when we make measurements on the individual qubit states? As we shall see, systems with two (or more) qubits have properties that are both weird and useful as computational resources.

Before we talk about quantum states, let's think about general systems with two (or more) parts. We know that quantum states carry information about probabilities of results of various measurements. From probability theory, we know that if two events are independent then the probability of event 1 *and* event 2 happening is given by the product of the probabilities of each of the events $P(1 \text{ AND } 2) = P(1)P(2)$. For example, if we have a system which consists of a standard six-sided die and a fair coin, the probability of getting a 5 on the die AND getting heads on the coin is

$$P(5 \text{ AND heads}) = P(5)P(\text{heads}) = 1/6 * 1/2 = 1/12. \tag{9.1}$$

CARDY: Yeah, I remember that from my high school statistics course. But isn't there some condition that the events must not be correlated or something like that?

ALICE: You are right, Cardy. We are assuming that the die and the coin tosses are independent. If the coin were glued to one side of the die, they would not be independent and Eq. (9.1) would not apply.

As we know, in quantum mechanics, probabilities are calculated from quantum state vectors. That suggests that if we have two "independent" qubits (we'll need to figure out what independent means for qubits), then the quantum state vector $|S\rangle$ for the system of two qubits ought to be the product of the quantum state vectors for each of the qubits:

$$|S\rangle = \underbrace{|A\rangle}_{\text{Alice's state}} \otimes \underbrace{|B\rangle}_{\text{Bob's state}} = |A\rangle|B\rangle, \tag{9.2}$$

Quantum Computing: From Alice to Bob. Alice Flarend and Bob Hilborn, Oxford University Press.
© Alice Flarend and Robert C. Hilborn (2022). DOI: 10.1093/oso/9780192857972.003.0009

where $|A\rangle$ is my state vector and $|B\rangle$ is Bob's state vector. The symbol $\otimes$ is used to indicate "multiplication" of two right vectors. That kind of product is called a "tensor product." We will often omit that symbol (as we have done on the right side of Eq. (9.2)) when the product form is obvious. In other situations, we will use the symbol to help guide the eye in interpreting equations.

Note that the qubits could be entirely different kinds of physical systems; I might use photon polarization and Bob might use neutron spin. Or I may use a photon that produces the sensation "blue light" in our eyes while Bob's photon produces the sensation "red light."

We also know that the quantum state vector of each qubit can be represented as a superposition of basis state vectors. So, if I use a linearly polarized photon qubit and Bob uses a spin-1/2 qubit of the type we looked at in Chapter 7, we can rewrite Eq. (9.2) as

$$|S\rangle = |A\rangle \otimes |B\rangle = \underbrace{[a_0|\text{hlp}\rangle + a_1|\text{vlp}\rangle]}_{\text{Alice's state}} \otimes \underbrace{[b_0|\uparrow\rangle + b_1|\downarrow\rangle]}_{\text{Bob's state}}. \tag{9.3}$$

We have used a superposition of linear polarization basis states for my state and a superposition of spin-up and spin-down states for Bob's. If the state vector for the system can be written in the simple product form shown in Eq. (9.3), the two qubits are independent in the sense that the probability of my observing a photon with vertical polarization is a_1^2 while the probability of Bob's observing an electron with spin up is b_0^2. Moreover, the probabilities for Bob's observations are independent of my observations and vice versa.

In most quantum computing applications, the two qubits in a two-qubit system are of the same physical type: for example, two photons, two electrons, two ions, or two superconducting circuits. From now on, we will assume, unless otherwise explicitly stated, that the qubits are of the same physical type.

BOB: What kind of state space does a two-qubit state occupy? Alice expresses her quantum state in terms of two basis state vectors while I may use a different pair of basis state vectors. That gives us a total of four basis state vectors, which tells us that two-qubit quantum states "live" in a four-dimensional state space. (You will pardon me if I don't draw a picture of that four-dimensional state space!) That space is called a product space because in a sense it contains the product of what Alice's state space contains and what my state space contains. Its dimensionality is the product of Alice's state space dimensionality and that of mine. In this case, $2 \times 2 = 4$.

There is a way (not very elegant, but it is straightforward) to see that we need a four-dimensional state space: We simply write down all the possible results of measurements on Alice's and my qubits. First, let's express the overall two-qubit state using the computational basis states, rather than the polarization and spin basis states in Eq. (9.3). There are four individual basis states:

$$|0\rangle_A, \ |1\rangle_A, \ |0\rangle_B, \text{ and } |1\rangle_B. \tag{9.4}$$

The subscripts tell us which of the qubits is associated with each state vector. Notice that we have four different possible states, two for each qubit. When Alice makes measurements on her states, she gets 0 or 1 (using our usual computational basis form) and I can get 0 or 1; so, together there are four possibilities: 00, 01, 10, and 11.

Table 9.1 Measurement results and system state vectors for a 2 × 2 product space.

Alice's result	Bob's result	System state vector
0	0	$\lvert 0\rangle_A \otimes \lvert 0\rangle_B = \lvert 0\rangle_A\lvert 0\rangle_B = \lvert 0_A 0_B\rangle$
0	1	$\lvert 0\rangle_A \otimes \lvert 1\rangle_B = \lvert 0\rangle_A\lvert 1\rangle_B = \lvert 0_A 1_B\rangle$
1	0	$\lvert 1\rangle_A \otimes \lvert 0\rangle_B = \lvert 1\rangle_A\lvert 0\rangle_B = \lvert 1_A 0_B\rangle$
1	1	$\lvert 1\rangle_A \otimes \lvert 1\rangle_B = \lvert 1\rangle_A\lvert 1\rangle_B = \lvert 1_A 1_B\rangle$

BOB: In the right-most column of Table 9.1, I have shown three different formats for the system state vectors that correspond to the measurement results listed in the first two columns. In the first format, the system state vectors are written as products of the individual state vectors (with labels A and B to distinguish Alice's from mine). The third, more succinct format uses a single state vector with two labels. I will always put Alice's label on the left and mine on the right. I should note that many authors do not use subscripts in this way and leave it to you to infer the appropriate basis states from the context. That is fine as long as we always agree to use the same order for the labels.

The crucial (and in some sense weird) point of this chapter is that there are system quantum states which describe qubits that are not independent. In those situations, what I observe depends on what Alice observes, and vice versa. Such states are called "entangled" states and play a fundamental role in QIS and QC. So, we will take some time to understand how to determine if a state of a two-qubit system is entangled (or not) and how we describe such an entangled state. Then, in the following chapters, we will see how entangled states can be prepared and manipulated for QIS and QC tasks.

9.2   General Two-Qubit States

ALICE: To explore the properties of multi-qubit states, we will start with a system of just two qubits. The generalization to more than two qubits is conceptually straightforward but is very messy to write out in detail.

To be specific, we will use two spin-1/2 qubits (though they might be different kinds of spin-1/2 particles, e.g. one electron and one neutron) and write the system basis states in a form like that in Eq. (9.2), namely $\lvert S\rangle = \lvert A\rangle\lvert B\rangle$. We will label my qubits A, and use B for Bob's. Later we will translate the formalism into the computational basis states $\lvert 0\rangle$ and $\lvert 1\rangle$.

For two spin-1/2 qubits there are four possible combinations of spin-up and spin-down for observations along a chosen direction. That means, as we mentioned in the previous section, we have a four-dimensional state space and four basis states. There are several different ways of writing the basis states; we shall use all of them as we proceed. Table 9.2 shows three of

Table 9.2 Four ways to write the basis state vectors for a two-qubit system.

Individual quantum state vectors in the spin basis	System vectors in the spin basis	System vectors in the computational basis	System column vectors
$\lvert\uparrow\rangle_A\lvert\uparrow\rangle_B$	$\lvert\uparrow_A\uparrow_B\rangle$	$\lvert00\rangle$	$\begin{pmatrix} 1 \\ 0 \\ 0 \\ 0 \end{pmatrix}$
$\lvert\uparrow\rangle_A\lvert\downarrow\rangle_B$	$\lvert\uparrow_A\downarrow_B\rangle$	$\lvert01\rangle$	$\begin{pmatrix} 0 \\ 1 \\ 0 \\ 0 \end{pmatrix}$
$\lvert\downarrow\rangle_A\lvert\uparrow\rangle_B$	$\lvert\downarrow_A\uparrow_B\rangle$	$\lvert10\rangle$	$\begin{pmatrix} 0 \\ 0 \\ 1 \\ 0 \end{pmatrix}$
TBC	TBC	TBC	TBC

Note: TBC means to be completed by the reader. See Try It 9.1.

the four basis states in four different formats. (You should fill in the last row of the Table. See Try It 9.1.)

In Table 9.2, we have used four formal representations: Going from left to right, the first three formats are the same as those in Table 9.1: a product of individual state vectors, a single state vector for the system of two qubits, and a computational basis right vector for the combined system. The right-most column introduces a set of four-entry column vectors whose entries are the projections of the state vectors along the four state-space directions. The subscripts, when used, remind us that one of the qubits belongs to Alice and the other to Bob. Note that the computational basis vectors in the third column are in numerical order (in the binary system) going from the top row down: $\lvert00\rangle$, $\lvert01\rangle$, $\lvert10\rangle$. That method of labeling states will be useful in describing multi-qubit systems with a large number of qubits. In the four-entry column vector representation of the states in Table 9.2, the top row column vector has a unit component along the first of the four state space directions and 0 along the others, the second row vector has unit length along the second state space dimension and 0 along the others, and so on. As you become more proficient at quantum thinking, you should be able to switch back and forth among these representations.

Try It 9.1

Complete the fourth row of Table 9.2.

ALICE: The four-entry column vectors on the far-right column of Table 9.2 are examples of the tensor product of the individual qubit column vectors. The general rule for the tensor product of two column vectors is:

$$\begin{pmatrix} a_1 \\ a_2 \end{pmatrix} \otimes \begin{pmatrix} b_1 \\ b_2 \end{pmatrix} = \begin{pmatrix} a_1 \begin{pmatrix} b_1 \\ b_2 \end{pmatrix} \\ a_2 \begin{pmatrix} b_1 \\ b_2 \end{pmatrix} \end{pmatrix} = \begin{pmatrix} a_1 b_1 \\ a_1 b_2 \\ a_2 b_1 \\ a_2 b_2 \end{pmatrix}. \tag{9.5}$$

Notice that, although at first glance it may look like it, this is not the same as multiplying two square matrices. You cannot multiply two 2×1 matrices using the matrix product rules. Tensor products are a different animal. See Appendix Quantum Toolkit for more on vector and matrix products.

9.3 Two-Qubit Basis Vectors are Orthogonal

BOB: Remember that in a two-dimensional (single qubit) state space, we wanted to have the basis state vectors perpendicular (orthogonal) to each other. That makes many of our calculations much easier to write out and to understand. It also reflects the fact that the basis states correspond to mutually exclusive measurement outcomes. We want to have that same orthogonality in our four-dimensional state space, though it is admittedly more difficult to picture that. We also need to remember that it is the state vectors that are perpendicular (in state space) and not the spins of Bob's and Alice's particles.

Try It 9.2

Use the multiplication rule introduced in Chapter 4 (row vector times a column vector) to show that the four basis state column vectors in column 3 of Table 9.2 are orthogonal to each other and also that they have unit length.

Try It 9.3

If we want to describe a spin-1/2 three-qubit state, how many basis states do we need? Hint: Write out all the possibilities of spin-up and spin-down.

Try It 9.4

Generalize these results to find how many basis states are needed for n spin-1/2 qubit states. Hint: For one qubit, we have a $2^1 = 2$-dimensional state space. For two qubits, we have $2^2 = 4$ dimensions.

While we are thinking about multi-qubit states, I want to show you some notation that will be used when we are dealing with many qubits. To be general, let's suppose that we are working with n qubits. Then we have 2^n basis states. It turns out to be useful to label those basis states and their state amplitudes with n binary bits, starting with $000\ldots0$ and ending with $111\ldots1$. Note that the numerically smallest label is $000\ldots0$ and the largest is $(111\ldots1)_2 = (2^n - 1)_{10}$ where we used subscripts to distinguish between binary and base 10 numbers.

In the QC and QIS literature, the symbol x is used to denote the labels. Using that convention, a general superposition state of the n-qubit system can be written as

$$|\psi\rangle = \sum_{x=000\ldots0}^{2^n-1} a_x |x\rangle. \tag{9.6}$$

We us upper-case Greek sigma (Σ) to indicate that we sum what follows with x ranging from 0 to $2^n - 1$. For example, for a two-qubit system, we have $2^n - 1 = 3$ and Eq. (9.6) becomes

$$|\psi\rangle = \sum_{x=0}^{11} a_x |x\rangle = a_{00} |00\rangle + a_{01} |01\rangle + a_{10} |10\rangle + a_{11} |11\rangle. \tag{9.7}$$

Try It 9.5

Write the equivalent of Eq. (9.7) for a three-qubit ($n = 3$) system.

The Σ notation is a bit messy when we are working with multiple qubits and states in QC and QIS applications; so, we have created our own more condensed notation:

$$|\psi\rangle = \sum_{x=000\ldots0}^{2^n-1} a_x |x\rangle = \{a_x |x\rangle\}. \tag{9.8}$$

The curly braces remind us that we are dealing with a set of amplitudes and states, and we implicitly remember that we need to sum over all the possible values of x for the n-qubit system. We will make use of that more compact notation in later chapters when we look at several QC and QI algorithms.

9.4 Single Qubits and Product States

ALICE: We should point out that product states are also used to describe the states of a single qubit (physical object) if that qubit has several distinct dynamical properties—properties that can be manipulated and change with time. In tech speak, such dynamical properties are called "degrees of freedom." A specific example will be helpful. Suppose the qubit is a photon. One of the dynamical properties is its linear polarization. We have seen how we can manipulate the linear polarization state by using polarizing filters, for example. But the photon has other dynamical properties: the direction or path in which it is traveling and its energy, which is proportional to the oscillation frequency of the corresponding classical wave. You may recall that a green photon has more energy than a red photon.

Taking those dynamical properties into account, we write the state $|S\rangle$ of the photon as

$$|S\rangle = |\text{polarization}\rangle \, |\text{direction of propagation}\rangle \, |\text{energy}\rangle . \tag{9.9}$$

Each part of the product state might be a superposition state, as we have seen for polarization. The overall state space for the qubit is then a tensor product state space.

If a particular dynamical property does not change in the situation we have set up, we can drop that part of the product state and ignore that part of the product state space. That is often, but not always, the case with the photon's energy.

Let's return to our old friend the polarizing beam splitter, shown in Figure 9.1. As we discussed in Chapter 4, the PBS splits an incoming light beam into two beams traveling in different directions. One of the outgoing beams is vlp light and the other is hlp. In that case the state for a photon that has interacted with the PBS (assumed to be a 50:50 beam splitter for simplicity's sake) is

$$|S\rangle = \frac{1}{\sqrt{2}} \, |\text{vlp}\rangle \otimes |\text{beam C}\rangle + \frac{1}{\sqrt{2}} \, |\text{hlp}\rangle \otimes |\text{beam D}\rangle . \tag{9.10}$$

Given this form, we know that, even if we don't directly measure the polarization mode, if we measure the photon's path and detect it in beam C, its polarization state was vlp. Similarly, with beam D and hlp polarization. We see that the PBS produces a correlation between the qubit's polarization state and its propagation direction state. In the next section, we will explore similar correlations when our system has more than one qubit.

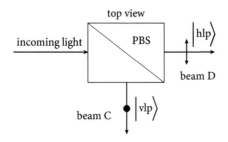

Fig. 9.1 A top view of a polarizing beam splitter (PBS).

9.5 Quantum Entanglement

ALICE: We will see in this section that if the state vector of the combined two-qubit system can be written as a product of the quantum state vectors of the initial subsystems (mine and Bob's) as illustrated in Eq. (9.2), then the probabilities of my observing spin-up or spin-down for my qubit are independent of what Bob observes for his qubit. You might ask if things could be otherwise. Hold onto your state space vectors because we are now going to show that there are system states for which Bob's and my observations are *not* independent. We will use the spin-up spin-down basis states to illustrate the issues, but you could use the computational basis states $|0\rangle$ and $|1\rangle$ as well.

To be quite general, we will write the system state vector for the two-qubit system as a superposition of the four basis states listed in Table 9.2. Three are listed in the second column of Table 9.2 with the fourth from the results of your Try It:

$$|S\rangle = c\,|\uparrow_A\uparrow_B\rangle + d\,|\uparrow_A\downarrow_B\rangle + e\,|\downarrow_A\uparrow_B\rangle + f\,|\downarrow_A\downarrow_B\rangle\,, \tag{9.11}$$

where $c, d, e,$ and f are numbers that satisfy $c^2 + d^2 + e^2 + f^2 = 1$; so, the probabilities add up to 1. We will show (spoiler alert!) that if $cf = de$ (the product of the amplitudes of the two outer terms is equal to the product of the amplitudes of the two inner terms) then the state vector can be written as shown in Eq. (9.3) as a product of the state vector for Bob's qubit and the state vector for my qubit. In that case, Bob's and my observations are independent. Remember that when we use "outer" and "inner" coefficients in this way, we need to use our standard ordering of the state labels with "outer" corresponding to $|00\rangle$ and $|11\rangle$ in the third column of Table 9.2 and "inner" corresponding to $|01\rangle$ and $|10\rangle$.

The derivation of the criterion is not too difficult. Essentially, we want to write the state vector in Eq. (9.11) as a product of two state vectors: one for me and one for Bob. In other words, we want to find the four amplitudes g, h, m, and n that satisfy

$$\begin{aligned}|S\rangle &= c|\uparrow_A\uparrow_B\rangle + d|\uparrow_A\downarrow_B\rangle + e|\downarrow_A\uparrow_B\rangle + f|\downarrow_A\downarrow_B\rangle \\ &= \underbrace{\{g|\uparrow_A\rangle + h|\downarrow_A\rangle\}}_{\text{Alice's qubit state}}\underbrace{\{m|\uparrow_B\rangle + n|\downarrow_B\rangle\}}_{\text{Bob's qubit state}}.\end{aligned} \tag{9.12}$$

Notice that the second line came about merely by factoring out my states as required for independent observations. Multiplying out the second line of Eq. (9.12) with the new variables and comparing that result to the first line, we find that we have four equations to satisfy:

$$gm = c \quad gn = d \quad hm = e \quad hn = f. \tag{9.13}$$

While this seems to get us nowhere, look carefully at the equations for c and d. They have a factor of g in common; so, dividing them will cancel out that variable. Similarly, e and f have a factor of h in common. This calculation yields $m/n = c/d$ and $m/n = e/f$. Further, both of those pairs have m/n in common so we have a criterion for independent states:

$$cf = de. \tag{9.14}$$

The mathematics can also be done using the other pairs of variables, c & e and d & f, leading to an identical result. Thus, our criterion for the state in Eq. (9.11) to be a product state is $cf = de$; the product of the two "outer" state amplitudes equals the product of the two "inner" state amplitudes.

Conversely—and this is the critical point—if $cf \neq de$, then the state vectors cannot be written as a simple product state and, as we shall see, my observations in general depend on what Bob observes and vice versa. The observations on the two qubits are not independent! In that case, the overall state vector of the system is called an *entangled state vector*. Or more succinctly, the system state is entangled.

This example shows that all entangled states are superposition states, but not all superposition states are entangled. The properties of entangled state vectors turn out to be rather weird (or at least very counter-intuitive) but they are exceedingly important for QIS and QC.

CARDY: I keep reading about entangled qubits, but you said that the state vector is entangled. Which is it?

ALICE: Another excellent observation, Cardy. Many people who work on QIS and QC say "Alice's qubit is entangled with Bob's qubit." In the language we have been using throughout this book, qubits are the physical systems (the photons, electrons, neutrons, circulating superconducting currents . . .) and the quantum states are the formal descriptions used to predict the outcomes of measurements on those qubits. From that point of view, it is better to say that we use an entangled state to describe the system. That entangled state contains all there is to say about the outcome of measurements on the system. We will return to this issue in Chapter 14 where we discuss some of the fundamental, almost philosophical, issues surrounding the relationship between qubits and quantum states. As we shall see in this chapter, it is not always possible to say that a qubit "has" a quantum state in the way a car has a headlight. A lot of conceptual confusion can be avoided by distinguishing between the concept of a qubit as a physical object and the concept of a quantum state. So, we will stick with saying "the system is described by an entangled state."

Try It 9.6

Below is a list of several two-qubit spin-1/2 states. Which ones are entangled states and which are unentangled states? Explain your reasoning.

$$|S_1\rangle = \frac{1}{\sqrt{2}}|\uparrow_A\uparrow_B\rangle - \frac{1}{\sqrt{2}}|\downarrow_A\downarrow_B\rangle$$

$$|S_2\rangle = \frac{\sqrt{3}}{2}|\uparrow_A\downarrow_B\rangle + \frac{1}{2}|\downarrow_A\uparrow_B\rangle$$

$$|S_3\rangle = \frac{\sqrt{3}}{2}|\uparrow_A\downarrow_B\rangle + \frac{1}{2}|\uparrow_A\uparrow_B\rangle$$

$$|S_4\rangle = \frac{\sqrt{3}}{2}|\uparrow_A\downarrow_B\rangle - \frac{1}{2}|\uparrow_A\uparrow_B\rangle$$

$$|S_5\rangle = \frac{1}{2\sqrt{2}}|\uparrow_A\uparrow_B\rangle + \frac{1}{2}|\uparrow_A\downarrow_B\rangle + \frac{1}{2}|\downarrow_A\uparrow_B\rangle + \frac{\sqrt{3}}{2\sqrt{2}}|\downarrow_A\downarrow_B\rangle$$

$$|S_6\rangle = \frac{1}{2}|\uparrow_A\uparrow_B\rangle + \frac{1}{2}|\uparrow_A\downarrow_B\rangle + \frac{1}{2}|\downarrow_A\uparrow_B\rangle + \frac{1}{2}|\downarrow_A\downarrow_B\rangle$$

$$|S_7\rangle = \frac{1}{2}|\uparrow_A\uparrow_B\rangle - \frac{1}{2}|\uparrow_A\downarrow_B\rangle + \frac{1}{2}|\downarrow_A\uparrow_B\rangle + \frac{1}{2}|\downarrow_A\downarrow_B\rangle. \qquad (9.15)$$

Partial answer: $|S_3\rangle$ is not an entangled state because it can be factored into a simple product state $|S_3\rangle = |\uparrow_A\rangle \left\{ \frac{\sqrt{3}}{2}|\downarrow_B\rangle + \frac{1}{2}|\uparrow_B\rangle \right\}$ with one factor for Alice and the other for Bob. We also see that the product of the two outer basis states' amplitudes $= 0$ and the product of the two inner basis states' amplitudes $= 0$. Since those products are equal, the state is not entangled.

BOB: The weirdness of entangled states is made more dramatic if we see what happens if Alice and I prepare our two-qubit system in an entangled state. Then Alice keeps one of the qubits while I transport the other qubit (gently!) to a distant location (say the other side of the galaxy) to make sure there is no hidden connection between the two objects.

Next, let's consider what happens if both Alice and I observe the spin orientation for our qubits. By the Born rule, the probabilities for the possible results are given by the squares of the basis state coefficients in Eq. (9.11). These probabilities are listed in Table 9.3.

So far, those results look reasonable based on what we have seen for single-qubit states. Life gets more interesting if Alice makes her observation first. As an example, suppose Alice makes an observation of the spin orientation of her qubit and finds that it is spin-up. (This will happen with probability $c^2 + d^2$ (the sum of the probabilities in rows one and two in Table 9.2).) Given that information, what are the probabilities for the results of my measurements?

To see what happens, let's use the general two-qubit quantum state in Eq. (9.11) with Alice's parts pulled to the left and mine expressed inside curly braces:

$$|S\rangle = \underbrace{|\uparrow_A\rangle}_{\text{Alice's}} \otimes \underbrace{\{c|\uparrow_B\rangle + d|\downarrow_B\rangle\}}_{\text{Bob's state, given Alice's}} + \underbrace{|\downarrow_A\rangle}_{\text{Alice's}} \otimes \underbrace{\{e|\uparrow_B\rangle + f|\downarrow_B\rangle\}}_{\text{Bob's state, given Alice's}}. \qquad (9.16)$$

Table 9.3 The probabilities for Alice's and Bob's observations from Eq. (9.11).

Alice's observation	Bob's observation	Probability for Alice AND Bob's observations
$\uparrow_A$	$\uparrow_B$	c^2
$\uparrow_A$	$\downarrow_B$	d^2
$\downarrow_A$	$\uparrow_B$	e^2
$\downarrow_A$	$\downarrow_B$	f^2

Try It 9.7

Write out Eq. (9.16) by hand starting from Eq. (9.11) to make sure you know how the terms are grouped.

BOB: We want to keep me normalized (or at least my quantum state vectors normalized), so we divide the terms in curly braces by the lengths of my states and then multiply Alice's state vector by that length:

$$|S\rangle = \underbrace{\sqrt{c^2 + d^2}\,|\uparrow_A\rangle}_{\text{Alice}} \otimes \underbrace{\left\{ \frac{c|\uparrow_B\rangle + d|\downarrow_B\rangle}{\sqrt{c^2 + d^2}} \right\}}_{\text{Bob}} + \underbrace{\sqrt{e^2 + f^2}\,|\downarrow_A\rangle}_{\text{Alice}} \otimes \underbrace{\left\{ \frac{e|\uparrow_B\rangle + f|\downarrow_B\rangle}{\sqrt{e^2 + f^2}} \right\}}_{\text{Bob}}.$$

$$(9.17)$$

ALICE: Let's unravel the rather messy-looking Eq. (9.17). Note that the coefficient of $|\uparrow_A\rangle$ is $\sqrt{c^2 + d^2}$ and the coefficient of $|\downarrow_A\rangle$ is $\sqrt{e^2 + f^2}$. These results tell us that when I make an observation, I will get spin-up with probability $c^2 + d^2$ and spin-down with probability $e^2 + f^2$. (As always, the probabilities are given by the *squares* of *my* state vector coefficients.) Note that my probabilities add to 1 because of our requirement that $c^2 + d^2 + e^2 + f^2 = 1$. Eq. (9.17) is a fairly complicated expression but it will be a key element in quantum computation.

Try It 9.8

Write out Eq. (9.17) by hand and explain to yourself or a friend what the various terms mean.

BOB: The rule that Alice invoked in working out the probabilities for a two-qubit state is called the generalized Born rule because it is (surprise!) a generalization of the Born rule for single-qubit states. Assuming that the measurement device states are the basis states in which the state vectors are prepared, we state the generalized Born rule as follows:

> **Generalized Born Rule**
>
> For a multi-qubit state, the probabilities of the outcomes of measurements on the (normalized) state of only one of the qubits are the squares of the amplitudes for the appropriate basis states for that qubit. After the measurement, the state of the system is the product of the basis state observed and the states associated with the other qubits.

CARDY: But what happens if we make measurements on all the qubits simultaneously? How do we figure out the probabilities for the joint measurement outcomes?

ALICE: Excellent question, Cardy. In that case, we go back to the state as expressed in Eq. (9.11). We use the squares of the appropriate amplitudes. For example, the probability that I will observe spin-up and Bob will observe spin-down is d^2, the square of the amplitude of the second state vector.

9.6   Generalized Born Rule and Entanglement

BOB: Let's explore the generalized Born rule and find the probabilities for my measurement results given that Alice has observed spin-up. After Alice's observation, we have new information, and that means that the appropriate quantum state vector for the overall system is, from Eq. (9.17),

$$|S \text{ after } \uparrow_A\rangle = \underbrace{|\uparrow_A\rangle}_{\text{Alice's state}} \otimes \underbrace{\frac{c|\uparrow_B\rangle + d|\downarrow_B\rangle}{\sqrt{c^2 + d^2}}}_{\text{Bob's state}}. \tag{9.18}$$

In other words, Alice's qubit is now described by the state $|\uparrow_A\rangle$ (the basis state associated with her measurement outcome), which we factored out in Eq. (9.18), and mine is described by the remaining part of that equation.

An important observation: After a measurement on one of the qubits, the system state is no longer an entangled state. For example, the system state in Eq. (9.18) is a simple product state, not an entangled state. Measurement crushes entanglement!

CARDY: That sounds like a sports headline for the English Premier League!

ALICE: It certainly does to a soccer fan. Maybe quantum physicists can have a league with those names for teams. Let's get back to our two-qubit analysis. Bob is getting anxious and decides to make a spin observation on his qubit. As we can see from Eq. (9.18) he should get spin-up with probability $c^2/(c^2 + d^2)$.

Try It 9.9

Explain why, under those same conditions, Bob will get spin-down with probability $d^2/(c^2 + d^2)$.

Try It 9.10

Show that Bob's probabilities add up to 1.

ALICE: Probabilities say nothing about an individual observation. For each individual state preparation–observation cycle, I will get either spin-up or spin-down and then Bob will get either spin-up or spin-down. It is only after many preparation–observation cycles that the numerical values of the probabilities emerge.

Now look specifically at the case in which I get spin-down for my qubit. Then the appropriate quantum state vector is

$$|S \text{ after } \downarrow_A\rangle = \underbrace{|\downarrow_A\rangle}_{\text{Alice's state}} \otimes \underbrace{\frac{e|\uparrow_B\rangle + f|\downarrow_B\rangle}{\sqrt{e^2 + f^2}}}_{\text{Bob's state}}. \tag{9.19}$$

Bob now has the probability $e^2/(e^2 + f^2)$ of getting spin-up and $f^2/(e^2 + f^2)$ of getting spin-down.

Try It 9.11

Show that Bob's probabilities add up to 1 when Alice gets spin-down.

ALICE: Let's summarize these results (Table 9.4).

CARDY: This is rather weird! It appears that the probabilities for Bob's observations depend on the results of Alice's observations even though Alice and Bob have their qubits on opposite sides of the galaxy. Since the state amplitudes c, d, e, and f could have any values (subject to the normalization constraint), the probabilities could all be different. Their observations are not independent. How could it be that what Alice observes affects the probabilities of what Bob observes? Is Alice doing something we don't know about?

ALICE: No way! I am an honest quantum mechanician. Let's see under what conditions the probabilities for Bob's measurements are the same for both of my observations. Bob's probabilities for getting spin-up are the same if

Table 9.4 The probabilities that Bob observes spin-up or spin-down given Alice's observations of spin-up or spin-down for the state given in Eq. (9.11).

Alice's observation	Probability for Bob's getting spin-up	Probability for Bob's getting spin-down
spin-up	$c^2/(c^2 + d^2)$	$d^2/(c^2 + d^2)$
spin-down	$e^2/(e^2 + f^2)$	$f^2/(e^2 + f^2)$

$$\frac{c^2}{c^2 + d^2} = \frac{e^2}{e^2 + f^2}. \tag{9.20}$$

Using just a little bit of algebra (cross-multiply, expand, and cancel equal terms, and then take the square root of both sides of the equation) shows that this condition can be written as

$$c^2 f^2 = d^2 e^2. \tag{9.21}$$

This result is consistent with the condition we asserted before in Eq. (9.14): The product of the two outer amplitudes is the same as the product of the two inner amplitudes (when we use the standard order of the state vectors). When Eq. (9.21) is satisfied, Bob's probabilities are independent of what Alice observes. You should note that Eq. (9.21) is less restrictive than Eq. (9.14). Take a look at the state vectors in Try It 9.5 and see if any have coefficients that satisfy Eq. (9.21) but not Eq. (9.14). Is the corresponding state entangled or not entangled?

> **Try It 9.12**
>
> Show that requiring equal probabilities for spin-down also gives the result stated in Eq. (9.21). Does that make sense?

ALICE: Our entanglement analysis should be comforting in the sense that there are conditions under which my and Bob's measurements are independent. Alas, that doesn't solve the problem of understanding the correlations between Bob's and my observations when we do have an entangled state. From a mathematical point of view, if you prepare a superposition state for a two-qubit system by randomly choosing state amplitudes, you will almost certainly end up with an entangled state for which Eqs. (9.14) and (9.21) are *not* satisfied. However, from a practical point of view, if I prepare my qubit in a quantum state here and Bob, in a distant location, prepares his qubit in a quantum state there, the overall system state will be a product state. To get an entangled system state, we need to act on both those qubits' states with some quantum gate or gates. That is both good news and bad news for QIS and QC. It is good because it tells us that it is relatively easy to produce specified entangled states by choosing the right quantum gates. It is bad because your qubit states might easily become entangled with the states of other qubits in the environment because of uncontrolled interactions, which consequently mess up the states you wanted to have for your QIS and QC algorithms.

Let's pause and summarize what we have learned so far: If $cf \neq de$, the system is described by an entangled state and the probabilities of Bob's measurement outcomes (in general) depend on my observations.

CARDY: Maybe there is somethings special about you, Alice, that leads to these weird results.

ALICE: Let's test that idea by seeing if Bob is special, in the quantum sense.

Try It 9.13

Go through two scenarios in which Bob makes the first observations. What are the probabilities for Alice to find spin-up and spin-down in those cases? Partial answer: When Bob gets spin-up, Alice gets spin-up with probability $c^2/(c^2 + e^2)$ and spin-down with probability $e^2/(c^2 + e^2)$.

Try It 9.14

Given the results of the previous Try It, show that the conditions for having an unentangled state are the same as those given in Eq. (9.21).

ALICE: There is nothing special about either of us when it comes to quantum measurements. Nevertheless, with an entangled state, the probabilities of getting spin-up or spin-down depend on what was observed previously for the other qubit. In essence, that is what is called a joint or conditional probability statement. Those joint probabilities tell us about correlations among the results, given the entangled state. The key issue is that these are correlations among "partial" measurements on the qubits individually in a situation for which the quantum state applies only to properties of the complete system. For an entangled system state, we don't have quantum states for the qubits individually. We will explore this issue in section 9.7.

This result applies to any two-qubit system, not just spin-1/2 systems. You can easily check that by replacing the spin-up, spin-down basis states with linear polarization states or with computational basis states $|0\rangle$ and $|1\rangle$.

Try It 9.15

Rewrite this section's arguments in terms of photon linear polarization states (horizontal and vertical polarization) and then with the computational basis states $|0\rangle$ and $|1\rangle$.

BOB: There is a deep metaphysical question here. How can Alice's observations affect my probabilities even though the qubits (and the measurement equipment) may be very far apart—so far that there is no time for a signal to get from Alice to me before I make my observations.

Some would say that the entangled system state has correlations built in and Alice's observations just invoke those correlations. That means that the qubits are not independent. In other words, they become mutually dependent if the state preparation process produces an entangled state. This mutual dependence is maintained even if the qubits are physically separated by large distances.

Try It 9.16

What happens if Alice and Bob make their observations simultaneously? Note: We implicitly treated that case in Table 9.2 and Alice explained the results at the end of section 9.5. Provide your own explanation of this situation.

Try It 9.17

Challenge. Show that the results of having Alice's measurements first, having Bob's measurements first, and having the two sets of measurements simultaneously are all consistent. Hint: If Alice measures first, from Eq. (9.17) the probability that she gets spin-up is c^2+d^2. The probability that Bob then gets spin-down is $d^2/(c^2+d^2)$. Hence the joint probability that Alice gets spin-up and Bob gets spin-down is the product of those two probabilities: $(c^2+d^2) \times d^2/(c^2+d^2) = d^2$. Work through the other three possible combinations of outcomes.

ALICE: When most people first encounter entanglement, they think—quite reasonably—that there must be some hidden mechanism that causes the correlations we have described. But any such mechanism, if it exists, must be very strange and would violate well-established principles. For example, as far as we know, no signal from one system to another can travel faster than the speed of light, $c = 3 \times 10^8$m/s. In 2017, a team of Chinese scientists (Yin et al. 2017) used light from a laser on an Earth satellite and two light detectors sensitive to the polarization of the light. The two detectors were separated by hundreds of kilometers on Earth and were set up to demonstrate correlations among photons in an entangled state. The spatial separations were such that no signal could have traveled between the receiving stations in time for one set of measurements to influence the other. Nevertheless, the correlations of observations built into the entangled state were observed.

9.7 Entanglement and the States of Just One of the Qubits

BOB: A bit of history: Erwin Schrödinger, whom we met in Chapter 4, invented the term entanglement. He noted that entanglement is *the* critical feature that distinguishes quantum mechanics from classical mechanics. Schrödinger's point is that if the two-qubit system has an entangled state, then it is not legitimate to say that Alice's qubit "has" a specific quantum state and my qubit "has" its own specific quantum state. If the system state is entangled, there is only the system state and not individual states of the individual qubits. The qubits are not independent and therefore cannot be said to have their own states.

Since this is such a key concept in quantum mechanics, let's see how it plays out. The general two-qubit state in Eq. (9.16) is a good starting point. If we think back to the beginning of this chapter, we wrote a two-qubit state as a product state $|S\rangle = |A\rangle |B\rangle$. If the system state can be factored in this way—a state vector for Alice's qubit multiplied by a state vector for my qubit—then, and only then, can we say that Alice's qubit is described by $|A\rangle$ and mine by $|B\rangle$. In the more general case shown in Eq. (9.16), we cannot factor the system state this way—a part associated with Alice's qubit and a part associated with my qubit. In other words, for an entangled two-qubit state, we don't have a definite state vector for Alice's qubit or a definite state vector for my qubit.

CARDY: That sounds pretty weird and worrisome. What does it mean for a qubit not to have a quantum state?

BOB: This might help: Another way to describe what is happening with an entangled state is to note that the results of measurements for one of the qubits depend on what happens to the other qubit. Remember that quantum states embody the information needed to predict the probabilities of those various results. In particular, the entangled state carries information about the correlations among the probabilities of the individual qubits' measurement outcomes. When correlations are present, we cannot assign each qubit its own state. We will take up the issue of correlated measurement outcomes in more detail in Chapter 14.

9.8 More on Entangled States

ALICE: There is a nice way of expressing the conditions for entangled states that allows us to extend the results to more than two qubits. These more general results will be useful when we discuss quantum error correction in Chapter 12. This is also a nice example of where a careful choice of mathematical notation makes the results easier to remember. Building on Eq. (9.11), let's make two generalizations: We will use computational basis states instead of spin states, and we change the coefficients from c, d, e, and f to symbols that are easier to associate with the basis states:

$$|\Psi\rangle = a_{00} |00\rangle + a_{01} |01\rangle + a_{10} |10\rangle + a_{11} |11\rangle. \tag{9.22}$$

CARDY: Aha! I see the pattern. You've labeled the coefficients with the symbols that name the basis states. How does that help with entangled states?

ALICE: Recall the condition expressed in Eq. (9.14) for a state not to be entangled. Using the notation in Eq. (9.22), we write the "not entangled" condition as

$$a_{00}a_{11} = a_{01}a_{10}. \tag{9.23}$$

If we write out the expression for a two-qubit simple product state (unentangled state) with amplitudes for each of the single qubit states labeled by a_0, a_1, b_0, and b_1 (notice that there is only one subscript for each of these amplitudes), we get

$$|\psi\rangle = \underbrace{(a_0|0\rangle + a_1|1\rangle)}_{\text{Alice's state}} \otimes \underbrace{(b_0|0\rangle + b_1|1\rangle)}_{\text{Bob's state}}$$
$$= a_0b_0|0\rangle \otimes |0\rangle + a_0b_1|0\rangle \otimes |1\rangle + a_1b_0|1\rangle \otimes |0\rangle + a_1b_1|1\rangle \oplus |1\rangle$$
$$= a_0b_0|00\rangle + a_0b_1|01\rangle + a_1b_0|10\rangle + a_1b_1|11\rangle. \tag{9.24}$$

In Eq. (9.24) we have left out the A and B subscripts on the basis states.

Comparing Eqs. (9.24) and (9.22), we see that single-qubit outermost product and middle product state amplitudes satisfy Eq. (9.23): $a_0b_0a_1b_1 = a_0b_1a_1b_0$. Each of the product state amplitudes appears once and only once on each side of the equation.

BOB: We can extend the results to handle three qubits, which means we have eight (2^3) basis states. The generalization of Eq. (9.22) is

$$|\Phi\rangle = a_{000}|000\rangle + a_{001}|001\rangle + a_{010}|010\rangle + a_{011}|011\rangle$$
$$+ a_{100}|100\rangle + a_{101}|101\rangle + a_{110}|110\rangle + a_{111}|111\rangle. \tag{9.25}$$

If you write the three-qubit product state corresponding to Eq. (9.24), you get

$$|\Phi\rangle = (\alpha_0|0\rangle + \alpha_1|1\rangle) \otimes (\beta_0|0\rangle + \beta_1|1\rangle) \otimes (\gamma_0|0\rangle + \gamma_1|1\rangle). \tag{9.26}$$

I won't multiply out all of the terms in Eq. (9.26), but you can see that the conditions to have the general three-qubit state expressible as a simple product state are

$$a_{000}a_{111} = a_{001}a_{110} = a_{011}a_{100} = a_{101}a_{010}. \tag{9.27}$$

If any of these conditions does *not* hold, then the state is an entangled state. Again, we see that if we choose the amplitudes randomly, we almost always get an entangled state. On the other hand, if we think about the practical implementation of quantum states on physically separated qubits, which are independent, you will need to act on an initial product state with appropriate quantum gates to produce a specific entangled state that you want for some QIS and QC purpose.

Try It 9.18

Show that there is a pattern among the 0s and 1s in the subscripts of each pair of coefficients in Eq. (9.27).

Try It 9.19

Multiply out the terms in Eq. (9.26) and verify that they satisfy the conditions in Eq. (9.27).

CARDY: But what if the conditions in Eq. (9.27) are almost satisfied but not quite? Is the state only a little entangled?

BOB: That is a cool question, Cardy, and one that is not addressed in most textbooks on quantum mechanics. Yes, we can have just a little bit of entanglement (no pun intended), and that will turn out to be an important concept when we deal with quantum error correction in Chapter 12. In more advanced books on quantum mechanics and QIS, there are fancier ways of describing the "degree of entanglement," but they would take us rather far afield. The following Challenge will show you how you might introduce a "small amount" of entanglement. In Chapter 14, we return to the question of providing a measure of the amount of entanglement.

Try It 9.20

Challenge. Suppose the condition in Eq. (9.23) is not quite satisfied. Figure out a way to write the state in Eq. (9.24) as a product state added to a smaller amplitude entangled state. Hint: There are many ways to do this. For example, we might use

$$|\psi\rangle = a_0 b_0 |00\rangle + a_0 b_1 |01\rangle + a_1 b_0 |10\rangle + a_1 b_1 |11\rangle + \delta \left(|00\rangle + |11\rangle\right).$$

The first four terms on the right of the previous equation represent a simple product state, as we know from Eq. (9.24). The last part of the expression is by itself an entangled state.

a. Explain how we know the last part of the state vector $\delta \left(|00\rangle + |11\rangle\right)$ is entangled.

b. Write out the condition among the state amplitudes to assure that the state is normalized.

Since the state vector has an entangled part, the overall state must be an entangled state, but we can change the amount of entanglement by changing δ.

c. Go through the measurement arguments in sections 9.4. and 9.5. to see how δ affects the results. We would expect that Bob's measurement results become less affected by Alice's measurement results as $\delta \to 0$.

9.9   Changing Basis States for Two-Qubit States

BOB: In Chapter 8 we learned how to express a general quantum state using a variety of basis states. That change of basis states was important in calculating the measurement outcome probabilities when the state preparation basis states are different from the measurement basis states. How does that work out for two-qubit states? You might also wonder if a change in basis states converts an entangled state into an unentangled state or vice versa.

In principle, the change-in-basis-states method is just what you might guess based on what we did in Chapter 8. Let's review the method. We are restricting ourselves to basis state changes that can be expressed in terms of a single angle θ_{AB}. In Chapter 15, we'll show you what to do in more general cases.

As before, we will write a single-qubit state in terms of two basis sets $|A_0\rangle$ and $|A_1\rangle$ and $|B_0\rangle$ and $|B_1\rangle$. We saw in Chapter 8 that we can write one set of basis states in terms of the other:

$$
\begin{aligned}
|A_0\rangle &= \cos\theta_{AB}\,|B_0\rangle + \sin\theta_{AB}\,|B_1\rangle \\
|A_1\rangle &= -\sin\theta_{AB}\,|B_0\rangle + \cos\theta_{AB}\,|B_1\rangle
\end{aligned}
\tag{9.28}
$$

and

$$
\begin{aligned}
|B_0\rangle &= \cos\theta_{AB}\,|A_0\rangle - \sin\theta_{AB}\,|A_1\rangle \\
|B_1\rangle &= \sin\theta_{AB}\,|A_0\rangle + \cos\theta_{AB}\,|A_1\rangle,
\end{aligned}
\tag{9.29}
$$

where θ_{AB} is the state space angle between $|A_0\rangle$ and $|B_0\rangle$ and we assume that $|A_0\rangle$ lies counterclockwise in state space by the angle θ_{AB} from $|B_0\rangle$. In a two-qubit state, we have products of the various individual qubit basis states. All we have to do is apply Eq. (9.28) or Eq. (9.29) to the appropriate right-vector basis states.

Rather than writing a general result, which turns out to be rather messy algebraically, let's do an example. Suppose we have two qubits α and β and we prepare them in a state

$$
|\psi\rangle = \frac{1}{\sqrt{2}}\left(|A_0\rangle_\alpha \otimes |A_0\rangle_\beta + |A_1\rangle_\alpha \otimes |A_1\rangle_\beta\right).
\tag{9.30}
$$

The subscripts indicate which state vector applies to which qubit. We want to change the basis states for qubit α from the A basis to the B basis. Let's break up the work into steps. First, we use the top line in Eq. (9.28) in the first part inside the parentheses in Eq. (9.30):

$$
|A_0\rangle_\alpha \otimes |A_0\rangle_\beta = \underbrace{\left[\cos\theta_{AB}|B_0\rangle_\alpha + \sin\theta_{AB}|B_1\rangle_\alpha\right]}_{|A_0\rangle_\alpha}\otimes|A_0\rangle_\beta
$$

$$
= |B_0\rangle_\alpha \otimes (\cos\theta_{AB}|A_0\rangle_\beta) + |B_1\rangle_\alpha \otimes (\sin\theta_{AB}|A_0\rangle_\beta),
\tag{9.31}
$$

where we used $(a\,|F\rangle) \otimes |G\rangle = |F\rangle \otimes (a\,|G\rangle)$, valid for any number a.

We now do the same thing for the second term inside the parentheses in Eq. (9.30),

$$|A_1\rangle_\alpha \otimes |A_1\rangle_\beta = \underbrace{\left[-\sin\theta_{AB}|B_0\rangle_\alpha + \cos\theta_{AB}|B_1\rangle_\alpha\right]}_{|A_1\rangle_\alpha} \otimes |A_1\rangle_\beta$$

$$= |B_0\rangle_\alpha \otimes (-\sin\theta_{AB}|A_1\rangle_\beta) + |B_1\rangle_\alpha \otimes (\cos\theta_{AB}|A_1\rangle_\beta). \qquad (9.32)$$

Finally, we add Eqs. (9.31) and (9.32) and then group terms multiplying $|B_0\rangle_\alpha$ and $|B_1\rangle_\alpha$:

$$|A_0\rangle_\alpha \otimes |A_0\rangle_\beta + |A_1\rangle_\alpha \otimes |A_1\rangle_\beta = |B_0\rangle_\alpha \otimes \underbrace{\left[\cos\theta_{AB}|A_0\rangle_\beta - \sin\theta_{AB}|A_1\rangle_\beta\right]}_{|B_0\rangle_\beta}$$

$$+ |B_1\rangle_\alpha \otimes \underbrace{(\sin\theta_{AB}|A_0\rangle_\beta + \cos\theta_{AB}|A_1\rangle_\beta)}_{|B_1\rangle_\beta}$$

$$= |B_0\rangle_\alpha \otimes |B_0\rangle_\beta + |B_1\rangle_\alpha \otimes |B_1\rangle_\beta. \qquad (9.33)$$

To get the final line in Eq. (9.33), we used Eq. (9.29) for the β qubit state.

CARDY: Eq. (9.33) is a strange result: The state seems to have the same form in any set of basis states.

BOB: Yes, that is unusual and certainly not true of entangled states in general (see Try It 9.20). We will make use of Eq. (9.33) in the following section.

> **Try It 9.21**
>
> Apply the change in basis states to the state $|\psi\rangle = \frac{1}{\sqrt{2}}\left(|A_0\rangle_\alpha \otimes |A_0\rangle_\beta - |A_1\rangle_\alpha \otimes |A_1\rangle_\beta\right)$.
> Does it maintain its form in different basis states the way the state in Eq. (9.30) does?
> Do the same for the state $|\psi\rangle = \frac{1}{\sqrt{2}}\left(|A_0\rangle_\alpha \otimes |A_1\rangle_\beta - |A_1\rangle_\alpha \otimes |A_0\rangle_\beta\right)$.

BOB: To wrap up this section, please remember that many authors do not include the tensor product symbol and the subscripts on the state vectors. That is, you may see a state written as $|\psi\rangle = (1/\sqrt{2})\left(|A_0\rangle|A_0\rangle + |A_1\rangle|A_1\rangle\right)$. It is up to you to keep the qubit states in the correct order.

9.10 Two Qubits and Three Measurement Bases

ALICE: We want to work through an example of entangled states that might be observed with two measurement devices, each of which has three possible measurement bases. This may sound like a rather contrived situation, but it has been used as an experimental test of some of the fundamentals of quantum mechanics and variations on this model show up in a variety of QIS and QC applications. We gave a brief introduction to the setup in Chapter 8.4.

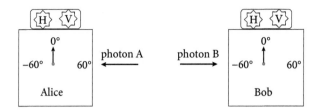

Fig. 9.2 Two photons interact with two different measurement devices, each of which has three choices for the orientation of the linear polarization detected by the devices. If "horizontal" polarization is detected for that measurement direction, the "H" light is lit on the device. If "vertical" polarization is observed, the "V" light is lit.

BOB: Figure 9.2 shows the set up. Two photons are launched toward two linear polarization measurement devices. Each device can be set to use a measurement basis along $0°$ degrees (relative to some direction), $60°$, or $-60°$ relative to that direction. Let's assume that the two photons are prepared in the following two-qubit state:

$$|\psi\rangle = \frac{1}{\sqrt{2}} \left(|\text{hlp}\rangle_{0°}|\text{hlp}\rangle_{0°} + |\text{vlp}\rangle_{0°}|\text{vlp}\rangle_{0°} \right). \tag{9.34}$$

Cardy, is that an entangled state or not?

CARDY: Let me think a minute . . .. Oh, I see. It is indeed an entangled state because the product of the outer amplitudes is ½ and the product of the inner amplitudes, which are not there, is 0. The products are not equal; so, by Eq. (9.14) the state is an entangled state.

BOB: Right on. Now we want to calculate the probabilities that Alice and I will record "H" or "V" for the three choices of measurement polarization directions. In particular, we want to find the fraction of the results that end up with Alice and I both recording H or both recording V.

CARDY: Why would you want to do that?

BOB: Good question, Cardy. It turns out this setup can be used to illustrate one of the most profound results in quantum mechanics: Bell's theorem, which we discuss in Chapter 14. For now, we will just use the setup to show how to change basis states and evaluate measurement probabilities for entangled states. Let's go through the various possibilities.

1. Both Alice and I use the $0°$ settings on our measurement devices. In this case we simply read off the squares of the appropriate coefficients in the state since it is already written in the $0°$ basis. We need the amplitude of the $|\text{hlp}\rangle_{0°}|\text{hlp}\rangle_{0°}$ term to get the probability of HH and the amplitude of the $|\text{vlp}\rangle_{0°}|\text{vlp}\rangle_{0°}$ term for the probability of VV. Each amplitude is $1/\sqrt{2}$. So, the probability is ½. Thus, the probability of getting the same result (either HH or VV) is 1. This is another example of the correlations that come along with entangled states.

2. Both Alice and I use $60°$ settings. There are two ways to figure this out. For the first, we follow the procedures from Chapter 8 giving the state in terms of the amplitudes in the $60°$ basis:

$$|\text{hlp}\rangle_{0°} = \frac{1}{2}|\text{hlp}\rangle_{60°} - \frac{\sqrt{3}}{2}|\text{vlp}\rangle_{60°}$$

$$|\text{vlp}\rangle_{0°} = \frac{\sqrt{3}}{2}|\text{hlp}\rangle_{60°} + \frac{1}{2}|\text{vlp}\rangle_{60°}. \tag{9.35}$$

We now substitute Eq. (9.35) in Eq. (9.34):

$$|\psi\rangle = \frac{1}{2}\left[\left(\frac{1}{2}|\text{hlp}\rangle_{60°} - \frac{\sqrt{3}}{2}|\text{vlp}\rangle_{60°}\right) \otimes \left(\frac{1}{2}|\text{hlp}\rangle_{60°} - \frac{\sqrt{3}}{2}|\text{vlp}\rangle_{60°}\right)\right]$$
$$+ \left[\left(\frac{\sqrt{3}}{2}|\text{hlp}\rangle_{60°} + \frac{1}{2}|\text{vlp}\rangle_{60°}\right) \otimes \left(\frac{\sqrt{3}}{2}|\text{hlp}\rangle_{60°} + \frac{1}{2}|\text{vlp}\rangle_{60°}\right)\right]$$
$$= \frac{1}{\sqrt{2}}\left[|\text{hlp}\rangle_{60°}|\text{hlp}\rangle_{60°} + |\text{vlp}\rangle_{60°}|\text{vlp}\rangle_{60°}\right]. \tag{9.36}$$

Note that the structure of the state is just like the $0°$ case we looked at before; so, we end up with the same results: Alice and I get the same results with probability 1.

The second method uses the general result in Eq. (9.33). We immediately recognize that when Alice and I choose the same basis set, no matter what its orientation is, we get identical results with probability 1. So, we know that holds for the choice of the $-60°$ basis as well.

3. Alice uses the $0°$ basis and I use the $60°$ settings.

This one is a bit trickier to analyze because the measurement device directions are not the same. But we proceed in the same way. We use Eq. (9.28) to change the basis set for my qubit's state:

$$|\psi\rangle = \frac{1}{\sqrt{2}}\left[\underbrace{|\text{hlp}\rangle_{0°}}_{\text{Alice}} \otimes \underbrace{\left(\frac{1}{2}|\text{hlp}\rangle_{60°} + \frac{\sqrt{3}}{2}|\text{vlp}\rangle_{60°}\right)}_{\text{Bob}} + \underbrace{|\text{vlp}\rangle_{0°}}_{\text{Alice}} \otimes \underbrace{\left(-\frac{\sqrt{3}}{2}|\text{hlp}\rangle_{60°} + \frac{1}{2}|\text{vlp}\rangle_{60°}\right)}_{\text{Bob}}\right]$$
$$= \frac{1}{\sqrt{2}}\left[\frac{1}{2}|\text{hlp}\rangle_{0°}|\text{hlp}\rangle_{60°} + \frac{1}{2}|\text{vlp}\rangle_{0°}|\text{vlp}\rangle_{60°}\right]$$
$$+ \frac{\sqrt{3}}{2}|\text{hlp}\rangle_{0°}|\text{vlp}\rangle_{60°} - \frac{\sqrt{3}}{2}|\text{vlp}\rangle_{0°}|\text{hlp}\rangle_{60°}. \tag{9.37}$$

From that, we read off that Alice and I will get HH with probability $\left(1/\left(2\sqrt{2}\right)\right)^2 = 1/8$ and VV with probability 1/8. So, for these settings, we will get matching results (VV or HH) with probability ¼. The results will be the same if we interchange the basis sets with Alice choosing $60°$ and I choose $0°$. We could also substitute $-60°$ for $60°$ and the results will be the same.

> **Try It 9.22**
>
> Convince yourself that Bob's claims about the various choices of basis sets are correct.

4. Alice uses the $-60°$ basis and I use the $60°$ settings. Proceeding with the appropriate substitutions, we get

$$|\psi\rangle = \frac{1}{\sqrt{2}}\left[\underbrace{\left(\frac{1}{2}|hlp\rangle_{-60°} - \frac{\sqrt{3}}{2}|vlp\rangle_{-60°}\right)}_{\text{Alice}} \otimes \underbrace{\left(\frac{1}{2}|hlp\rangle_{60°} - \frac{\sqrt{3}}{2}|vlp\rangle_{60°}\right)}_{\text{Bob}}\right]$$

$$+ \left[\underbrace{\left(+\frac{\sqrt{3}}{2}|hlp\rangle_{-60°} + \frac{1}{2}|vlp\rangle_{-60°}\right)}_{\text{Alice}} \otimes \underbrace{\left(-\frac{\sqrt{3}}{2}|hlp\rangle_{60°} + \frac{1}{2}|vlp\rangle_{60°}\right)}_{\text{Bob}}\right]$$

$$= -\frac{\sqrt{1}}{2\sqrt{2}}\left[|hlp\rangle_{-60°}|hlp\rangle_{60°} + |vlp\rangle_{-60°}|vlp\rangle_{60°}\right]$$

$$+ \frac{\sqrt{3}}{2\sqrt{2}}|vlp\rangle_{-60°}|hlp\rangle_{60°} - \frac{\sqrt{3}}{2\sqrt{2}}|hlp\rangle_{-60°}|vlp\rangle_{60°}. \tag{9.38}$$

Again, we are looking for the cases in which Alice and I both get H or both get V. The probability that we both get H is $\left(1/\left(2\sqrt{2}\right)\right)^2 = 1/8$, with the identical result for V. Hence, the probability for getting any identical result is ¼.

ALICE: Since Bob and I are equivalent, at least as far as quantum mechanics is concerned, we will get the same results if I choose the $60°$ basis and Bob chooses $-60°$.

We can now pull together all these results to find out how often (on average) Alice and I get the same results, namely HH or VV. Table 9.5 lists the fractions of the runs that yield the same results.

Since there are nine settings, the overall average fraction of the runs that give the same measurement results is $4.5/9 = 0.5$. The bottom line is that Alice and I will get the same measurement results 50% of the time if we choose equally among the three basis settings.

Table 9.5 Settings $0°$, $60°$, and $-60°$ and the fraction of measurements with those settings for which Alice and Bob get the same results.

Settings	$0°\ 0°$	$60°\ 60°$	$-60°\ -60°$	$0°\ 60°$	$60°\ 0°$	$0°\ -60°$	$-60°\ 0°$	$60°\ -60°$	$-60°\ -60°$
Fraction of same results	1	1	1	1/4	1/4	1/4	1/4	1/4	1/4

CARDY: That was a lot of math. And you didn't work through every combination in detail. I hope all that effort will turn out to be useful.

ALICE: Yes, it was a lot of math, but as you may have noticed it was 90% algebra with multiplication and addition. The remaining and important 10% was reading off the measurement result probabilities from the algebra. You certainly deserve a lot of credit because you were perseverant in sticking with us through all that. When we get to Chapter 14, you will see why it was important to do this math with different measurement basis sets for multi-qubit states.

Try It 9.23

Work through at least one of the remaining settings cases to verify that Bob's claims about how they worked out are correct.

 CHAPTER SUMMARY

- Most quantum applications involve systems of more than one qubit. Multi-qubit states can be built up as superpositions of products of single-qubit states. For n qubits we need 2^n basis states.

- The product of two state vectors is called a tensor product and is often denoted with the symbol $\otimes$. For example, $|\psi\rangle = |A\rangle \otimes |B\rangle$, or without the symbol $|\psi\rangle = |A\rangle\,|B\rangle$.

- Entangled states cannot be factored into simple product states because there is a correlation between the states.

- You can use quantum gates acting on multi-qubit systems to produce entangled states. Qubit interactions with their environment can wipe out the entangled state you produced with quantum gates.

- All entangled states are superposition states, but not all superposition states are entangled states.

- We can distinguish entangled from non-entangled states mathematically by a relationship among the state amplitudes.

- Entangled states imply correlations among the probabilities of the measurements carried out on the individual qubits. It is important to note that the correlations in the measurement outcomes are not due to an interaction that occurs between the qubits during the measurement.

- Measurements kill off entanglement. After an observation of one of the qubits, that qubit's state is no longer entangled with the state of the other qubits.

- Entangled states do not assign separate states to each qubit. Only the system can be said to have a state.

- We can exchange basis states for multi-qubit states by using the methods developed in Chapter 8 on each of the qubit states individually.

FURTHER READING

A survey of earlier experiments that demonstrated entanglement distribution and tests of Bell's theorem with special attention to the loopholes that have been closed over the years:

Alain Aspect, "Closing the Door on Einstein and Bohr's Quantum Debate." https://physics.aps.org/articles/v8/123.

The 2017 entangled polarization photon experiment:

J. Yin, Y. Cao, Y.-H. Li, S.-K. Liao, L. Zhang, J.-G. Ren, W.-Q. Cai, W.-Y. Liu, B. Li, H. Dai, G.-B. Li, Q.-M. Lu, Y.-H. Gong, Y. Xu, S.-L. Li, F.-Z. Li, Y.-Y. Yin, Z.-Q. Jiang, M. Li, J.-J. Jia, G. Ren, D. He, Y.-L. Zhou, X.-X. Zhang, N. Wang, X. Chang, Z.-C. Zhu, N.-L. Liu, Y.-A. Chen, C.-Y. Lu, R. Shu, C.-Z. Peng, J.-Y. Wang, and J.-W. Pan, "Satellite-based entanglement distribution over 1200 kilometers." *Science* **356** 1140–1144 (2017).

10 Quantum Circuits and Multi-Qubit Applications

Therefore, send not to know
For whom the bell tolls,
It tolls for thee.

John Donne, "For Whom the Bell Tolls" (1624)

10.1 Two-Qubit Gates

BOB: We are now ready to learn how to carry out operations on entangled and other two-qubit quantum states. We will consider only gates that have two inputs and two outputs. As we noted in Chapter 6, such gates can be reversible. That means we can use the gates to act on states to produce new states and then apply them again to revert back to where we started, if so needed.

10.2 CNOT Gates

ALICE: Let's start with the CNOT gate, introduced for classical logic gates in Chapters 2 and 3. The quantum circuit schematic is shown in Figure 10.1 which shows what we might expect for a quantum CNOT gate based on what we know about a classical-computing CNOT gate. We will see that the correspondence is correct if the states $|A\rangle$ and $|B\rangle$ are the computational basis states $|0\rangle$ and $|1\rangle$, but the results are more interesting (and more useful) for more general quantum states. Recall that the upper-qubit line is called the control qubit and the lower-qubit line is the target or controlled qubit.

First, we need to translate from classical bits 0 and 1 to quantum state vectors $|0\rangle$ and $|1\rangle$. Recall that the logic table for the classical CNOT gate is (Table 10.1):

Table 10.2 shows the corresponding table for a quantum CNOT gate acting on the computational basis states.

When a CNOT gate processes (operates on) the computational basis states, the table looks just like that for the classical computing CNOT logic gate.

CARDY: Is there a matrix representation of the CNOT gate?

ALICE: There sure is. Because we have four possible basis states, the matrix is a 4×4 matrix (four rows and four columns):

Quantum Computing: From Alice to Bob. Alice Flarend and Bob Hilborn, Oxford University Press.
© Alice Flarend and Robert C. Hilborn (2022). DOI: 10.1093/oso/9780192857972.003.0010

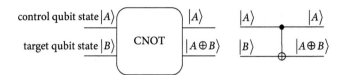

Fig. 10.1 A quantum controlled-NOT gate. On the left, a gate box representation. On the right is the commonly used circuit representation. $\oplus$ means XOR (see Chapter 2).

Table 10.1 CNOT truth table for classical bits A and B.

Input		Output	
A	B	A	$A \oplus B$
0	0	0	0
0	1	0	1
1	0	1	1
1	1	1	0

Table 10.2 The truth table for a quantum CNOT gate acting on computational basis states.

Input		Output	
A	B	A	$A \oplus B$
$|0\rangle$	$|0\rangle$	$|0\rangle$	$|0\rangle$
$|0\rangle$	$|1\rangle$	$|0\rangle$	$|1\rangle$
$|1\rangle$	$|0\rangle$	$|1\rangle$	$|1\rangle$
$|1\rangle$	$|1\rangle$	$|1\rangle$	$|0\rangle$

$$\text{CNOT} \Leftrightarrow \begin{pmatrix} 1 & 0 & 0 & 0 \\ 0 & 1 & 0 & 0 \\ 0 & 0 & 0 & 1 \\ 0 & 0 & 1 & 0 \end{pmatrix} \tag{10.1}$$

Try It 10.1

In Chapter 6, we stated four conditions the matrix associated with quantum gates should satisfy. Check that the CNOT matrix satisfies those conditions.

ALICE: We should check that this matrix does what it is supposed to do. As an example, let's try out the fourth entry in Table 10.2 with the matrix operating on the column vector for the state $|1_A\rangle\,|1_B\rangle$

$$\begin{pmatrix} 1 & 0 & 0 & 0 \\ 0 & 1 & 0 & 0 \\ 0 & 0 & 0 & 1 \\ 0 & 0 & 1 & 0 \end{pmatrix} \begin{pmatrix} 0 \\ 0 \\ 0 \\ 1 \end{pmatrix} = \begin{pmatrix} 0 \\ 0 \\ 1 \\ 0 \end{pmatrix}, \tag{10.2}$$

which is just what we want: The control state is $|1_A\rangle$ and the target input state is $|1_B\rangle$. On the right side of Eq. (10.2), upper output state is $|1_A\rangle$, unchanged from the input, and the target output state is $|0_B\rangle$.

CARDY: Slow down, please. I am getting confused about our basis state notation. It seems that you went quickly from vectors in Table 10.2 to column vectors in Eq. (10.2).

ALICE: Okay, Cardy. Let's spell that out. You may remember from the previous chapter that the system computational basis states for a two-qubit system are $|00\rangle, |01\rangle, |10\rangle$, and $|11\rangle$. The column vector on the left side of Eq. (10.2) corresponds to $|11\rangle$. When CNOT acts on that state, Table 10.2 indicates that the result is $|10\rangle$, which corresponds to the column vector on the right of Eq. (10.2). Notice that the first "1" label does not change because it represents the control bit.

CARDY: Thanks. That helped. But I will need to practice translating among those various representations.

Try It 10.2

Check the other entries in the table using the same method. This is a good exercise in matrix multiplication.

BOB: Life gets more interesting if we consider more complicated input quantum state vectors. Let's assume that the input states are

$$|A\rangle = \frac{1}{\sqrt{2}}\,|0_A\rangle + \frac{1}{\sqrt{2}}\,|1_A\rangle$$

$$|B\rangle = |0_B\rangle, \tag{10.3}$$

where $|A\rangle$ is Alice's control state and $|B\rangle$ is Bob's target state.

Using the state product rule, we see that the overall initial ("input") state of the system is

$$|S_{input}\rangle = \frac{1}{\sqrt{2}}\,\{|0_A\rangle + |1_A\rangle\}\,\,|0_B\rangle = \frac{1}{\sqrt{2}}\,|0_A\rangle\,|0_B\rangle + \frac{1}{\sqrt{2}}\,|1_A\rangle\,|0_B\rangle. \tag{10.4}$$

Cardy, you should be able to tell very quickly if that input state is an entangled state or not. Which is it?

CARDY: Oh, yeah. I need the product of the outer coefficients to be different from the product of the inner coefficients to have an entangled state. I will write this out

$$c\,|0_A\rangle\,|0_B\rangle + d\,|0_A\rangle\,|1_B\rangle + e\,|1_A\rangle\,|0_B\rangle + f\,|1_A\rangle\,|1_B\rangle$$
$$= \frac{1}{\sqrt{2}}\,|0_A\rangle\,|0_B\rangle + 0\,|0_A\rangle\,|1_B\rangle + \frac{1}{\sqrt{2}}\,|1_A\rangle\,|0_B\rangle + 0\,|1_A\rangle\,|1_B\rangle. \tag{10.5}$$

This one seems to have $cf = 0$ and $de = 0$ for those products. So, it is not an entangled state.

BOB: Excellent work, Cardy. Now let's write the output after the CNOT gate acts on that state. We can see from Table 10.2 that the first term on the right side of Eq. (10.4) is not changed under the action of the CNOT gate, but the second term becomes $\frac{1}{\sqrt{2}}|1_A\rangle\,|1_B\rangle$. Thus, the output state of the system is

$$|S_{out}\rangle = \frac{1}{\sqrt{2}}\left\{|0_A\rangle + |1_A\rangle\right\}|0_B\rangle = \frac{1}{\sqrt{2}}\,|0_A\rangle\,|0_B\rangle + \frac{1}{\sqrt{2}}\,|1_A\rangle\,|1_B\rangle. \tag{10.6}$$

Surprise! This state is an entangled state. Note, as usual for an entangled state, the output state cannot be written as a product of an Alice state and a Bob state. The CNOT gate can act on an unentangled quantum state vector and produce an entangled state vector.

Try It 10.3

Check that Eq. (10.6) does indeed describe an entangled state.

BOB: The first quantum CNOT gate was implemented by David Wineland and his team at the National Institute of Standards and Technology in Boulder, Colorado in 1995. The group used trapped ions as qubits. Wineland went on to win the Nobel Prize in Physics in 2012.

To see how the CNOT turns an unentangled state into an entangled one, let's look at the action of CNOT on the basis states written in the combined right vector form. For example, we write $|0_A\rangle\,|0_B\rangle = |00\rangle$. Then, for a general two-qubit input state, we may write:

$$\text{CNOT}\left\{c\,|00\rangle + d\,|01\rangle + e\,|10\rangle + f\,|11\rangle\right\}$$
$$= c\,|00\rangle + d\,|01\rangle + e\,|11\rangle + f\,|10\rangle. \tag{10.7}$$

The e and f terms are interchanged under the operation of the CNOT gate.

Try It 10.4

Use the matrix form of the CNOT gate in Eqs. (10.1) and (10.2) to check the result given in Eq. (10.7). Hint: Write the column vector in Eq. (10.2) with entries c, d, e, and f.

CARDY: Yeah, that seems to work. Is it okay to just switch the last two terms and not worry about all that matrix multiplication each time?

ALICE: Yes, Cardy, that is fine, but you need to be sure the basis states are expressed in the standard order with my state on the left and Bob's on the right with the labels increasing in numerical order (00, 01, 10, 11) (in binary form). As long as you do that, the shortcut method is just fine.

10.3 Quantum Circuits (Quantum Gate Arrays)

ALICE: Now let's look how we combine the CNOT gate with other operations on quantum states.

Figure 10.2 shows a simple quantum computational circuit. These circuits are read from left to right as we did before. Usually, we start with all the qubits (four in this case) in the computational basis state $|0\rangle$. This is not necessary, but it is the most common choice. The states are then acted upon by the various gates. Finally, on the far right, the resulting states meet up with measurement devices (usually in the computational basis) and the results are recorded on the output wires of the measurement gates.

Looking closer at the circuit, you can see the top-row qubit is a control for a CNOT gate with the second-row qubit serving as the target. The second-row qubit is also a target for a CNOT gate with the third-row qubit as the control.

CARDY: I see that the fourth-row qubit serves as control, while the third-row qubit is a target. Is that right?

ALICE: You are definitely getting the hang of this, Cardy. I should point out that while most current QCs use this gate-array model, there are other models for quantum computers. For example, there is "measurement-based quantum computation" and "topological quantum computation." For our purposes, we don't need to know about those other models because at least in principle they can be translated into the quantum circuit model (and vice versa). We will describe some of these other models in Chapter 16.

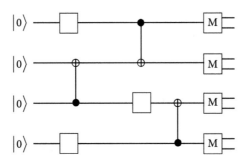

Fig. 10.2 A typical (but simple) four-qubit quantum computational circuit. The input states are on the left and are acted upon by various gates as you proceed to the right across the diagram. The open squares represent unspecified gates. The boxes on the far right are measurement devices.

10.4 **The Bell Circuit**

ALICE: The following circuit—a combination of the Hadamard gate from Chapter 6, acting on a single qubit, and the CNOT gate acting on two qubits—is named after John Stewart Bell (1928–90). Bell's theories had (and still have) a major impact on our understanding of the conceptual foundations of quantum mechanics and on quantum computing.

Figure 10.3 shows a Bell circuit (BC) composed of a Hadamard gate and a CNOT gate. We can also represent the Bell circuit as a gate box as shown on the right in Figure 10.3. This gate plays a big role in various quantum algorithms.

To understand what the Bell circuit does, recall that the Hadamard gate converts a computational basis state into a superposition state. For example, with $|A\rangle = |0\rangle$, we know that

$$H\,|0\rangle = \frac{1}{\sqrt{2}}\,|0\rangle + \frac{1}{\sqrt{2}}\,|1\rangle. \tag{10.8}$$

In the Bell circuit, that superposition state is the control input for the CNOT gate. If $|B\rangle = |0\rangle$, the output of the BC is

$$|\psi\rangle = \text{CNOT}\,[H\,|0\rangle] = \frac{1}{\sqrt{2}}\,|00\rangle + \frac{1}{\sqrt{2}}\,|11\rangle, \tag{10.9}$$

which we recognize as an entangled state. We could also write this as

$$BC\,|00\rangle = \frac{1}{\sqrt{2}}\,|00\rangle + \frac{1}{\sqrt{2}}\,|11\rangle = |\text{Bell}_{00}\rangle = |\beta_{00}\rangle = |\Phi^{+}\rangle. \tag{10.10}$$

The resulting state is called a Bell state (the right side of Eq. (10.10) gives three different notations for the Bell state). We will see the Bell state and its cousins (see Try It 10.5) quite frequently in what follows. The BC and its variants show up in many QI and QC situations because they are gates that produce entangled states from non-entangled states. We cannot stress enough that entanglement plays a huge role in many QIS and QC applications.

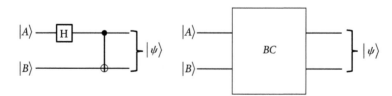

Fig. 10.3 The Bell circuit (BC). The large curly brace on the right side reminds us that the output of the BC is an entangled state for the *system* of two output qubits; it cannot be associated with either of the output qubit lines alone.

Try It 10.5

Show that the following operations are implemented by the Bell circuit:

$$BC\,|01\rangle = \frac{1}{\sqrt{2}}\,|01\rangle + \frac{1}{\sqrt{2}}\,|10\rangle = |\text{Bell}_{01}\rangle = |\beta_{01}\rangle = |\Psi^{+}\rangle$$

$$BC\,|10\rangle = \frac{1}{\sqrt{2}}\,|00\rangle - \frac{1}{\sqrt{2}}\,|11\rangle = |\text{Bell}_{10}\rangle = |\beta_{10}\rangle = |\Phi^{-}\rangle$$

$$BC\,|11\rangle = \frac{1}{\sqrt{2}}\,|01\rangle - \frac{1}{\sqrt{2}}\,|10\rangle = |\text{Bell}_{11}\rangle = |\beta_{11}\rangle = |\Psi^{-}\rangle \qquad (10.11)$$

Show that the Bell states are normalized and orthogonal.

Try It. 10.6

Challenge. Is the Bell circuit reversible? Hint: Apply BC to the state on the right side of Eq. (10.10). Section 10.5 explains why this may not work the way you expect.

Using Matrices for the Bell Circuit

ALICE: If we want to use matrices to represent the Bell circuit, we need 4×4 matrices and four-element column vectors because the circuit acts on two qubits. In particular, we need the matrix representation for the CNOT gate (Eq. (10.1)) and the matrix that represents the Hadamard gate acting on Alice's qubit while nothing happens to Bob's. The latter matrix is the tensor product of two matrices: one the Hadamard gate acting on Alice's qubit and the identity operator acting on Bob's; so, we need to know how to form the tensor product of two matrices.

For any operators, F and G, each acting on a single-qubit state, each operator may be represented by a 2×2 matrix. How do we combine those matrices to represent the operator $F \otimes G$? The general product rule for 2×2 matrices is called the Kronecker product, the "natural" generalization of the tensor product rule for state vectors. An operator F and its matrix elements are given by

$$F \Rightarrow \begin{pmatrix} F_{11} & F_{12} \\ F_{21} & F_{22} \end{pmatrix}, \qquad (10.12)$$

where the subscripts refer to the row and column numbers $F_{\text{row column}}$. Using that notation, we write the product of two operators as

$$F \otimes G \Rightarrow \begin{pmatrix} F_{11} \otimes G & F_{12} \otimes G \\ F_{21} \otimes G & F_{22} \otimes G \end{pmatrix} = \begin{pmatrix} F_{11}G_{11} & F_{11}G_{12} & F_{12}G_{11} & F_{12}G_{12} \\ F_{11}G_{21} & F_{11}G_{22} & F_{12}G_{21} & F_{12}G_{22} \\ F_{21}G_{11} & F_{21}G_{12} & F_{22}G_{11} & F_{22}G_{12} \\ F_{21}G_{21} & F_{21}G_{22} & F_{22}G_{21} & F_{22}G_{22} \end{pmatrix}. \quad (10.13)$$

Let's work this out for the product $H \otimes I$. We know the matrices representing H and I

$$H \Rightarrow \frac{1}{\sqrt{2}} \begin{pmatrix} 1 & 1 \\ 1 & -1 \end{pmatrix} \text{ and } I \Rightarrow \begin{pmatrix} 1 & 0 \\ 0 & 1 \end{pmatrix}. \quad (10.14)$$

Using those matrices in Eq. (10.13), we find

$$H \otimes I = \frac{1}{\sqrt{2}} \begin{pmatrix} 1 & 0 & 1 & 0 \\ 0 & 1 & 0 & 1 \\ 1 & 0 & -1 & 0 \\ 0 & 1 & 0 & -1 \end{pmatrix}. \quad (10.15)$$

Let's now apply those results to the Bell circuit with the input state $|00\rangle$. We want to evaluate the combination $CNOT[(H \otimes I)|00\rangle]$. For the $H \otimes I$ part we get

$$\frac{1}{\sqrt{2}} \begin{pmatrix} 1 & 0 & 1 & 0 \\ 0 & 1 & 0 & 1 \\ 1 & 0 & -1 & 0 \\ 0 & 1 & 0 & -1 \end{pmatrix} \begin{pmatrix} 1 \\ 0 \\ 0 \\ 0 \end{pmatrix} = \frac{1}{\sqrt{2}} \begin{pmatrix} 1 \\ 0 \\ 1 \\ 0 \end{pmatrix}. \quad (10.16)$$

Now we apply the CNOT matrix to the column vector on the right in Eq. (10.16)

$$\frac{1}{\sqrt{2}} \begin{pmatrix} 1 & 0 & 0 & 0 \\ 0 & 1 & 0 & 0 \\ 0 & 0 & 0 & 1 \\ 0 & 0 & 1 & 0 \end{pmatrix} \begin{pmatrix} 1 \\ 0 \\ 1 \\ 0 \end{pmatrix} = \frac{1}{\sqrt{2}} \begin{pmatrix} 1 \\ 0 \\ 0 \\ 1 \end{pmatrix} = \frac{1}{\sqrt{2}} \begin{pmatrix} 1 \\ 0 \\ 0 \\ 0 \end{pmatrix} + \frac{1}{\sqrt{2}} \begin{pmatrix} 0 \\ 0 \\ 0 \\ 1 \end{pmatrix}, \quad (10.17)$$

which is the matrix and column vector equivalent of Eq. (10.10).

CARDY: Ah! I see that the amplitudes for the lower two states in the column vector have been switched.

10.5   The Reverse Bell Circuit

BOB: The Bell circuit, like most of the gates we will be using, is reversible in the sense that if we use the output of the BC as input to first the CNOT gate and then send the upper qubit in Figure 10.3 through the Hadamard gate, we end up with the original set of states. We will call such a gate the reverse Bell circuit (RBC).

Figure 10.4 shows the circuit diagram for the RBC. As promised, it is just the reverse of the BC shown in Figure 10.3.

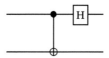

Fig. 10.4 The circuit diagram for the reverse Bell circuit.

BOB: The RBC acting on the output state in Eq. (10.10) begins with the CNOT:

$$\text{RBC}\,[\text{BC}\,|00\rangle] = \text{CNOT}\left\{\frac{1}{\sqrt{2}}\,|00\rangle + \frac{1}{\sqrt{2}}\,|11\rangle\right\} = \frac{1}{\sqrt{2}}\,|00\rangle + \frac{1}{\sqrt{2}}\,|10\rangle. \tag{10.18}$$

Next the Hadamard works only on the upper registry, which is shown in Figure 10.4 with the "H" box being only in the top line:

$$H\left\{\frac{1}{\sqrt{2}}\,|00\rangle + \frac{1}{\sqrt{2}}\,|10\rangle\right\} = \frac{1}{\sqrt{2}}\,(|00\rangle + |01\rangle + |00\rangle - |01\rangle) = |00\rangle = |0\rangle\,|0\rangle.$$

CARDY: Something big happened. We seem to have lost the entangled superposition state.

BOB: Yes, the Hadamard gate replaces the initial input state with a superposition state. The second application of the Hadamard gate reduces the superposition back to a simple product state. Both the Bell circuit and the reverse Bell circuit are quite useful for QI and QC.

Try It 10.7

Apply the RBC to each of the superposition states in the previous Try It. Does the RBC successfully return the original input states? The Bell circuit is reversible, but you need to reverse the order of the H and CNOT gates,

10.6   **Entanglement Swapping**

ALICE: In Chapter 9, we found that entanglement is a rather strange quantum phenomenon with many counterintuitive notions. Here is an example that sets up an entangled state for two qubits that are always far apart and have no direct interaction with each other! No matter how strange it seems, this kind of entanglement is useful in several quantum cryptographic protocols.

We start with four qubits: A, B, C, and D. Qubits A and B are prepared in an entangled state while C and D are also prepared in an entangled state but one that is completely independent of the AB entangled state. The qubits go off in different directions. See Figure 10.5. We then arrange to carry out a measurement on qubits C and D. With the right choice of measurement basis states, we find that qubits A and D end up in an entangled state even though they have

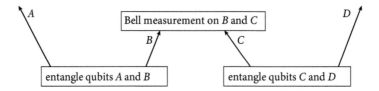

Fig. 10.5 A diagram of the entanglement swapping process. Qubits *A* and *B* are produced in an entangled state. Qubits *C* and *D* are in an independent entangled state. Qubits *B* and *C* are subject to a Bell measurement. The result is that *A* and *D* are left in an entangled state.

never met up with each other and have not interacted directly. We say that the procedure has swapped the entanglement from the *AB* and *CD* systems for entanglement in the *AD* system.

CARDY: That sounds totally weird. How do we know that qubits *A* and *D* are in fact in an entangled state?

ALICE: Good question. As before, having qubits in an entangled state means that measurements on one of the qubits are correlated with measurements on the other qubit.

Let's work through an example of how entanglement swapping works. Suppose the qubits are photons and we prepare the pairs of qubits in the Bell state $|\text{Bell}_{00}\rangle$ for horizontal (hlp) and vertical (vlp) polarization (see Eq. (10.10)).

$$|AB\rangle = \frac{1}{\sqrt{2}} \{|\text{hlp}\rangle_A |\text{hlp}\rangle_B + |\text{vlp}\rangle_A |\text{vlp}\rangle_B\}$$

$$|CD\rangle = \frac{1}{\sqrt{2}} \{|\text{hlp}\rangle_C |\text{hlp}\rangle_D + |\text{vlp}\rangle_C |\text{vlp}\rangle_D\}. \tag{10.19}$$

Since *AB* and *CD* are prepared independently, the state for the overall *ABCD* system is a product of two two-qubit states:

$$|ABCD\rangle = |AB\rangle \, |CD\rangle. \tag{10.20}$$

Qubits *B* and *C* proceed to a measurement device whose measurement basis states are the four Bell states (Eq. (10.10) and Try It 10.5). Later we'll talk briefly about how to build such a device for polarized photons.

To express what happens when the qubits *B* and *C* interact with the Bell state measurement device, we need to express the system state Eq. (10.20) as a superposition of Bell states for the *BC* qubit combination. To do that, we use Eq. (10.19) in (10.20) and expand the products of state vectors:

$$|ABCD\rangle = \frac{1}{2} \overbrace{\{|\text{hlp}\rangle_A |\text{hlp}\rangle_B + |\text{vlp}\rangle_A |\text{vlp}\rangle_B\}}^{AB \text{ entangled state}} \overbrace{\{|\text{hlp}\rangle_C |\text{hlp}\rangle_D + |\text{vlp}\rangle_C |\text{vlp}\rangle_D\}}^{CD \text{ entangled state}}$$

$$= \frac{1}{2} |\text{hlp}\rangle_A \underbrace{|\text{hlp}\rangle_B |\text{hlp}\rangle_C} |\text{hlp}\rangle_D + \frac{1}{2} |\text{hlp}\rangle_A \underbrace{|\text{hlp}\rangle_B |\text{vlp}\rangle_C} |\text{vlp}\rangle_D$$

$$+ \frac{1}{2} |\text{vlp}\rangle_A \underbrace{|\text{vlp}\rangle_B |\text{hlp}\rangle_C} |\text{hlp}\rangle_D + \frac{1}{2} |\text{vlp}\rangle_A \underbrace{|\text{vlp}\rangle_B |\text{vlp}\rangle_C} |\text{vlp}\rangle_D. \tag{10.21}$$

We have grouped terms for the B and C qubits because they are the ones that will be subject to Bell measurements. Using Eqs. (10.10) and (10.11), the four underlined terms in Eq. (10.21) can be written in terms of the Bell states:

$$|\text{hlp}\rangle_B |\text{hlp}\rangle_C = \frac{1}{\sqrt{2}} |\text{Bell}_{00}\rangle_{BC} + \frac{1}{\sqrt{2}} |\text{Bell}_{10}\rangle_{BC}$$

$$|\text{hlp}\rangle_B |\text{vlp}\rangle_C = \frac{1}{\sqrt{2}} |\text{Bell}_{01}\rangle_{BC} + \frac{1}{\sqrt{2}} |\text{Bell}_{11}\rangle_{BC}$$

$$|\text{vlp}\rangle_B |\text{hlp}\rangle_C = \frac{1}{\sqrt{2}} |\text{Bell}_{01}\rangle_{BC} - \frac{1}{\sqrt{2}} |\text{Bell}_{11}\rangle_{BC}$$

$$|\text{vlp}\rangle_B |\text{vlp}\rangle_C = \frac{1}{\sqrt{2}} |\text{Bell}_{00}\rangle_{BC} - \frac{1}{\sqrt{2}} |\text{Bell}_{10}\rangle_{BC}. \tag{10.22}$$

Using Eq. (10.22) in Eq. (10.21) and grouping the Bell states gives

$$|ABCD\rangle = \frac{1}{2} |\text{Bell}_{00}\rangle_{BC} \underbrace{\frac{1}{2} \{ |\text{hlp}\rangle_A |\text{hlp}\rangle_D + |\text{vlp}\rangle_A |\text{vlp}\rangle_D \}}_{|\text{Bell}_{00}\rangle_{AD}}$$

$$+ \frac{1}{2} |\text{Bell}_{10}\rangle_{BC} \underbrace{\frac{1}{2} \{ |\text{hlp}\rangle_A |\text{hlp}\rangle_D - |\text{vlp}\rangle_A |\text{vlp}\rangle_D \}}_{|\text{Bell}_{10}\rangle_{AD}}$$

$$+ \frac{1}{2} |\text{Bell}_{01}\rangle_{BC} \underbrace{\frac{1}{2} \{ |\text{hlp}\rangle_A |\text{vlp}\rangle_D + |\text{vlp}\rangle_A |\text{hlp}\rangle_D \}}_{|\text{Bell}_{01}\rangle_{AD}}$$

$$+ \frac{1}{2} |\text{Bell}_{11}\rangle_{BC} \underbrace{\frac{1}{2} \{ |\text{hlp}\rangle_A |\text{vlp}\rangle_D - |\text{vlp}\rangle_A |\text{hlp}\rangle_D \}}_{|\text{Bell}_{11}\rangle_{AD}}. \tag{10.23}$$

Implementing the AD Bell states indicated in Eq. (10.23) yields the final result:

$$|ABCD\rangle = \frac{1}{2} |\text{Bell}_{00}\rangle_{BC} |\text{Bell}_{00}\rangle_{AD} + \frac{1}{2} |\text{Bell}_{10}\rangle_{BC} |\text{Bell}_{10}\rangle_{AD}$$

$$+ \frac{1}{2} |\text{Bell}_{01}\rangle_{BC} |\text{Bell}_{01}\rangle_{AD} + \frac{1}{2} |\text{Bell}_{11}\rangle_{BC} |\text{Bell}_{11}\rangle_{AD}. \tag{10.24}$$

CARDY: That's interesting. The entire state can be written as a sum of products of Bell states.

ALICE: Yes. That result is an example showing that the Bell states can indeed be used as basis states for a two-qubit system.

Now let's see what happens if we subject the BC qubits to a measurement device whose basis states are the Bell states. This is an example of a measurement basis that is not tied to a simple physical property like spin-up or spin-down. Note that each of the possible Bell states appears in Eq. (10.24).

CARDY: Ah, yes. I see that the amplitudes for the various Bell states are all the same. So, we have a 25% probability of getting any one of them.

ALICE: Exactly right. Suppose we carry out the Bell state measurement and get $|\text{Bell}_{01}\rangle_{BC}$. What is the state used to describe the system after that measurement?

CARDY: It should be the measurement basis state multiplied by the corresponding state for the AD qubits in Eq. (10.24). So, we should have $|\text{Bell}_{01}\rangle_{BC}|\text{Bell}_{01}\rangle_{AD}$. Wait ... that is an entangled state because it is a Bell state. That is weird because A and D were never entangled before.

ALICE; Superb! That result indeed tells us that the state for the AD qubits after the measurement is $|\text{Bell}_{01}\rangle_{AD}$, an entangled state! We start with qubits A and B in an entangled state while C and D are in an independent entangled state. After the process, qubits A and D are described by an entangled state. We have swapped the entanglement from AB and CD to AD even though the A and D qubits have never directly interacted. That is truly strange. Nonetheless, if we carry out further measurements on qubits A and D, we find that those measurement outcomes are correlated, just as we would expect for an entangled state.

BOB: How are Bell state measurements carried out? It turns out that you can set up a combination of beam splitters, polarizing materials, and photon detectors of the type we have described before. Photons described by each of the four Bell states will trigger a different combination of detectors (see Further Reading (Pan, 1998), (Jin, 2015), and (Kirby, 2016).) From the pattern of the detector responses, we can deduce the Bell state that resulted from the measurement. These experiments were first carried out in 1998, and since then variations on those experiments have been used in many quantum communications protocols.

10.7   No-Cloning Theorem

ALICE: As we dig deeper into quantum information, quantum cryptography, and quantum computing, we will discover that if we could prepare a qubit in a general quantum state without our knowing what that state is, lots of things in QIS would be much easier. At the same time we would lose much of the utility of qubits and quantum states. The notion of duplicating a general quantum state is called cloning. The formal statement of the impossibility of doing so is called the no-cloning theorem, which we mentioned in Chapter 5.

First, let's specify what we would like to accomplish by cloning: We want to prepare a second qubit in (or more accurately, described by) a quantum state that is exactly the same state associated with a previously received qubit.

CARDY: Isn't cloning easy if you know the state coefficients for the original state? You use standard state preparation procedures, which we have already learned, to prepare the second qubit in that state.

ALICE: Cloning is an interesting task only if we do not know what the first qubit's quantum state is. Specifically, that means we do not know the state coefficients. As you noted, if we knew the state coefficients and had sufficiently clever technical equipment, we could prepare the second qubit in a state identical to the state of the first qubit. But the question is: Can we clone the qubit's state if we don't know the first qubit's state coefficients? As we saw in Chapter 5, acting on a qubit with a measurement device gives us at best limited knowledge about the original state coefficients because the state associated with the qubit may be a superposition state. In general, we would have to use a series of observations, perhaps with two or more measurement basis choices, on a string of identically prepared qubits to get more information about the state. Such a procedure is called quantum tomography, in analogy with medical tomography.

As an example, let's think about a spin-1/2 qubit prepared in one of two superposition states

$$|S_{\pm}\rangle = \frac{1}{\sqrt{2}} \left\{ |\uparrow\rangle \pm |\downarrow\rangle \right\}. \tag{10.25}$$

If we observe the spin-orientation with a spin-up, spin-down measurement device, 50% of the time we get spin-up and 50% of the time we get spin-down with either of the superposition states. But with a single observation, we would only get one or the other result. And if we do not know the qubit was in a superposition state, we may think it had previously been in the basis state we measured. A single observation does not give us enough information to clone the state. Can we devise a cloning procedure that requires just one operation on the original qubit? That process, according to standard quantum mechanics, is impossible. Let's see why.

BOB: Here's the general argument: Let's suppose that the original qubit's state is represented by

$$|\psi\rangle = a_0 |A_0\rangle + a_1 |A_1\rangle \tag{10.26}$$

in Alice's A basis. We would like to build a quantum gate that can operate on that state and produce a second qubit described by that same state while keeping the first qubit and its state intact. We also need to keep in mind that the only gates we want to use are reversible gates; so that means if we want two qubits out, we must put in two qubits. One of the input qubits will have the state described in Eq. (10.26)—the one we want to duplicate—while the other can be one of the basis states. It turns out that it doesn't matter what that other state is; so, we choose something simple like $|0\rangle$. If you could create a cloning gate circuit C, it would look like the Figure 10.6 which shows what we would want a cloning gate to do if we could build one. We start with two input qubits. One is described by the state vector $|\psi\rangle$ that we would like to clone. Cloning means we want to produce another qubit in that same state; so, we end up with two qubits described by the same quantum state vector.

Let's work out what happens with three different input states for Alice's qubit.

1. Suppose $|\psi\rangle = |A_0\rangle$. We want the action of the cloning gate to be

$$C \left\{ |A_0\rangle \, |0\rangle \right\} = |A_0\rangle \, |A_0\rangle. \tag{10.27}$$

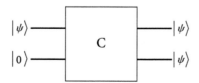

Fig. 10.6 A pictorial representation of what a cloning gate C circuit would look like and what it is supposed to do.

2. Suppose $|\psi\rangle = |A_1\rangle$. Then we want

$$C\left\{|A_1\rangle\,|0\rangle\right\} = |A_1\rangle\,|A_1\rangle. \tag{10.28}$$

3. Suppose the state is given by Eq. (10.26). Then we want to get

$$C\left\{a_0\,|A_0\rangle + a_1\,|A_1\rangle\right\}|0\rangle = \left\{a_0\,|A_0\rangle + a_1\,|A_1\rangle\right\}\left\{a_0\,|A_0\rangle + a_1\,|A_1\rangle\right\}. \tag{10.29}$$

That is, in all three cases, we end up with two qubits each described by the same quantum state. So far, no obvious problems. But let's look at Eq. (10.29) in two different ways. First, let's multiply out the right side of Eq. (10.29):

$$C\left\{a_0\,|A_0\rangle + a_1\,|A_1\rangle\right\}|0\rangle = a_0^2\,|A_0\rangle\,|A_0\rangle + a_0 a_1\,|A_0\rangle\,|A_1\rangle + a_1 a_0\,|A_1\rangle\,|A_0\rangle$$
$$+ a_1^2\,|A_1\rangle\,|A_1\rangle. \tag{10.30}$$

Try It 10.8

Carry out the multiplication expansion to arrive at Eq. (10.30).

Now let's expand the left side of Eq. (10.29) knowing that the cloning gate, like all the other gates we are working with, is a linear operator (gate) which can be distributed to each part of a sum of state vectors. Applying that rule and referring to Eqs. (10.27) and (10.28) gives us

$$C\left\{a_0\,|A_0\rangle + a_1\,|A_1\rangle\right\}|0\rangle = C\left\{a_0\,|A_0\rangle\,|0\rangle\right\} + C\left\{a_1\,|A_1\rangle\,|0\rangle\right\}$$
$$= a_0\,|A_0\rangle\,|A_0\rangle + a_1\,|A_1\rangle\,|A_1\rangle. \tag{10.31}$$

Now we are in trouble! We see immediately that Eqs. (10.30) and (10.31) cannot both be true in general because Eq. (10.30) has products of the state coefficients while Eq. (10.31) involves the state coefficients only to the first power. That means that we have a contradiction that follows from our assumption that we can "build" a cloning operation for quantum states. Therefore, the rules of logic tell us that we cannot build a device that meets our specifications.

CARDY: Not so fast. I think I see that if either $a_0 = 1$ and $a_1 = 0$, or vice versa, then the two equations are compatible.

BOB: You're absolutely right! You have noticed the exception that most quantum computing experts overlook or, at least, don't mention. In the situation you specified, the quantum state is actually one of my basis states and not a state of superposition. If that is the case, the laws of quantum mechanics allow for cloning. But the no-cloning theorem says that for a general quantum superposition state—one in which your two sets of numerical values do not hold—cloning is not possible.

The no-cloning theorem prevents us from using the quantum equivalent of bit fan-out, which is a key element in classical computing architecture. On the positive side, it prevents eavesdroppers from reading the quantum state of a qubit you are sending to Bob and replacing the intercepted qubit with another in the same quantum state. Remember how the lack of cloning foiled the eavesdropping attempt described at the end of Chapter 8. The eavesdropper could not clone a photon's linear polarization state and that led to our being able to detect the presence of the eavesdropper.

CARDY: Perhaps I could get around the theorem by using a different input state for the lower state on the input side of Figure 10.6.

Try It 10.9

Follow up on Cardy's suggestion: use $|1\rangle$ as the lower state in Figure 10.6. Also, try a superposition state in that input. Does the No-Cloning Theorem still hold?

10.1 Superdense Coding

ALICE: I think we are now ready to show Cardy a quantum information application that demonstrates an advantage over classical computing. Some QIS people call this "superdense coding," which sounds fantastic. It's great rhetoric but the reality, though impressive, is more modest. Perhaps dense coding would be a better name, which is what we will call it.

The basic idea behind dense coding is that you can send two classical bits with just one *single* qubit if you are clever about preparing the qubit in an entangled state and then having appropriate gates ready at the receiving end to process that qubit. It turns out that this method is a nice example of secure quantum communication. But it works only if the message sender has already sent you a properly prepared qubit.

I want to send Bob two classical bits. The four possibilities are 00, 01, 10, or 11. I will show that I can do this by sending Bob just one qubit if we do some preliminary work. First, I prepare two qubits C and D in the entangled Bell state:

$$|\psi\rangle = |\text{Bell}_{00}\rangle = \frac{1}{\sqrt{2}}|0_C 0_D\rangle + \frac{1}{\sqrt{2}}|1_C 1_D\rangle. \tag{10.32}$$

I keep one of the qubits and send the other to Bob.

CARDY: Wait! Doesn't this violate the no-cloning theorem? You have two qubits described by the same state.

ALICE: Excellent question that raises an important issue. The entangled state in Eq. (10.32) is the state of the two-qubit *system*. We can't assign a state to either of the qubits individually. All we have done is move one of the system parts (one of the qubits) from my location to Bob's location. The system state stays the same. We don't violate the no-cloning theorem because we did not duplicate the entangled state.

Let's continue the argument: I now act on my qubit with one of the four gates we introduced in Chapter 6—the identity gate I, the X gate, the Y gate, or the Z gate—depending on which of the pair of classical bits I want Bob to receive. Here is how this works. If I want to send 00, I apply the identity gate, which of course is equivalent to doing nothing to the two-qubit state. I then launch my qubit to Bob. When Bob receives the new qubit, he has an RBC act on the two-qubit system: the qubit he had already received and the qubit I just sent him. As we saw before in Eq. (10.18), the output of the RBC when acting on the two states in Eq. (10.32) is the state $|0_C0_D\rangle = |0_C\rangle |0_D\rangle$. Bob then measures each of the qubits using the computational basis and gets 0 and 0, just what I wanted him to see.

Now let's try another case. I want to send 0 and 1 to Bob. I prepare the same two-qubit state described in Eq. (10.32) and, as before, send one qubit (D) to Bob and keep the other (C). I apply the X gate to my qubit C after which the two-qubit system state is

$$X_C \left[\frac{1}{\sqrt{2}} |0_C0_D\rangle + \frac{1}{\sqrt{2}} |1_C1_D\rangle \right] = \frac{1}{\sqrt{2}} |1_C0_D\rangle + \frac{1}{\sqrt{2}} |0_C1_D\rangle. \tag{10.33}$$

Note that the X gate has flipped just the parts of the system state associated with my qubit.

I launch my qubit to Bob. Then he acts on it and the one he had received previously with the RBC:

$$\text{RBC}\frac{1}{\sqrt{2}} \{|1_C0_D\rangle + |0_C1_D\rangle\} = \frac{H_i}{\sqrt{2}} [\text{CNOT} \{|1_C0_D\rangle + |0_C1_D\rangle\}]$$
$$= \frac{H_i}{\sqrt{2}} \{|1_C1_D\rangle + |0_C1_D\rangle\}. \tag{10.34}$$

Remember that C is the control bit and D is the target bit for the CNOT gate.

Next, the Hadamard gate acts on the C part of the entangled state. Let's recall the results from Chapter 6:

$$H|0\rangle = \frac{1}{\sqrt{2}} [|0\rangle + |1\rangle]$$
$$H|1\rangle = \frac{1}{\sqrt{2}} [|0\rangle - |1\rangle]. \tag{10.35}$$

Using these results in Eq. (10.34) yields

$$\text{RBC}\frac{1}{\sqrt{2}} \{|1_C0_D\rangle + |0_C1_D\rangle\} = \frac{H_i}{\sqrt{2}} \{|1_C1_D\rangle + |0_C1_D\rangle\}$$
$$= \frac{1}{2} \{|0_C\rangle - |1_C\rangle\} |1_D\rangle + \frac{1}{2} \{|0_C\rangle - |1_C\rangle\} |1_D\rangle$$
$$= |0_C\rangle |1_D\rangle. \tag{10.36}$$

Bob then measures those qubits with the computational basis and gets 0 and 1. Success!

Try It 10.10

Work through the remaining two cases: sending 10 and 11. The answers are given below.
Answer for bits 10: To send 1 and 0, Alice applies the Z gate to her qubit and the two-qubit state becomes

$$\frac{1}{\sqrt{2}}\left|00\right\rangle - \frac{1}{\sqrt{2}}\left|11\right\rangle.$$

The RBC produces $\left|1\right\rangle\left|0\right\rangle$, which Bob then measures and gets 1 and 0.
Answer for bits 11: To send 1 and 1, Alice applies the Y gate to her qubit and the two-qubit state becomes

$$\frac{1}{\sqrt{2}}\left|01\right\rangle - \frac{1}{\sqrt{2}}\left|10\right\rangle.$$

Application of the RBC to the two qubits gives $\left|1\right\rangle\left|1\right\rangle$ from which Bob measures 1 and 1.

BOB: This scheme is a nice example of secure quantum communication since I need to have both qubits to decode the information Alice is sending me. As we saw before, if someone were eavesdropping and tried to intercept Alice's "modified" qubit, the scheme would not work because they would have to measure the qubit to get information about it and, in the process, modify the state. If the eavesdropper does not access my qubit, there will be nothing to decode.

CARDY: Pretty cool. You do seem to get two bits by sending one qubit, but with a bunch of other operations required to get the result.

ALICE: Yes, there are preparatory steps you need to take to make this work, but this example illustrates the general point that for some computational or information problems, an appropriately designed quantum algorithm has an advantage over the corresponding classical algorithm.

Try It 10.11

Challenge: Show that the scheme still works if the initial state is any two-qubit entangled state, not necessarily a Bell state.

10.9 Quantum State Teleportation

BOB: Another neat application is, in essence, the reverse of dense coding: Alice can send two classical bits that will allow Bob to prepare a qubit in a state that Alice has chosen. Note that she is not sending a qubit to Bob, just the two classical bits, and Bob will use those bits to prepare his qubit in the desired state.

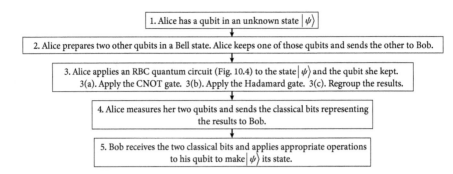

Fig. 10.7 A flow chart for the quantum state teleportation process.

CARDY: Didn't *Star Trek* have teleportation?

BOB: Yes, but we are teleporting a quantum state, not an object—neither a qubit or Commander Spock. Alas, there is no physical teleportation. That's why we are calling this quantum *state* teleportation.

CARDY: Another problem: This sounds like a violation of the no-cloning theorem! Aren't we in trouble again?

ALICE: Quantum state teleportation does not violate the no-cloning theorem because, as we shall see, at any particular time, there is only one qubit with the specified state description. We have teleported only the state information without teleporting the qubit. This is another example of why it is important to distinguish between the qubit—a physical system—and a state—an abstract description that contains information about the results of measurements.

ALICE: Here is how we, Alice and Bob, implement quantum state teleportation. I have a qubit described by a state $|\psi\rangle$. I don't need to know what that state is. I want to send Bob two classical bits that will enable him to prepare one of his qubits in the state $|\psi\rangle$. At no time do I send him the amplitudes for $|\psi\rangle$. As before, I make use of entanglement to make the process work. There are several steps to the argument, but we have seen most of these operations before.

BOB: Let's go through each step in detail. Figure 10.7 shows a flow chart of the process.

Step 1

Alice has a qubit whose state is given by

$$|\psi\rangle = a\,|0\rangle + b\,|1\rangle. \tag{10.37}$$

Alice does not need to know the numerical values of a and b.

Step 2

Alice prepares a two-qubit system in the by-now-familiar entangled Bell state

$$|AB\rangle = |\text{Bell}_{00}\rangle = \frac{1}{\sqrt{2}}\,|00\rangle + \frac{1}{\sqrt{2}}\,|11\rangle. \tag{10.38}$$

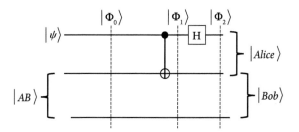

Fig. 10.8 The quantum circuit for quantum state teleportation. The upper wire is the control wire for the CNOT gate. The lower two wires represent the two qubits that are part of the entangled Bell state pair shared by Alice and Bob. The output on the lower right is the state of Bob's qubits after the operation. The output on the upper right is the state of Alice's qubits after the operation. The vertical dashed lines indicate the locations of the states described in the text.

As customary, Alice keeps one of the qubits and sends the other to me. So, at this point, Alice has two qubits: one from Step 1 described by the state in Eq. (10.37) and the other from Step 2, which is her part of the two-qubit system described by Eq. (10.38).

Step 3

To find the appropriate classical bits to send to me so that I can prepare my qubit in the state $|\psi\rangle$, Alice subjects the state $|\psi\rangle$ and the entangled Bell state to a reverse Bell circuit (RBC) with an additional circuit "wire." See Figure 10.8.

The state vector for the total system initial state on the left in Figure 10.8 is

$$|\Phi_0\rangle = |\psi\rangle |AB\rangle = \{a\,|0\rangle + b\,|1\rangle\} \left\{ \frac{1}{\sqrt{2}}\,|00\rangle + \frac{1}{\sqrt{2}}\,|11\rangle \right\}$$

$$= \frac{a}{\sqrt{2}}\,|00\rangle\,|0\rangle + \frac{a}{\sqrt{2}}\,|01\rangle\,|1\rangle + \frac{b}{\sqrt{2}}\,|10\rangle\,|0\rangle + \frac{b}{\sqrt{2}}\,|11\rangle\,|1\rangle. \qquad (10.39)$$

BOB: Let's unpack what we have written. The first line of Eq. (10.39) is just the usual product of state vectors. Since it is a simple product, not a superposition, the overall system state is *not* an entangle state. The second line of Eq. (10.39) repackages the state vectors. The double entry states, such as $|01\rangle$ are the state vectors that will be affected by the CNOT gate.

Step 3(a)

The CNOT gate will now be used to switch the labels on the last two terms of the second line of Eq. (10.39). The resulting system state in Figure 10.8 (just before the Hadamard gate) will be

$$|\Phi_1\rangle = \frac{a}{\sqrt{2}} \{|0\rangle\,|0\rangle\,|0\rangle \; + \; |0\rangle\,|1\rangle\,|1\rangle\} + \frac{b}{\sqrt{2}} \{|1\rangle\,|1\rangle\,|0\rangle \; + \; |1\rangle\,|0\rangle\,|1\rangle\}. \qquad (10.40)$$

> **Try It 10.12**
>
> Work out the details that lead to Eq. (10.40).

Step 3(b)

Let's recall what the Hadamard gate does when acting on the computational basis states:

$$H \left|0\right\rangle = \frac{1}{\sqrt{2}} \left|0\right\rangle + \frac{1}{\sqrt{2}} \left|1\right\rangle$$

$$H \left|1\right\rangle = \frac{1}{\sqrt{2}} \left|0\right\rangle - \frac{1}{\sqrt{2}} \left|1\right\rangle. \tag{10.41}$$

Noting that the Hadamard gate is acting only on the left-most state vector in each of the terms in Eq. (10.40), we arrive at the rather messy looking result

$$H \left|\Phi_1\right\rangle = \left|\Phi_2\right\rangle = \frac{a}{2} \left\{ \left|0\right\rangle \left|0\right\rangle \left|0\right\rangle + \left|1\right\rangle \left|0\right\rangle \left|0\right\rangle + \left|0\right\rangle \left|1\right\rangle \left|1\right\rangle + \left|1\right\rangle \left|1\right\rangle \left|1\right\rangle \right\}$$

$$+ \frac{b}{2} \left\{ \left|0\right\rangle \left|1\right\rangle \left|0\right\rangle - \left|1\right\rangle \left|1\right\rangle \left|0\right\rangle + \left|0\right\rangle \left|0\right\rangle \left|1\right\rangle - \left|1\right\rangle \left|0\right\rangle \left|1\right\rangle \right\}. \tag{10.42}$$

You can decipher what happened because we have grouped the terms in the following way: The first two sets of state vectors inside the curly braces multiplied by a are the ones that came from applying the Hadamard gate to the first term in the curly braces multiplied by a in Eq.(10.40). (Remember that the Hadamard gate transforms a basis state into a superposition state.) The second term comes from applying H to the second term in Eq. (10.40).

> **Try It 10.13**
>
> Work through the details in going from Eq. (10.40) to Eq. (10.42).

Step 3(c)

We regroup the terms in Eq. (10.42) in the following way:

$$2 \left|\Phi_2\right\rangle = \left|00\right\rangle \left\{ a \left|0\right\rangle + b \left|1\right\rangle \right\} + \left|01\right\rangle \left\{ a \left|1\right\rangle + b \left|0\right\rangle \right\}$$

$$+ \left|10\right\rangle \left\{ a \left|0\right\rangle - b \left|1\right\rangle \right\} + \left|11\right\rangle \left\{ a \left|1\right\rangle - b \left|0\right\rangle \right\}. \tag{10.43}$$

Note that the right side of Eq. (10.43) is a combination of the computational basis states and, in curly braces, variations on the state (Eq. (10.37)) Alice wants me to reproduce for my qubit.

Step 4

Alice now measures the two qubits in her possession in the computational basis. Remember that one qubit has the unknown state $|\psi\rangle$ and the other is one of the qubits in the Bell state she prepared. The possible measurement results are (translated in the usual way to classical bits) 00, 01, 10, and 11. Table 10.3 indicates the list of possible results of Alice's measurement, and the appropriate state vector for the qubit in my possession. Remember that one of Alice's qubits is entangled with mine.

Step 5

After these manipulations, we see that my qubit is described by a state similar to but not the same as the one Alice wants to teleport unless Alice's results are 0,0 as shown in the first line of Table 10.3. Alice sends me the two classical bits she obtained from her measurement. Then I operate on my qubit to get its state to be the one Alice wants to transport. For example, if Alice measurement results are 0,1, she sends me those two bits (in that order). After receiving the two bits 0,1, I know I must operate on my qubit to get the desired state. I have figured out that if I get 0,1, my qubit's state is the one in the third column, second row of Table 10.3. I operate on my qubit with the Pauli X gate (matrix), which interchanges the basis state vector labels. That is, it will change $a|1\rangle + b|0\rangle$ to $a|0\rangle + b|1\rangle$, just what I want! The required gate operations are listed in Table 10.4

Try It 10.14

Use the properties of the Pauli gates (matrices) to verify that the gate operations listed in Table 10.4, when applied to Bob's qubit's state after Alice's measurements, will give the desired teleportation state.

Table 10.3 The results of Alice's measurements and Bob's qubit state after step 4 of the quantum state teleportation algorithm.

Alice's measurement results in classical bits	Alice's two-qubit state after the measurement	Bob's qubit's state after Alice's measurement
0,0	$|0\rangle\,|0\rangle$	$a|0\rangle + b|1\rangle$
0,1	$|0\rangle\,|1\rangle$	$a|1\rangle + b|0\rangle$
1,0	$|1\rangle\,|0\rangle$	$a|0\rangle - b|1\rangle$
1,1	$|1\rangle\,|1\rangle$	$a|1\rangle - b|1\rangle$

Table 10.4 The gate Bob applies to his qubit depends on the two classical bits that Alice sends to him.

Alice sends	The gate that Bob applies to his qubit
0,0	I
0,1	X
1,0	Z
1,1	Y

CARDY: My head feels like it has been teleported, with my brain left behind! That was a lot of work. Why didn't Alice just use the same preparation that led to the state $|\psi\rangle$ and then ship that qubit to Bob? That would have been a lot simpler.

ALICE: The problem is that in general I don't know how the qubit whose state is to be teleported was prepared. If I did, your method would work just fine. But it is limited to those cases where I know all the details of the state $|\psi\rangle$. The quantum state teleportation method used here works whether I know the numerical values of a and b or not. Once again, superposition and entanglement make the process work.

It is important to note that at the end of the teleportation process, Bob does not know the state amplitudes of $|\psi\rangle$. All he knows is that he followed the quantum rules and those guarantee that his qubit is now described by $|\psi\rangle$.

CARDY: But what if Bob wants to check that the state teleported is the same as the state Alice started with?

BOB: Good question! To check, Alice would need to use a series of prepare–measure cycles on her state $|\psi\rangle$ to find what it predicts for the probabilities of measurement outcomes. I would need to do the same after the teleportation process.

Try It 10.15

Suppose that in Alice's "unknown" state $|\psi\rangle = a\,|0\rangle + b\,|1\rangle$, we have $a = 1/2$ and $b = -\sqrt{3}/2$. Go through the steps of the teleportation process and show that Bob is able to produce the state $|\psi\rangle$.

ALICE: Some authors claim that Bob's qubit "instantaneously jumps into one of the four states" listed on the right in Table 10.3 after I make my measurement, and thusly quantum state teleportation allows for instantaneous transmission of information over long distances. The flaw in that argument is that Bob's qubit state does not change until he receives the two classical bits and applies the appropriate gate and that takes time. There is no "instantaneous" transfer of information.

CARDY: But if Bob's qubit state changes immediately after Alice makes her measurement, couldn't Bob measure the state of his qubit before the classical bits arrive; so we would have instantaneous transmission of state information?

Try It 10.16

Is Cardy right? If Cardy is wrong, what is wrong with that line of reasoning?

BOB: We should emphasize one more time that quantum state teleportation does not involve the transport of any physical object (qubit) once the originally entangled qubits are set up, with Alice having one and Bob having the other. It is only the abstract quantum state information describing the qubit's state that is transported.

10.10 **SWAP Gates**

ALICE: Before moving on to more quantum algorithms, we want to introduce one more two-qubit gate: the SWAP gate. As the name implies, the SWAP gate interchanges something between two qubits. That something is their quantum states. Although the algorithms we discuss in the remainder of this book do not require SWAP gates, they are found in many other quantum algorithms so we thought you should be introduced to them. They also provide another nice example of using matrices to represent quantum gates.

Since we are dealing with two qubits, the basis states can be taken to be the four-entry column vectors in Table 9.2. We need a 4×4 matrix to represent operations on those vectors. The SWAP operator that interchanges the states $|01\rangle$ and $|10\rangle$ has the following matrix representation in that column vector basis:

$$\text{SWAP}_{10} \Leftrightarrow \begin{pmatrix} 1 & 0 & 0 & 0 \\ 0 & 0 & 1 & 0 \\ 0 & 1 & 0 & 0 \\ 0 & 0 & 0 & 1 \end{pmatrix}. \tag{10.44}$$

Try It 10.17

Use the column vector representation of the state $|01\rangle$ to show that $\text{SWAP}_{10} |01\rangle = |10\rangle$ and $\text{SWAP}_{10} |10\rangle = |01\rangle$. What happens if you apply the SWAP gate to the states $|00\rangle$ and $|11\rangle$? Explain why those results make sense.

CHAPTER SUMMARY

- The quantum CNOT gate is a two-qubit gate, which applies a NOT gate to one input state depending upon the state of the other input state. CNOT can be used to create an entangled state.

- The Bell circuit consists of a Hadamard gate acting on the control state followed by a CNOT, produces an entangled state. The reverse Bell circuit (RBC) has the gates in the opposite order, producing a product state from an entangled state.

- The no-cloning theorem states it is impossible to prepare a second qubit described by a quantum state that is exactly the same as the state associated with a previously prepared qubit. A general quantum state is a superposition state. One observation does not provide enough information to produce a second superposition state identical to the first.

- Superdense coding allows for two classical bits to be sent with a single qubit by leveraging an entangled Bell state. One qubit is kept by the sender and the other is sent to the receiver. The sender uses a gate on the first qubit, which is then sent to the receiver. The receiver applies an RBC to that qubit and measures the two qubits, yielding the desired classical bits.

- Quantum state teleportation allows a receiver to replicate a quantum state in a distant location using two classical bits and a qubit prepared as part of an entangled state. Only information is transported, not qubits. At no time does the process result in two qubits described by the same state; so, the no-cloning theorem is not violated. The sender applies quantum gates to the qubit they kept and the qubit whose state is to be teleported. That combination is measured, resulting in two classical bits, which are sent to the receiver. The classical bits tell the receiver what gates to apply to their qubit to produce the desired state. Neither the sender nor the receiver needs to know the details of the teleported state.

FURTHER READING

On Bell measurements: Brian T. Kirby, Siddhartha Santra, Vladimir S. Malinovsky, and Michael Brodsky, "The utility of entanglement swapping in quantum communications." *Proc. SPIE 9980, Quantum Communications and Quantum Imaging XIV*, 99800L (13 September 2016); https://doi.org/10.1117/12.2237760.

On entanglement swapping: Rui-Bo Jin, Masahiro Takeoka, Ryosuke Shimizu, and Masahide Sasaki, "Highly efficient entanglement swapping and teleportation at telecom wavelength." *Nature Scientific Reports* **5**, 9333 (2015).

Jian-Wei Pan, Dik Bouwmeester, Harald Weinfurter, and Anton Zeilinger, "Experimental Entanglement Swapping: Entangling Photons That Never Interacted." *Phys. Rev. Lett.* **80**, 3891–3894 (1998).

On entanglement swapping used in quantum communications: Brian P. Williams, Ronald J. Sadlier, and Travis S. Humble, "Superdense Coding over Optical Fiber Links with Complete Bell-State Measurements." *Phys. Rev. Lett.* **118**, 050501 (2017).

11 Quantum Computing Algorithms

> Contrariwise, continued Tweedledee, if it was so, it might be; and if it were so, it would be; but as it isn't, it ain't. That's logic.
>
> Lewis Carroll, *Through the Looking Glass (1872)*

11.1 Welcome to the Quantum Algorithm Zoo

ALICE: We are finally ready to tackle some actual quantum computing algorithms. As we shall see, quantum computing algorithms are not just classical computing algorithms modified to run on a quantum computer. Quantum algorithms are new ways of thinking about problems and are designed to exploit the properties of quantum states, quantum gates, and measurement devices.

BOB: We will start with a few of the standard quantum algorithms, which were invented decades ago, primarily to show how we can use the features of quantum states, particularly entangled superposition states, to implement algorithms more quickly or more efficiently than we can with classical computers. Many of these algorithms are, in a sense, designed to solve "made up" problems—problems selected because they are amenable to quantum algorithms and show some "advantage" over classical computational algorithms. There are many slightly different versions of these algorithms. We will use versions that are fairly simple to understand.

ALICE: The development of quantum algorithms is still in its infancy. Links to almost all the known quantum algorithms can be found on the website Quantum Algorithm Zoo at https://quantumalgorithmzoo.org. Many of the algorithms deal with fairly esoteric topics in mathematics, but in a later chapter, we will explore one that allows us to find the energy states of atoms and molecules. I read that many of the companies that have built working quantum computers are hoping that smart people like Cardy will develop algorithms that allow those computers to do something useful, but which had not been possible in a practical way even with classical supercomputers.

CARDY: I'm ready to help out, but first I have to learn the basics of quantum algorithms.

BOB: As Alice noted, all the quantum computing algorithms we will look at have a common structure. As illustrated in Figure 11.1, you start by preparing a set of qubits in some initial state, usually one of the computational basis states. You then act on those qubits' states with various quantum gates. At the end, the resulting qubits interact with measurement devices, again usually employing the computational basis states. You then interpret the results of those measurements to answer the question the algorithm is designed to address. For now, we will

Quantum Computing: From Alice to Bob. Alice Flarend and Bob Hilborn, Oxford University Press.
© Alice Flarend and Robert C. Hilborn (2022). DOI: 10.1093/oso/9780192857972.003.0011

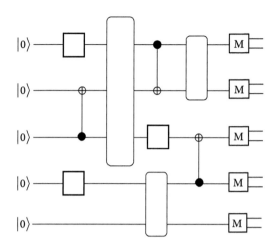

Fig. 11.1 A schematic of a typical quantum circuit (gate array) designed to implement a quantum computational algorithm. The blank square boxes are single qubit gates. The rectangular symbols indicate multi-qubit gates. The boxes labeled M are measurement devices. The other symbols are CNOT gates.

focus on the algorithms themselves. In Chapter 15, we will describe how you translate those structures into commands that the quantum computer will execute—a quantum computer program.

11.2 The Deutsch Algorithm

ALICE: In 1985, David Deutsch introduced one of the first quantum algorithms to show an advantage over classical computing algorithms. Many others, including Deutsch himself, subsequently modified the algorithm to make it more efficient and understandable.

Deutsch's algorithm asks a specific question about mathematical functions for which the independent variable (the input) is a binary digit, that is, 0 or 1, and the dependent variable (the output) is 0 or 1. Since there are only two possible input values and two possible output values, we have four possible functions. Customarily these are labeled f_0, f_1, f_2, and f_3. Of course, you could use any four names you would like. We don't need to worry about the details of how the functions turn the inputs into the outputs. All we need to do is specify the output for each of the two possible inputs 0 and 1. The following table displays the results.

The functions are divided into two classes: (1) constant functions and (2) balanced functions. Constant means that the output is the same no matter what the input is. The other two give 0 for 50% of the inputs and 1 for the other 50% and are called balanced functions. For a one-qubit system, all non-constant functions are balanced, as shown in Table 11.1. For a two-qubit system, we might have a function that gives $f(00) = 0$ and $f(01) = f(10) = f(11) = 1$. That function is non-constant, but not balanced. We will restrict ourselves to constant and balanced functions.

BOB: Deutsch posed the following problem. Suppose someone gives you a box with an input and an output. Built into the box is one of the four functions given in Table 11.1. How many

Table 11.1 Constant and balanced single-qubit binary functions.

Input $= 0$	Input $= 1$	Type of function
$f_0(0) = 0$	$f_0(1) = 0$	Constant
$f_1(0) = 0$	$f_1(1) = 1$	Balanced
$f_2(0) = 1$	$f_2(1) = 0$	Balanced
$f_3(0) = 1$	$f_3(1) = 1$	Constant

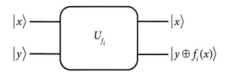

Fig. 11.2 The quantum gate that implements one of the four functions f_i, $i = 0, 1, 2$, or 3, listed in Table 11.1. See the text for the definitions of x and y in this context.

times do you need to evaluate the function—by providing a number at the input and reading the output—to tell if the function is in the "constant" group or the "balanced" group?

CARDY: The answer seems pretty easy: I just try 0 at the input and record the output. I then try 1 at the input and record the output. If the two outputs are the same, I have a constant function, and if they are different I have a balanced function. In fact, I could tell you which of the four functions is in the box once I have those results.

ALICE: Excellent reasoning, Cardy, and quite correct for classical thinking: You need two evaluations of the function box, with two different inputs, to answer the constant vs. balanced question. What Deutsch did was to prove that, with appropriate quantum gates, you need evaluate the function only once, not twice, to get the answer.

To make this work, we will take advantage of the superposition form of qubit states. We will need a quantum gate with two inputs and two outputs as shown in Figure 11.2. The variables are x, y, and $y \oplus f_i(x)$, each of which are either 0 or 1. Remember that we want to use reversible gates and those require the same number of input rows on the left as we have output rows on the right. In Figure 11.2, the upper row (sometimes called the *upper register*) state appears unchanged on the output side (on the right) while the lower row (the *lower register*) state is changed as indicated in the figure.

CARDY: I recall seeing the circle symbol with the plus sign in Chapter 2 and briefly in Chapter 10. Please remind me what it means.

ALICE: That symbol means "addition modulo 2" or "exclusive OR," also known as XOR. Just think of adding two classical binary digits and dropping any "carry" term if the result exceeds 1 as we showed in Chapter 2. Table 11.2 repeats the XOR truth table, so we can have the results in front of us.

CARDY: I think I see what is happening. If both inputs are the same, then the output is 0. But if the two inputs are different, the output is 1. An example of how the functions fit in would be a big help.

Table 11.2 The exclusive OR (XOR) truth table

Operation	Result
$0 \oplus 0$	0
$0 \oplus 1$	1
$1 \oplus 0$	1
$1 \oplus 1$	0

ALICE: OK. Let's suppose that we have the function f_1 inside the gate box in Figure 11.2 and that $x = 0$ and $y = 1$. From Table 11.1, we see that $f_1(0) = 0$. From Table 11.2, we read off $1 \oplus 0 = 1$; so, $y \oplus f_1(0) = 1$ and the lower register output state is $|1\rangle$.

This might be a good time to remind ourselves of a feature of the way we have written the lower register output state $|y \oplus f(x)\rangle$. We have previously emphasized that what appears inside the state vector symbol is just a name to label the state. However, for many quantum algorithms, it is useful to have that name be a numerical value that is generated from other numerical values, here y and $f(x)$. We saw something like that when we labeled the four two-qubit state vectors as $|00\rangle$, $|01\rangle$, $|10\rangle$, and $|11\rangle$. The state labels increase in binary numerical order as we read from left to right.

CARDY: Phew! That seems like a lot of work to figure out all the possibilities. It doesn't seem more efficient that just evaluating the function first with 0 and then again with 1.

ALICE: Later we will show you how to build the gate to do this automatically and it won't seem like that much work anymore. For now, let's assume that we have the function gate working properly. It turns out that we can write down the four possibilities for the results of acting on the computational basis states, leaving out the unnecessary details of the functions. We'll let f denote any one of the four functions. We'll also use the tensor product symbol $\otimes$ to separate the state vectors and make the equations easier to read.

If the gate U_f acts on the initial system product state $|x\rangle \otimes |y\rangle = |0\rangle \otimes |0\rangle$, we get

$$U_f\left[|x\rangle \otimes |y\rangle\right] = U_f\left[|0\rangle \otimes |0\rangle\right] = |0\rangle \otimes |y \oplus f(x)\rangle$$
$$= |0\rangle \otimes |0 \oplus f(0)\rangle = |0\rangle \otimes |f(0)\rangle. \tag{11.1}$$

This works because $0 \oplus f(0) = f(0)$ no matter what $f(0)$ is: If $f(0) = 0$, we have $0 \oplus 0 = 0 = f(0)$. If $f(0) = 1$, then we have $0 \oplus f(0) = 0 \oplus 1 = 1 = f(0)$. Those are the only possibilities.

CARDY: I needed to pay close attention to the symbols. The $\oplus$ and the $\otimes$ look very alike, but they mean different things. The $\oplus$ returns either a 0 or a 1, while the $\otimes$ returns the product of the state vectors.

BOB: You should also keep track of where the 0s come from in this example. In the example, both x and y are 0, but that will not always be the case. Sometimes, for example, x will be 1 and you will be evaluating $f(1)$ instead of $f(0)$. And then there are 0s in the example from the $0 \oplus 0 = 0$, which comes from evaluating $y \oplus f(0)$. The binary value is the same but the calculation is different. Knowing how the calculation works leads to understanding the algorithm.

ALICE: Good advice, Cardy. We just did a single case of $|0\rangle \otimes |0\rangle$. Let's make it look simpler for other cases by introducing the notation $g(y) = 1 \oplus f(y)$. All the results from Eq. (11.1) can be summarized by:

$$U_f \left[|0\rangle \otimes |0\rangle \right] = |0\rangle \otimes |0 \oplus f(0)\rangle = |0\rangle \otimes |f(0)\rangle$$
$$U_f \left[|0\rangle \otimes |1\rangle \right] = |0\rangle \otimes |1 \oplus f(0)\rangle = |0\rangle \otimes |g(0)\rangle$$
$$U_f \left[|1\rangle \otimes |0\rangle \right] = |1\rangle \otimes |0 \oplus f(1)\rangle = |1\rangle \otimes |f(1)\rangle$$
$$U_f \left[|1\rangle \otimes |1\rangle \right] = |1\rangle \otimes |1 \oplus f(1)\rangle = |1\rangle \otimes |g(1)\rangle \qquad (11.2)$$

for any of the four possible functions in Table 11.1.

Try It 11.1

Check the results in Eq. (11.2) by using Table 11.2.

CARDY: It doesn't seem that we have made much progress. In fact, it seems that we would still need to use two sets of input states to decide if f is constant or not.

BOB: Good observation. What Deutsch was able to show—and this is typical in almost all quantum algorithms—is that if we use superposition states in place of the basis states as inputs to the quantum gate, the "magic" of quantum mechanics allows us to answer the question with just one application of the function gate.

ALICE: Figure 11.3 shows the overall quantum circuit (gate array) for the Deutsch algorithm. In region 1, two Hadamard gates convert the initial states into superposition states—a general and crucial strategy in most quantum algorithms. In region 2, the function gate acts on those states to produce the output states. The upper row output state is acted upon by another Hadamard gate in region 3. And finally, in region 4, the measurement device tells us whether the upper row output state is $|0\rangle$ or $|1\rangle$.

Deutsch algorithm

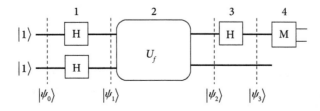

Fig. 11.3 The quantum circuit for the Deutsch algorithm. The vertical dashed lines indicate where the various state labels apply.

The measurement step is another general feature of quantum algorithms. We read out the answer (or something logically linked to the answer) using measurement devices. For the Deutsch algorithm, the states have been cleverly set up so that measurement result $|0\rangle$ in the upper register indicates that the function f was constant while $|1\rangle$ in the upper register tells us that the function was balanced. The cleverness of the algorithm is in making that association work properly. We will now go through the details to show how it works.

Let's assume that we prepare the circuit input state as the product of the two input states $|\psi_0\rangle = |1\rangle |1\rangle$. See Figure 11.3 to see where the state labels apply.

CARDY: I thought you said earlier that we like to start with the $|0\rangle |0\rangle$ states.

BOB: Many people do prefer that method. If you start with $|0\rangle |0\rangle$, you simply apply the quantum X gate (the equivalent of a NOT gate) to each of the states to get $|1\rangle |1\rangle$. To keep the diagrams as simple as possible, let's assume we have done that.

ALICE: Next, we use our old friend the Hadamard gate on each of the input states to generate superposition states in both the upper register labeled A and the lower register labeled B. Recall that the action of the Hadamard gate on the computational basis state $|1\rangle$ is given by

$$H|1\rangle = \frac{1}{\sqrt{2}}\left(|0\rangle - |1\rangle\right). \tag{11.3}$$

The resulting system state is expressed as

$$|\psi_1\rangle = H_A \otimes H_B |1\rangle_A |1\rangle_B = \frac{(|0\rangle_A - |1\rangle_A)}{\sqrt{2}} \otimes \frac{(|0\rangle_B - |1\rangle_B)}{\sqrt{2}}$$

$$= \frac{1}{2}\left(|0\rangle_A|0\rangle_B - |0\rangle_A|1\rangle_B - |1\rangle_A|0\rangle_B + |1\rangle_A|1\rangle_B\right). \tag{11.4}$$

We have used the tensor product symbol $\otimes$ between the two Hadamard gate symbols to remind us that one of them acts on the A register state and the other acts on the B register state. Note that Eq. (11.4) contains all four possible states in our two-qubit system. This is sometimes called "quantum parallelism": we are operating on all possible states for the system simultaneously. Sounds awesome. We'll come back to this issue later.

Try It 11.2

Is the state in Eq. (11.4) an entangled state? Hint: Review Eq. (9.14).

Now we are ready to have state $|\psi_1\rangle$ interact with the function gate in region 2 of Figure 11.3. The function gate operates on the state given in Eq. (11.4). We have already listed the pieces needed in Eqs. (11.1) and (11.2).

Let's pull all those pieces together by writing out the $|\psi_2\rangle$ state after the function gate operation. We will drop the A and B subscripts and remember that the states on the left of a product are in the upper register, while those on the right are in the lower register.

$$|\psi_2\rangle = U_f\left(\mathrm{H} \otimes \mathrm{H}|1\rangle|1\rangle\right)$$

$$= \frac{1}{2}\left(\underbrace{|0\rangle}_{\text{upper}} \otimes \underbrace{|f(0)\rangle}_{\text{lower}} - \underbrace{|0\rangle}_{\text{upper}} \otimes \underbrace{|g(0)\rangle}_{\text{lower}} - \underbrace{|1\rangle}_{\text{upper}} \otimes \underbrace{|f(1)\rangle}_{\text{lower}} + \underbrace{|1\rangle}_{\text{upper}} \otimes \underbrace{|g(1)\rangle}_{\text{lower}}\right). \quad (11.5)$$

The "trick" is to write the resulting state separately for constant and balanced functions. For constant functions we have, by definition,

$$f(0) = f(1) \text{ and } g(0) = g(1). \quad (11.6)$$

For balanced functions, we have

$$f(0) \neq f(1) \text{ and } g(0) \neq g(1). \quad (11.7)$$

Furthermore, for the balanced functions, we get

$$f(1) = g(0) \text{ and } g(1) = f(0). \quad (11.8)$$

CARDY: Slow down! A lot just went by.

BOB: Okay, let's check one possibility using Tables 11.1 and 11.2. For the balanced function, we have

$$f_1(0) = 0 \text{ and } g_1(0) = 1 \oplus f_1(0) = 1. \quad (11.9)$$

But we also have $g(1) = 1 \oplus f_1(1) = 0$. So, we see that $g_1(0) \neq g_1(1)$, $f_1(1) = g_1(0)$, and $g_1(1) = f_1(0)$. You can easily check the other results by writing out all the possible cases.

Try It 11.3

Verify the relationships stated in Alice's response "trick" paragraph.

CARDY: Okay. Everything seemed to work out properly.

ALICE: Let's go on and first assume that f is a constant function. Then we replace $f(1)$ and $g(1)$ in Eq. (11.5) with $f(0)$ and $g(0)$, respectively. Carrying out those substitutions and doing some rearrangement yields

$$|\text{constant}\rangle = \underbrace{\frac{1}{2}(|0\rangle - |1\rangle)}_{\text{upper}} \otimes \underbrace{(|f(0)\rangle - |g(0)\rangle}_{\text{lower}} = \frac{1}{2}(|0\rangle - |1\rangle) \otimes |F_f\rangle, \qquad (11.10)$$

where we introduced $|f(0)\rangle - |g(0)\rangle = |F_f\rangle$ to simplify the notation further and to make the results more obvious. Following the same procedure for balanced fs gives us

$$|\text{balanced}\rangle = \underbrace{\frac{1}{2}(|0\rangle + |1\rangle)}_{\text{upper}} \otimes \underbrace{|F_f\rangle}_{\text{lower}}. \qquad (11.11)$$

Try It 11.4

Check the details in arriving at Eqs. (11.10) and (11.11).

BOB: Cardy, what is the difference between those two equations?

CARDY: It seems that the difference is in the sign that appears in the superposition state associated with the upper register output.

ALICE: Excellent observation! How can we use that difference to answer the Deutsch question? There is a lot going on in this algorithm and for most of us it will be quite a cognitive overload to keep all of this in mind at once. The best strategy is to follow along with the steps. Once we are done with the algorithm, you should then go back through the steps to see how the pieces come together to answer the original question.

BOB: We now employ another "trick": Have the upper register output state interact with another Hadamard gate. Remember that a Hadamard gate acting on a superposition state "undoes" the superposition and produces a simple basis state.

$$\text{H}\frac{1}{\sqrt{2}}(|0\rangle + |1\rangle) = |0\rangle \text{ and } \text{H}\frac{1}{\sqrt{2}}(|0\rangle - |1\rangle) = |1\rangle. \qquad (11.12)$$

So, in this case, we get

$$|\psi_3 \text{ constant}\rangle = \text{H}|\text{constant}\rangle = \frac{1}{\sqrt{2}}|1\rangle \otimes |F_f\rangle$$

$$|\psi_3 \text{ balanced}\rangle = \text{H}|\text{balanced}\rangle = \frac{1}{\sqrt{2}}|0\rangle \otimes |F_f\rangle. \qquad (11.13)$$

The final step is to have the upper register state after the Hadamard operation interact with a measurement device set up to observe $|0\rangle$ or $|1\rangle$. Here is the grand conclusion: If the observation tells us that the upper row output state was $|1\rangle$ then, according to Eq. (11.13), we must have had a constant function f in the function box. If the observation tells us that the output state was $|0\rangle$, then the function f was in the balanced category. We have the answer. Sound the trumpets!

CARDY: I follow all the steps and you seemed to have answered the question with only one function evaluation as promised. But I don't think I could have pulled that off myself. How would I know what "tricks" to use?

ALICE: Well, that is true of building many proofs and algorithms in mathematics and computer science. It often takes several attempts to get a workable algorithm, and then more work is needed to clarify and simplify the algorithm (where possible). This was your first quantum algorithm, so it is understandable if you don't yet see a pattern to how they work.

BOB: Let's go over the algorithm, keeping Figure 11.3 in front of us. You recall that the first two Hadamard gates converted the initial states into superposition states as described in Eq. (11.4). That is a general and crucial strategy in most quantum algorithms. The function gate then acts on those states to produce the output states shown in Eq. (11.5). Next, the upper register state is acted upon by another Hadamard gate. That Hadamard gate produces a system product state as described in Eq. (11.13). And finally, the measurement device allows us to determine whether the upper row output state is $|0\rangle$ or $|1\rangle$.

As Alice mentioned, that last step is another general feature of quantum algorithms. We read out the answer (or something logically linked to the answer) using measurement devices. For the Deutsch algorithm, the states have been cleverly set up so that a measurement result $|1\rangle$ in the upper register is associated with the function f being constant while $|0\rangle$ in the upper register tells us that the function was balanced. The cleverness of the algorithm is in making that association work properly.

Note in Eq. (11.13) the output state of the lower register is $|F_f\rangle$ for both measurement outcomes; so, we don't learn anything else about the function. For example, if it were constant, we don't know whether it was f_0 or f_3.

CARDY: I am worried that if we make a measurement on the upper register state, won't that affect the lower-row state and mess up the result?

ALICE: We saw how that works for entangled quantum states; so, you could reframe your question as "Are the states in the Deutsch algorithm entangled states?"

CARDY: Ah! I see it! The states at the end are product states, not entangled states. So, I needn't worry about measurements on one part affecting the state of the other.

ALICE: Before we look at some further details of the Deutsch algorithm, I want to alert you to some terminology, which is often used for quantum algorithms that make use of one or more function gates. Those function gates are often called "oracles" and having states interact with the oracle is called *querying the oracle*. For the Deutsch algorithm, we would say that the algorithm requires only one query to the oracle while the corresponding classical algorithm requires, as Cardy said, two queries. So, if your definition of "advantage" is requiring fewer oracle queries, then the Deutsch algorithm has an advantage over the classical algorithm.

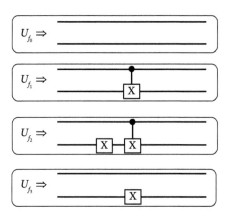

Fig. 11.4 Circuit diagrams for the four f function gates.

11.3 Implementing U_f with Simple Quantum Gates

BOB: This section is not directly needed for what follows, but we think it will help you to realize that the function box U_f could be constructed directly from just a few of the quantum gates we described earlier. The Figure 11.4 (after Mermin, 2007, p. 42) shows you the circuit diagrams. Each circuit consists of two input wires and two output wires. A qubit state is associated with each wire. For example, in the top part of Figure 11.4 each wire could represent either $|0\rangle$ or $|1\rangle$. Since those wires connect directly to the output, the states do not change.

In the second function circuit, the lower register state is subject to a controlled-NOT (CNOT) gate, drawn as a controlled X gate, with the upper register state acting as the control state. In that case, if the upper register state is $|0\rangle$, then the lower register state output is the same as the lower register state input. If the upper register state is $|1\rangle$, the lower register state is "flipped." For example, $|1\rangle$ becomes $|0\rangle$. That set of operations is just what is needed to implement function f_1 from Table 11.1.

> **Try It 11.5**
>
> Using the results from Chapter 6, verify that these circuits perform as promised.

11.4 The Simon Search Algorithm

ALICE: As a second example of a classic quantum algorithm, we look at the Simon algorithm, first developed in the 1990s. While it has some general structural similarities to the Deutsch algorithm, it is also more complicated; so, hang on to your gates. We will follow the analysis given in Bernhardt's *Quantum Computing for Everyone*, but with several additional explanations and examples to make the steps clearer. Once again, the strategy is to make use of superposition and

Table 11.3 Secret code example with $s = 01$

x	s	$x \oplus s$
00	01	01
01	01	00
10	01	11
11	01	10

entangled quantum states to create an algorithm that is more efficient than the corresponding classical algorithm.

The issue that the Simon algorithm addresses is constructing a method to find a "secret code" s that satisfies $f(x) = f(x \oplus s)$, where f is a function that takes a binary string x (a sequence of 0s and 1s) and generates a binary string for its output. Our condition for the secret code is that f generates the same output sequence when either x or $x \oplus s$ is the input. This means the function gives the same output for two different inputs. We exclude the string $s = 000\ldots$ since $x = x \oplus 000\ldots$ for any x and then obviously $f(x) = f(x)$. We want this to be an "if and only if" statement; so, if y is another binary sequence, we have $f(x) \neq f(y)$ if $y \neq x \oplus s$.

Let's illustrate the XOR addition of two sequences of 0s and 1s representing x and s. For example, we might have $x = 011100111$ and $y = 110001101$ as two binary strings of length 9. $\oplus$ means that we add the corresponding bits but drop any carry bit. Starting on the left with the first bits of both strings, we have $0 \oplus 1 = 1$, followed by $1 \oplus 1 = 0$, and so on. We then find that $x \oplus y = 101101010$. In short, if the two corresponding bits in the sequences are the same the output is a 0 and if they are different the output is 1. Since we are carrying out this operation bit by bit, we could also start on the right end of the sequence.

To be sure of what we are trying to accomplish with the Simon search algorithm, let's try the method with a two-bit sequence and $s = 01$. There are only four possible two-bit sequences, so we can easily write down all the possibilities (Table 11.3).

ALICE: Based on the first and last columns of the table, we see that for the secret code $s = 01$ to meet the $f(x) = f(x \oplus s)$ requirement, the function must satisfy

$$f(00) = f(01)$$
$$f(10) = f(11). \tag{11.14}$$

Note that these are the only two possibilities for two-bit binary numbers.

For the Simon algorithm, we want to be general and use binary strings of length n for both the input and output of the function f. We don't need to know anything about the function except that you get the same binary string output (e.g. $11011100011\ldots$) if you add (bitwise) the string associated with s to the original sequence x.

CARDY: It isn't obvious to me that this will work if we don't know explicitly what f produces for each of the inputs.

BOB: That is an important point. We saw a similar situation with the Deutsch algorithm: We did not need to specify the function to have an algorithm that works. The goal is to be as general as possible, and here that means we do not require a specific function for the algorithm.

ALICE: The construction strategy is much like that used in the Deutsch algorithm. We let the states interact with quantum gates in such a way that a measurement of one of the output states gives us information that will allow us to find the secret code. It will turn out that the quantum gates don't give us the code s directly, but they will produce equations that we can solve using classical computational techniques to find the code.

Let's first think about this from a classical computing point of view. We could try find s by brute force: Just choose an input sequence x at random and evaluate $f(x)$. Then, try another input sequence and keep repeating the process until we find a match. How many trials do we have to run? For example, for a string of length $n = 3$, the number of sequences is $2^3 = 8$ and for $n = 20$, we have $2^{20} = 1048576$ sequences.

Fortunately, we don't need to go through every possible sequence. First, we might get lucky and happen upon a match early on. At worst, we need to go through one more than half the total, that is $2^{n-1} + 1$, to be guaranteed to find a match. Note that the number increases exponentially with n, so it gets big really fast as n increases. We will see that the Simon algorithm is likely to give an answer with many fewer inputs, a number that grows only linearly with n. The Simon algorithm was the first to show that using a quantum computing algorithm can in principle give a tremendous advantage over classical computation.

CARDY: What did you mean by saying that the Simon algorithm is *likely* to give an advantage? That sounds rather strange.

ALICE: Yes, strange indeed! We shall see that probability plays an important role in the Simon algorithm. It also shows up in many other quantum algorithms. We will talk about this more after we have seen how the Simon algorithm works.

Try It 11.6

Use $n = 2$ and convince yourself that with the classical algorithm described in the previous paragraphs you need to try two input numbers to find s.

BOB: Now the quantum method. First, we need to build a function box U_f of the type we used for the Deutsch algorithm, except in this case we have only one function to worry about. That seems to make it easier, but there is a catch. The complication is that we want to be able to handle bit sequences with n bits. That requires n input states for the 0s and 1s that specify x and n input states to link to the n bit sequences that are generated by f. That means we need a function box with $2n$ states on the input side and $2n$ on the output side.

CARDY: We need the same number on the input and output sides so the function gate could be represented by an orthogonal (square) matrix. Right?

ALICE: You are remembering correctly, Cardy.

ALICE: In Figure 11.5, we have drawn the Simon function gate. On the left are two registers of n input states each with horizontal lines connecting to the box. On the right are two registers,

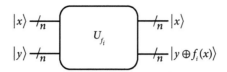

Fig. 11.5 The circuit diagram for the function gate for the Simon algorithm. The horizontal line with the diagonal slash $/_n$ and subscript n means we have n lines in that register. This figure is just the n-qubit version of Figure 11.2.

each with n output states. The state names $(x, y, y \oplus f(x))$ each represent an n binary bit sequence.

If we used just the bare function gate, we would not have made any progress over a classical algorithm because to make sure we find the functions that match, we would have to evaluate the function with up to $2^{n-1} + 1$ different inputs. However, we can again invoke the magic of quantum states by using a superposition state as the input to the top register connection to the function gate. In a sense we have the function gate operate on all the possible basis states at once. Then by making measurements on the upper register of output states, we will get information that will help us find s.

CARDY: I am beginning to appreciate the power of using superposition states!

ALICE: Remember that the standard quantum computational way of generating superposition states that include all the basis states is to operate with the Hadamard gate. For a single qubit, as we have seen, the Hadamard gate generates the equal-amplitude superposition states, as illustrated in Eq. (11.3).

BOB: How do we generalize this to multi-qubit states? Symbolically, we can write down a Hadamard gate that acts on n qubit states as $\mathrm{H}^{\otimes n}$ where $\otimes$ reminds us that the basis states are tensor product states. For $n = 2$, the result is

$$\mathrm{H}^{\otimes 2} \left|00\right\rangle = \frac{1}{2} \left(+ \left|00\right\rangle + \left|01\right\rangle + \left|10\right\rangle + \left|11\right\rangle \right)$$

$$\mathrm{H}^{\otimes 2} \left|01\right\rangle = \frac{1}{2} \left(+ \left|00\right\rangle - \left|01\right\rangle + \left|10\right\rangle - \left|11\right\rangle \right)$$

$$\mathrm{H}^{\otimes 2} \left|10\right\rangle = \frac{1}{2} \left(+ \left|00\right\rangle + \left|01\right\rangle - \left|10\right\rangle - \left|11\right\rangle \right)$$

$$\mathrm{H}^{\otimes 2} \left|11\right\rangle = \frac{1}{2} \left(+ \left|00\right\rangle - \left|01\right\rangle - \left|10\right\rangle + \left|11\right\rangle \right). \tag{11.15}$$

ALICE: We will show you how to derive these specific results once we have gone through the complete algorithm. But, if you look carefully at Eq. (11.15), you will see there is a pattern to the plus and minus signs: There are all plusses across the top row and down the first column, where the amplitudes for all the $\left|00\right\rangle$ states are $+1$. The minus signs are distributed the same way for row two and column two, and so on. Many of those features generalize to the n qubit case.

BOB: Now let's get back to the Simon algorithm. The quantum circuit (gate array) is shown in Figure 11.6. We start with the system state $\left|\psi_0\right\rangle$ on the far left. That state is a product of two sets of basis states.

$$\left|\psi_0\right\rangle = \left|0\right\rangle_n \left|0\right\rangle_n. \tag{11.16}$$

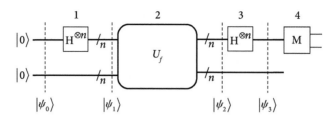

Fig. 11.6 The quantum circuit (gate array) for the n-qubit Simon algorithm. The system state $|\psi\rangle$ is indicated at four different positions indicated by the vertical dashed lines in the diagram. Although only one measurement device is shown at the top right of the diagram, there are actually n measurement devices. Each measurement device interacts with one qubit.

Here, we used the subscripts n on the right vectors to remind us that we are dealing with n-bit states. As before, we put the upper-register states on the left of each product, while the lower-register states go on the right.

As we read across the quantum circuit from left to right, the input states in the upper register in region 1 are acted upon by the Hadamard gate $H^{\otimes n}$ producing superposition states that connect to the function box. On the output side in region 3, the states in the upper register are acted upon another Hadamard gate $H^{\otimes n}$ and then at the end in region 4 interact with an n qubit measurement device.

It will be helpful to work this out step by step for $n = 2$, which is simple enough that we can write down all the steps, but complex enough to show how the algorithm works.

For two-qubit states, the input states on the far left of Figure 11.6 in both the upper and lower registers are $|00\rangle$. From Eq. (11.15), we see that after the upper-register Hadamard gate acts on the input state, the state $|\psi_1\rangle$ for the system (upper register and lower register) will be

$$|\psi_1\rangle = H^{\otimes 2}|00\rangle|00\rangle = \frac{1}{2}\left(\underbrace{|00\rangle}_{\text{upper}} \otimes \underbrace{|00\rangle}_{\text{lower}} + |01\rangle \otimes |00\rangle + |10\rangle \otimes |00\rangle + |11\rangle \otimes |00\rangle\right).$$

$$(11.17)$$

Note that as promised we have the upper-register states on the left side of each of the products and the lower-register states, all $|00\rangle$, on the right. If you prefer, you may add subscripts to remind you which group the states belong to.

In region 2, the system state then interacts with the function gate. Since the function gate (like all of our quantum gates) is linear in the states, we can act separately on each term in Eq. (11.17) and then add up the results. For example, when we act on the second term, we have $x = 01$ and $y = 00$; so, the lower-register output state is $|00 \oplus f(01)\rangle = |f(01)\rangle$ because the upper-register $x = 01$. Assembling the results, we find that the system state after the function box is

$$|\psi_2\rangle = U_f\left(H^{\otimes 2}|00\rangle\right)|00\rangle$$

$$= \frac{1}{2}\left(\underbrace{|00\rangle}_{\text{upper}} \otimes \underbrace{|f(00)\rangle}_{\text{lower}} + |01\rangle \otimes |f(01)\rangle + |10\rangle \otimes |f(10)\rangle + |11\rangle \otimes |f(11)\rangle\right).$$

$$(11.18)$$

Note that Eq. (11.18) contains the function f evaluated with each of the binary sequences associated with the two-qubit system; so in that sense we are handling all of the possibilities at once. This is another example of the power of using superposition states.

CARDY: Ah ha! I have spotted another entangled state in Eq. (11.18). Am I correct?

ALICE: Yes, that is an entangled state. As usual, entangled states play critical roles in the quantum algorithm.

Try It 11.7

Check that the state in Eq. (11.18) is an entangled state.

We are almost done with the algorithm! In region 3 the top-register qubits interact with the Hadamard gate to give us (using the results stated in Eq. (11.15)) the rather complex-looking state

$$|\psi_3\rangle = \frac{1}{4}\left(\begin{array}{l} \underbrace{(+|00\rangle + |01\rangle + |10\rangle + |11\rangle)}_{\text{upper}} \otimes \underbrace{|f(00)\rangle}_{\text{lower}} \\ + (+|00\rangle - |01\rangle + |10\rangle - |11\rangle) \otimes |f(01)\rangle \\ + (+|00\rangle + |01\rangle - |10\rangle - |11\rangle) \otimes |f(10)\rangle \\ + (+|00\rangle - |01\rangle - |10\rangle + |11\rangle) \otimes |f(11)\rangle \end{array}\right). \qquad (11.19)$$

Try It 11.8

Use Eq. (11.15) in Eq. (11.18) to get Eq. (11.19).

We can now rearrange Eq. (11.19), collecting terms associated with each of the upper-register states. The result is

$$|\psi_3\rangle = \frac{1}{4}\left(\begin{array}{l} \underbrace{|00\rangle}_{\text{upper}} \otimes \underbrace{(|f(00)\rangle + |f(01)\rangle + |f(10)\rangle + |f(11)\rangle)}_{\text{lower}} \\ + |01\rangle \otimes (|f(00)\rangle - |f(01)\rangle + |f(10)\rangle - |f(11)\rangle) \\ + |10\rangle \otimes (|f(00)\rangle + |f(01)\rangle - |f(10)\rangle - |f(11)\rangle) \\ + |11\rangle \otimes (|f(00)\rangle - |f(01)\rangle - |f(10)\rangle + |f(11)\rangle) \end{array}\right). \qquad (11.20)$$

BOB: Note that Eq. (11.20) has all four possible function arguments with various combinations of plus signs and minus signs. Since we are looking for situations in which some of the function inputs give equal results $f(x) = f(x \oplus s)$, we expect that the terms in several rows might add up to 0, thus removing those upper register states from the system state. Then when we make measurements on the upper-register qubits in region 4, the associated basis states never appear as results of the measurements. Though we haven't yet got to the details, we can guess that the measurement information results can then be used to find the secret code s. Each of the n measurement devices will provide two output numbers, which are the labels of the basis states.

But we are still not yet at the final answer. We will show that if the measurement results $(x_o x_1)$ satisfy a mathematical relationship with the code sequence $s = (s_0 s_1)$, namely,

$$(x_0 \times s_0) \oplus (x_1 \times s_1) \equiv x \cdot s = 0, \tag{11.21}$$

where $\times$ means standard multiplication, then the corresponding row of functions does *not* add up to zero. The expression in Eq. (11.21) is called the "dot product" between the two binary sequences because it is similar to the definition of the ordinary vector dot product as a sum of the products of the corresponding components:

$$\vec{A} \cdot \vec{B} = A_x B_x + A_y B_y + A_z B_z, \tag{11.22}$$

with the ordinary plus sign replaced by bitwise XOR $\oplus$. It is easy to convince yourself that the result of $x \cdot y$ is either 0 or 1 no matter what the length of the bit sequences is because the bit values are either 0 or 1.

One word of warning: The dot product of two binary sequences is not the same as a vector dot product. In fact, you can have a situation in which $x \cdot x = 0$. This occurs if x has an even number of 1s. It does *not* mean the x is orthogonal to itself!

Try It 11.9

Write out $x \cdot y$ in detail for the seven-bit sequences $x = 1010111$ and $y = 1110000$.

CARDY: I hope you are going to show me where that came from, as well.

ALICE: You bet. But let's finish off the algorithm. We are finished with the quantum part of the algorithm and now use some classical algorithms to wrap up the solution. Eq. (11.21) is a linear equation for the two unknowns s_0 and s_1. You may remember from algebra that for two unknowns you generally need two equations to get a definite solution. For the Simon algorithm, since we are excluding $s = 0$ and since the possible values for the solutions are just 0 and 1, it turns out that we need only one equation for the two-qubit situation. For n bits, we will need $n - 1$ equations. Details aside, the crucial point is that from the measurement outcomes we can generate enough equations to solve for the binary bits in the code.

In the n bit case, we will have a system of $n - 1$ simultaneous linear equations to solve. There are lots of tools for solving that system of equations. So, you can carry out that part of the algorithm using classical computational methods. In Chapter 15, we talk about when there might be an advantage to solving linear equations on a quantum computer. But currently, classical computers do that task quite efficiently.

BOB: Let's work through the solution for a two-qubit example to make sure we know how the Simon procedure works. Suppose when we measure the upper-register output states, we find that the results are either 00 or 11. We know that 00 doesn't give us any information about s_0 and s_1; so, we use 11 in Eq. (11.21) with $(1 \times s_0) \oplus (1 \times s_1) = s_0 \oplus s_1 = 0$ from which we have $s_0 s_1 = 00$ or 11. But, remember that 00 is not allowed; so, we conclude that $s = 11$. We are done! In this case, the solution of the linear equations did not require much work.

Let's summarize what we have done. In both the Deutsch algorithm and the Simon algorithm, we cleverly set up the state of the system so that entangled states, appropriately manipulated, will allow measurements that give us information from which we could calculate what we wanted to find, for example the Simon secret code s. Quantum state entanglement was the key feature because entangled states have correlations among the measurements associated with the various qubits of the system. It is those correlations that let the quantum magic produce the desired result.

ALICE: But there was one important difference between the Deutsch algorithm and the Simon algorithm. In the Deutsch algorithm, once we have the function gate set up, it takes only one measurement observation to find out if the function is constant or balanced. On the other hand, in the Simon algorithm, because the measurement outcomes are random, it is possible that we might have repeated observations of the 0 basis states, and they give us no useful information for the Simon question. Or we may get the same measurement results repeatedly—just like getting a string of heads when we repeatedly flip a coin. That set of repeating results doesn't give us new information, that is, new equations to help solve for the bits in the code s. You can estimate the probability that you will get sufficient information after a certain number of measurements and, indeed, as the number of measurements increases you are highly likely to get enough information to answer the Simon question. But there is no guarantee. In principle you could repeat the preparation–measurement cycle many times and still not get a definite answer. That is why we may say only that it is highly likely that the Simon algorithm will give an answer with fewer measurements than the classical computation; we are not guaranteed that will happen.

CARDY: That is indeed strange, but as I have come to appreciate, lots of things in quantum mechanics are strange. That is the way nature seems to work.

11.5 Multi-Qubit Hadamard Gates

CARDY: I think I understand how the Simon algorithm works. It certainly involved a lot of symbolic reasoning. Will you now explain how the Hadamard gates work, where the minus signs go, and why the binary sequence plays a role in finding s?

ALICE: Let's tackle the Hadamard gate question first. Since multi-qubit Hadamard gates appear frequently in QIS and QC applications, we will use this section to explain how to construct multi-qubit Hadamard gates. Then we will show how the $x \cdot s = 0$ condition arises in the Simon algorithm. We won't need these details in subsequent chapters, so you should feel free to skip this section. But going through the analysis gives you some good background in quantum gates and state manipulation that may prove to be useful in more complex QIS and QC applications.

The matrix representation of a multi-qubit Hadamard gate is built by arranging the single-qubit Hadamard matrices into larger matrices (more rows and columns). These more complicated structures carry out the equivalent operations on the multi-qubit states. Let's start by reminding ourselves of the one-qubit Hadamard matrix:

$$H^{\otimes 1} = H \Rightarrow \begin{pmatrix} 1 & 1 \\ 1 & -1 \end{pmatrix}. \tag{11.23}$$

For two-qubit states, the Hadamard gate for the system is a 4×4 matrix built using four one-qubit Hadamard matrices:

$$H^{\otimes 2} = \begin{pmatrix} H^{\otimes 1} & H^{\otimes 1} \\ H^{\otimes 1} & -H^{\otimes 1} \end{pmatrix} \Rightarrow \begin{pmatrix} \begin{pmatrix} 1 & 1 \\ 1 & -1 \end{pmatrix} & \begin{pmatrix} 1 & 1 \\ 1 & -1 \end{pmatrix} \\ \begin{pmatrix} 1 & 1 \\ 1 & -1 \end{pmatrix} & \begin{pmatrix} -1 & -1 \\ -1 & 1 \end{pmatrix} \end{pmatrix} = \begin{pmatrix} 1 & 1 & 1 & 1 \\ 1 & -1 & 1 & -1 \\ 1 & 1 & -1 & -1 \\ 1 & -1 & -1 & 1 \end{pmatrix}. \tag{11.24}$$

CARDY: That's cool! It looks just like the pattern of plusses and minuses we had in Eqs. (11.19) and (11.20).

ALICE: Exactly right. It is a good exercise to apply that Hadamard matrix to each of the two-qubit basis states we learned about in Chapter 9, here shown as column vectors:

$$\begin{pmatrix} 1 \\ 0 \\ 0 \\ 0 \end{pmatrix}, \begin{pmatrix} 0 \\ 1 \\ 0 \\ 0 \end{pmatrix}, \begin{pmatrix} 0 \\ 0 \\ 1 \\ 0 \end{pmatrix}, \begin{pmatrix} 0 \\ 0 \\ 0 \\ 1 \end{pmatrix} \tag{11.25}$$

to make sure it produces the superposition states in Eq. (11.19).

Try It 11.10

Work through the matrix multiplications that Alice suggested.

BOB: For a three-qubit system, since $2^3 = 8$, the Hadamard matrix needs eight rows and eight columns; so, we construct it from four two-qubit Hadamard matrices:

$$H^{\otimes 3} = \begin{pmatrix} H^{\otimes 2} & H^{\otimes 2} \\ H^{\otimes 2} & -H^{\otimes 2} \end{pmatrix}. \tag{11.26}$$

You can see that the process of building these is very straightforward.

CARDY: Okay. Now I see where those Hadamard matrices come from and how they are built from the one-qubit Hadamard matrix. But I'm still worried about getting the minus signs in the right places. I will try this for a three-qubit system.

Try It 11.11

Write out the complete three-qubit Hadamard matrix. Hint: Follow the same procedure that was used in going from Eq. (11.23) to Eq. (11.24). The answer is in Eq. (11.27).

$$H^{\otimes 3} \Rightarrow \begin{pmatrix} 1 & 1 & 1 & 1 & 1 & 1 & 1 & 1 \\ 1 & -1 & 1 & -1 & 1 & -1 & 1 & -1 \\ 1 & 1 & -1 & -1 & 1 & 1 & -1 & -1 \\ 1 & -1 & -1 & 1 & 1 & -1 & -1 & 1 \\ 1 & 1 & 1 & 1 & -1 & -1 & -1 & -1 \\ 1 & -1 & 1 & -1 & -1 & 1 & -1 & 1 \\ 1 & 1 & -1 & -1 & -1 & -1 & 1 & 1 \\ 1 & -1 & -1 & 1 & -1 & 1 & 1 & -1 \end{pmatrix}. \tag{11.27}$$

ALICE: Great! Now let's tackle the issue of using the dot product $x \cdot s$ to put the minus signs in the correct places in the Hadamard matrices and how that leads us to the code s. This is an interesting example of how changing notation and rearranging mathematical expressions lead to interesting conclusions. The key notion is that $x \cdot y = 0$ or 1 for any n-bit binary sequences and that $(-1)^{x \cdot y} = 1$ if $x \cdot y = 0$ and $(-1)^{x \cdot y} = -1$ if $x \cdot y = 1$.

To illustrate the general idea, let's return to the single-qubit Hadamard gate and apply it to the basis states $|0\rangle$ and $|1\rangle$. We will put the two results together in one equation by using the variable x, which will be either 0 or 1 (the only two possibilities!):

$$H|x\rangle = \frac{1}{\sqrt{2}} \left(|0\rangle + (-1)^x |1\rangle \right) = \sum_{y=0}^{1} (-1)^{x \times y} |y\rangle = \left\{ (-1)^{x \times y} |y\rangle \right\}. \tag{11.28}$$

$x \times y$ is ordinary multiplication. Note that including $(-1)^x$ and $(-1)^{x \times y}$ takes care of getting the correct signs when we apply the Hadamard gate. The last part of the equation makes use of our curly brace notation introduced in Chapter 9 to replace summation signs.

In Eq. (11.28), the sum over y contains only two terms, and you should check for yourself that the signs come out correctly. Of course, if we used only one-qubit Hadamard gates, the condensed notation would not be useful. However, when we want to generalize to n-qubits, the fancier notation makes the general expression much easier to write down and to manipulate.

For n-qubit systems, the action of $H^{\otimes n}$ on one of the x basis states is

$$H^{\otimes n}|x\rangle_n = \sum_{y=0}^{2^n-1} (-1)^{x\cdot y}|y\rangle_n = \left\{(-1)^{x\cdot y}|y\rangle_n\right\}. \tag{11.29}$$

Now we will use that notation to rewrite the Simon algorithm states. After the first Hadamard gate in Figure 11.6, the system quantum state is

$$|\psi_1\rangle = \frac{1}{2^{n/2}}\left(\sum_{x=0}^{2^n-1}|x\rangle_n\right) \otimes |0\rangle_n = \left\{1/2^{n/2}\,|x\rangle\right\} \otimes |0\rangle_n. \tag{11.30}$$

In the last term, we again used our curly bracket notation.

Let's take a close look at Eq. (11.30). Note that we have labeled the superposition states by the n-bit x values, which run from $00\ldots0$ to $11\ldots1$—all the possible n-bit values. This is another example of using numerical values to label states—something that is not necessary, but it turns out to be useful.

BOB: The next step is to have the system state interact with the function gate U_f, after which the system state is

$$|\psi_2\rangle = \frac{1}{2}\sum_{x=0}^{2^n-1}|x\rangle_n|f(x)\rangle_n = \frac{1}{2}\left\{|x\rangle_n|f(x)\rangle_n\right\}. \tag{11.31}$$

Now the upper-register part of the system state interacts with the second Hadamard gate to give

$$|\psi_3\rangle = \frac{1}{2^n}\sum_x\left(\sum_y(-1)^{x\cdot y}|y\rangle_n\right) \otimes |f(x)\rangle_n$$

$$= \frac{1}{2^n}\left\{(-1)^{x\cdot y}\,|y\rangle\,|f(x)\rangle_n\right\}. \tag{11.32}$$

To unclutter the notation, we have left off the details of the limits of the sums. Both run over the full range of the n bit numbers $00\ldots0$ through $11\ldots1$. Since we have two state labels (x and y) inside the curly braces, we need to remember to sum over both variables.

CARDY: Can you convince me that $(-1)^{x\cdot y}$ is the right thing to use in those equations?

BOB: Here is the argument: We replaced the $(-1)^{x\times y}$ term in Eq. (11.28), where x and y were single-bit numbers (0 or 1), with $(-1)^{x\cdot y}$, where x and y now are each n-bit numbers. In other words, we replaced the standard multiplication indicated by $\times$ with the dot product multiplication. We have already shown that the procedure works for $n=2$. But let's look at this from a different point of view that allows us to handle larger n cases easily. The key idea is that we can use x and y to label the rows and columns of a matrix. If we then use $(-1)^{x\cdot y}$ for the matrix elements labeled by row and column, we automatically get the right plus or minus sign in the Hadamard matrix. Here is what it looks like for $n = 2$:

$$
\begin{array}{c}
\begin{array}{cccc}
y=00 & y=01 & y=10 & y=11
\end{array} \\
\begin{array}{c}
x=00 \\
x=01 \\
x=10 \\
x=11
\end{array}
\begin{pmatrix}
(-1)^{x\cdot y} & (-1)^{x\cdot y} & (-1)^{x\cdot y} & (-1)^{x\cdot y} \\
(-1)^{x\cdot y} & (-1)^{x\cdot y} & (-1)^{x\cdot y} & (-1)^{x\cdot y} \\
(-1)^{x\cdot y} & (-1)^{x\cdot y} & (-1)^{x\cdot y} & (-1)^{x\cdot y} \\
(-1)^{x\cdot y} & (-1)^{x\cdot y} & (-1)^{x\cdot y} & (-1)^{x\cdot y}
\end{pmatrix}
=
\begin{pmatrix}
1 & 1 & 1 & 1 \\
1 & -1 & 1 & -1 \\
1 & 1 & -1 & -1 \\
1 & -1 & -1 & 1
\end{pmatrix}.
\end{array}
\tag{11.33}
$$

CARDY: That seems to work!

BOB: For the $n = 3$ Hadamard gate matrix in Eq. (11.27) in the previous Try It, you can use the same method to find where the "correct" minus signs go. The rows and columns of that matrix can be labeled by the eight basis state labels corresponding to the values of x (the three-bit sequence):

$$000 \quad 001 \quad 010 \quad 011 \quad 100 \quad 101 \quad 110 \quad 111. \tag{11.34}$$

You should also note that these numbers run from 0 (on the left) to $2^n - 1 = 7$ (on the right). For an 8×8 matrix, there are 64 combinations of row labels x and column labels y. You should check that $(-1)^{x\cdot y}$ gives the sign that matches the ones in the explicitly constructed three-qubit Hadamard matrix. The first row and first column should be easy.

CARDY: Let me think about that for a minute. Oh, yeah. For the first row $x = 000$ and $(000) \cdot y = 0$ for all the columns. And for the first column $y = 000$ and $x \cdot (000) = 0$ for all the rows.

ALICE: That is superb, Cardy. The others are easy enough to write down and work out the dot products to check against the results given in Eq. (11.27). Now let's move on with the rest of the Simon algorithm.

Just as we did in going from Eq. (11.19) to Eq. (11.20), we rearrange Eq. (11.32) to get

$$
\begin{aligned}
|\psi_3\rangle &= \frac{1}{2^n} \sum_y |y\rangle_n \left(\sum_x (-1)^{x\cdot y} |f(x)\rangle_n \right) \\
&= \frac{1}{2^n} \left\{ (-1)^{x\cdot y} |y\rangle_n |f(x)\rangle \right\}.
\end{aligned}
\tag{11.35}
$$

The final step in the Simon circuit is the action of the n measurement devices on the upper-register state in region 4 of Figure 11.6. The measurement device is designed to observe the basis states $|y\rangle_n$. Each measurement of the n-qubit upper register will yield a single value y. The probability of getting a specific value of y depends, of course, on the square of the term in parentheses in Eq. (11.35), which is in effect the amplitude for that state. So, let's play with that term to see how it gives us information about the code sequence s.

For the Simon algorithm, remember that we want to find the secret code s that satisfies $f(x) = f(x \oplus s) = z$, where z is one of the n bit binary numbers. For each x_1 in the set $\{x\}$

there is one and only one x_2 in the set such that $f(x_1) = f(x_2) = z$, as I specified in the second paragraph of section 11.4. Using that notation, we rewrite the parenthetical terms in Eq. (11.35) as

$$\sum_z \left[(-1)^{x_1 \cdot y} + (-1)^{x_2 \cdot y} \right] |z\rangle . \tag{11.36}$$

Since $x_1 \oplus s = x_2$, the square bracket term in Eq. (11.36) is

$$\left[(-1)^{x_1 \cdot y} + (-1)^{(x_1 \oplus s) \cdot y} \right] = (-1)^{x_1 \cdot y} \left[1 + (-1)^{y \cdot s} \right] . \tag{11.37}$$

For the dot product as we have defined it, the only two possibilities for $y \cdot s$ are 0 (even) and 1 (odd). So, for any y that appears as the result of a measurement, we will have

$$y \cdot s = 0 = (y_0 \times s_0) \oplus (y_1 \times s_1) . \tag{11.38}$$

As we mentioned in the previous section, a series of measurement results will yield several such equations, which in principle can be solved by standard algebraic methods to find s.

CARDY: I am tired after all this thinking. Figuring out these algorithms is hard work. Even though it is mostly addition and multiplication, there are a lot of states to keep track of and the action of the quantum gates mixes things up.

ALICE: That's not too surprising, Cardy. It is always challenging to make sense of new notation and new ways of manipulating symbols. Of course, quantum mechanics itself has its conceptual challenges. But if we follow the quantum rules carefully, we get results that agree with experiments. That is quite amazing.

In some ways, the current state of quantum algorithms is what classical computer programming was like before high-level languages like Fortran, C, Python and so on were developed. I am sure that within a decade or so, we will have similar high-level languages for quantum computing.

In the next chapter we will turn to some quantum algorithms that address more obviously interesting applications.

 CHAPTER SUMMARY

- The two quantum algorithms discussed in this chapter illustrate that the use of superposition and entanglement may make quantum computing more effective than classical computing by providing a method to have the analysis work on all possible states at once. This does not mean that the calculation is done repetitively on each possible state, but rather the superposition of states interacts with quantum gates in a way that results in a single measurement having a high probability of containing an answer. As usual, that probability is given by the square of the coefficient of the state. Because there is a small but nonzero probability of a wrong answer—due to the probabilistic nature of measurements on superposition states—you may need to run an algorithm several times. But with a clever quantum algorithm, the process can be less computationally intensive than the corresponding classical algorithm. Almost all quantum algorithms make use of entanglement to allow measurements on one part of the system's state to give information about another part of that state.

- The Deutsch algorithm analyzes whether a function is constant (always returns the same output regardless of input) or balanced (output depends on input). Table 11.1 shows the four possible outcomes for single-qubit inputs. A classical algorithm would need to evaluate the function twice to determine whether it is constant or balanced. The Deutsch algorithm can determine the type of function with only a single evaluation of the function.

- The Deutsch algorithm circuit, like those for many other quantum algorithms, has an upper register and a lower register. The first step is to put both registers into states of superposition using a Hadamard gate. Next the function gate acts on the two superposition states and gives a different output depending on whether the function is constant or balanced. Finally, the upper register is acted upon again by a Hadamard gate to reverse the state of superposition into a single basis state. That state then interacts with a measurement device and the measurement result can indicate whether the function was constant if the measurement yielded a 1 or balanced if the measurement result was 0.

- The Simon search algorithm finds a string s (the "secret code") that satisfies $f(x) = f(x \oplus s)$. The function f need not be specified.

- The approach is similar to the Deutsch algorithm in that a Hadamard gate is used to put both inputs into states of superposition, but is different because there is only one function in the Simon algorithm. The superposition exiting the function gate and entering a second Hadamard gate yields a matrix of results with a pattern of signs that assures some of the results will cancel out and not appear in the final result, which is then sent to a measurement device that yields a string that is mathematically related to the secret code, but is itself not the actual secret code.

- To determine the individual bits of the secret code, a classical algorithm can be used to solve this system of $n - 1$ equations. The advantage of the Simon quantum algorithm is that, as the number of bits in the secret code grows, the number of times the quantum algorithm needs to be run in order get an answer grows linearly, whereas for a classical algorithm, it grows exponentially.

 FURTHER READING

Chris Bernhardt, *Quantum Computing for Everyone* (MIT Press, Cambridge, MA, 2019). Chapter 8 has a nice but compact introduction to the Simon algorithm.

N. David Mermin, *Quantum Computer Science: An Introduction* (Cambridge University Press, Cambridge, 2007). Page 42 shows the simple circuits that implement the Deutsch functions.

Yohan Vianna, Mariana R. Barros, and Malena Ho-Meyal, "Classical realization of the quantum Deutsch algorithm." *American Journal of Physics* 86, 914 (2018). This paper provides a good introduction to the Deutsch algorithm and shows how to use lenses and diode lasers to build a classical analog that implements the quantum algorithm.

12 More Quantum Algorithms

Let us record the atoms as they fall upon the mind..

Virginia Woolf, *The Common Reader* (1925)

12.1 Introduction

ALICE: We are now ready to tackle two more quantum algorithms, both of which have advantages over the corresponding classical algorithms. We'll also look at how error correction, a critical component of information science, is handled for quantum communications. We will show too how quantum computing can be used to find energies and shapes of complex molecules—a crucial step in finding new drugs and enhanced materials.

In the following chapter, we will explore the famous Shor factoring algorithm. Because the math behind some of these algorithms is rather advanced, we will provide just enough detail so you will be able to understand the problems being solved and the general flow of the quantum algorithm.

12.2 Grover Search Algorithm

ALICE: The first algorithm we will tackle is the Grover search algorithm, which is a quantum algorithm that speeds up searching through databases. Lov Grover developed the algorithm in 1996. Although there are now many improved variations on the original algorithm, we'll take a close look at how the original works because it uses a strategy that many other quantum algorithms use.

First, let's think about a classical search question. Suppose we have N cards, only one of which has a special purple dot on it. If we spread the cards out face down on a table, how many cards do we need to turn over to find the one with the purple dot? Well, we might be lucky and find it on the first try. More likely, we will have to turn over many cards before finding the one with the purple dot. In the worst case, we might turn over $N - 1$ cards without seeing the purple dot. If that happens, we know that the dot must be on the remaining card, but we still turn it over to make sure that no one has messed with the cards. It turns out that *on average*, we need to flip just more than half of the cards to find the purple dot. The average is 2 for 3 cards, 2.5 for 4 cards, and 50.5 for 100 cards.

Quantum Computing: From Alice to Bob. Alice Flarend and Bob Hilborn, Oxford University Press.
© Alice Flarend and Robert C. Hilborn (2022). DOI: 10.1093/oso/9780192857972.003.0012

For the quantum version, we search for a particular quantum state. For a system with n qubits, we have, as we know, $N = 2^n$ possible states. Suppose we want to find just one state, equivalent to the card with the purple dot. To set up the search mechanism, we label the states $|x\rangle$ with the binary number x ranging from 0 through $2^n - 1$.

We also need to "tag" the "target" state, so the search mechanism has something to look for. Let's assume that there is a function $f(x)$ defined such that $f(x)=0$ unless $x=w$, where $|w\rangle$ is the target state and someone has arranged for $f(w) = 1$. Another way of putting this is that we give each of the basis states the amplitude $(-1)^{f(x)}/\sqrt{2^n}$. So, all the amplitudes have same magnitude. All will be positive except for the amplitude of the target state $|w\rangle$, which is negative. That amplitude is the quantum equivalent of the purple dot in our card example.

Try It 12.1

Show that $(-1)^{f(x)}/\sqrt{2^n}$ satisfies the positive and negative amplitude criteria stated in the previous paragraph.

ALICE: We want to develop an algorithm that starts with a superposition state with equal amplitudes for all the basis states and then manipulates the amplitudes to increase the one associated with $|w\rangle$ while decreasing the other amplitudes. This increases the probability of observing the target state. When we carry out a measurement on the superposition state, we are much more likely to observe $|w\rangle$ than any of the other basis states. For a system with N states, Grover found that the number of iterations of the algorithm needed to identify the target state increases as $\sqrt{N}$, while, as we saw, the number of card "flips" for the classical algorithm increases proportional to N.

The Grover algorithm, like the Simon algorithm, is probabilistic. All we can calculate is the probability of getting the target state when we make a measurement on the superposition state. But we will see that the probability can be very close to 1 under many circumstances.

To see how the Grover algorithm works, we will use a geometric picture of a two-dimensional slice of the N-dimensional state space. This geometric picture makes it relatively easy to understand how the algorithm works, and it allows us to prove many interesting results about the algorithm, including the probability that the measurement will give the target state. Later we show how to translate the geometric approach into a quantum operator method.

Geometric Approach

For this geometric picture, we use a state space diagram with one axis along the $|w\rangle$ direction, the target state we want to find, and the other perpendicular to $|w\rangle$. That perpendicular direction represents a superposition of all the *other* basis states, each of which individually is perpendicular to $|w\rangle$. Let's call it $|\psi_{\text{other}}\rangle$. In the diagram, we will also include the

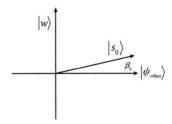

Fig. 12.1 Two-dimensional state space projection of the "target" state $|w\rangle$, the superposition of all the other basis states $|\psi_{\text{other}}\rangle$, and the initial equal-amplitude superposition state $|s_0\rangle$. β_0 is the angle between $|\psi_{\text{other}}\rangle$ and $|s_0\rangle$.

equal-amplitude superposition state

$$|s_0\rangle = \tfrac{1}{\sqrt{2^n}} \left(|00...0\rangle + |00...1\rangle + \ldots + |11...1\rangle \right)$$
$$= \tfrac{1}{\sqrt{2^n}} \sum_{x=0}^{2^n-1} |x\rangle = \left\{ 1/\sqrt{2^n} \, |x\rangle \right\}, \tag{12.1}$$

where we made use of the curly-brace state notation introduced in Chapter 9 to avoid the summation symbol.

CARDY: But how do we know in advance what $|w\rangle$ is? You said that $|w\rangle$ is what we want to find?

BOB: Exactly the right question! We just have to know that someone tagged the target state by making its amplitude negative. We don't need to know the specific state that was tagged. The beauty of the algorithm is that it "finds" that state automatically.

The Grover algorithm starts with $|s_0\rangle$, the equal-amplitude superposition of all the basis states in Eq. (12.1). In the geometric picture shown in Figure 12.1, you will note that, by construction, the angle between $|s_0\rangle$ and $|\psi_{\text{other}}\rangle$ is just a bit larger than $0°$. The algorithm simply magnifies that slight difference to get the subsequent superposition state more aligned with $|w\rangle$.

It looks like it should be easy to get the superposition aligned with the $|w\rangle$ direction; all we need to do is rotate $|s_0\rangle$ towards the $|w\rangle$ direction. But the problem is that the states are actually in an N-dimensional state space and it takes many parameters to specify rotations in such a space. For example, for two dimensions we need one angle; for three dimensions two angles; for four dimensions three angles, etc. The matrix that implements the rotation operation gets correspondingly more complicated. So, if we can't figure out the rotations, what can we do to move the system state vector towards the $|w\rangle$ direction?

The "trick" is to build the rotation as a product of reflections. The first reflection acts on the superposition state $|s_0\rangle$ and reflects it with respect to the $|\psi_{\text{other}}\rangle$ direction. This has the effect of multiplying the amplitude associated with $|w\rangle$ in the superposition $|s_0\rangle$ by -1. That step then sets up all the subsequent steps. See Figure 12.2. The second step reflects the new state $|s_1\rangle$ with respect to the original $|s_0\rangle$ state direction to produce the state $|s_2\rangle$ shown in Figure 12.3.

CARDY: Ah! I see geometrically that those two operations bring the superposition state closer to the $|w\rangle$ direction.

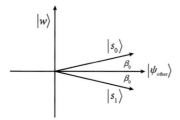

Fig. 12.2 The state space projection from Figure 12.1 with $|s_1\rangle$, the reflection of the state vector $|s_0\rangle$ with respect to the $|\psi_{\text{other}}\rangle$ axis.

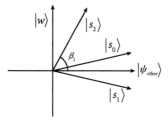

Fig. 12.3 $|s_2\rangle$ is the result of reflecting $|s_1\rangle$ with respect to the $|s_0\rangle$ direction. β_1 is the angle between $|\psi_{\text{other}}\rangle$ and $|s_2\rangle$.

ALICE: Yes, you have noticed that the reflections are equivalent to a rotation in state space. By the way, the connection between reflections and rotations is frequently used in computer graphics.

Quantum Operator Approach

ALICE: We now show how to carry out those reflections formally using quantum operators. The geometry gives us a clue. The first step is reflection with respect to $|\psi_{\text{other}}\rangle$. This corresponds to changing the sign of the amplitude for all states that are perpendicular to $|\psi_{\text{other}}\rangle$, which is just $|w\rangle$ because we have defined $|\psi_{\text{other}}\rangle$ that way. So, the first step makes the amplitude of $|w\rangle$ negative and produces the superposition state $|s_1\rangle$.

How do we find the part of $|s_0\rangle$ that lies along $|w\rangle$? To find that component, we express the state vector $|s_0\rangle$ as the sum of a part $|s_0\rangle_\parallel$ parallel to $|w\rangle$ and a part $|s_0\rangle_\perp$ perpendicular to $|w\rangle$ and hence parallel to $|\psi_{\text{other}}\rangle$:

$$|s_0\rangle = |s_0\rangle_\parallel + |s_0\rangle_\perp = |w\rangle \underbrace{\langle w|s_0\rangle}_{\substack{\text{project} \\ \text{along } |w\rangle}} + |\psi_{\text{other}}\rangle \underbrace{\langle \psi_{\text{other}}|s_0\rangle}_{\substack{\text{project} \\ \text{along } |\psi_{\text{other}}\rangle}} . \qquad (12.2)$$

To understand Eq. (12.2), recall that the Dirac bracket $\langle w|s_0\rangle$ expresses the projection of the state vector $|s_0\rangle$ along the direction of the state vector $|w\rangle$. Then all we need to do is change the sign of the part that is parallel to our desired state. This gives $|s_1\rangle$:

$$|s_1\rangle = -|w\rangle \langle w|\, s_0\rangle + |\psi_{\text{other}}\rangle \langle \psi_{\text{other}}|\, s_0\rangle. \qquad (12.3)$$

BOB: The quantum operator that carries out this reflection will be called $R^{(\psi_{\text{other}})}$

$$\begin{aligned} R^{(\psi_{\text{other}})} &= -\,|w\rangle\,\langle w| \;+\; |\psi_{\text{other}}\rangle\,\langle\psi_{\text{other}}| \\ &= 2\,|\psi_{\text{other}}\rangle\,\langle\psi_{\text{other}}| - I \\ &= 2P_{\psi_{\text{other}}} - I\,, \end{aligned}$$

(12.4)

where we used the identity operator $I \;=\; |w\rangle\,\langle w| + |\psi_{\text{other}}\rangle\,\langle\psi_{\text{other}}|$. We have also defined a short-hand notation $P_{\psi_{\text{other}}}$ for the projection operator $|\psi_{\text{other}}\rangle\,\langle\psi_{\text{other}}|$. The second and third lines in Eq. (12.4) will give us another useful geometrical interpretation, as we shall see shortly.

Try It 12.2

Use the projection operator form of the identity operator and show how that converts the first line in Eq. (12.4) to the second line.

CARDY: Got it! The use of that operator definitely makes the formula more manageable and it's easier to see what is going on with the reflection.

Try It 12.3

The rotation by reflection also works for regular vectors. Draw a diagram, like the one below, with a regular three-dimensional vector $\vec{A} = A_x\hat{x} + A_y\hat{y} + A_z\hat{z}$ with $A_x = 3$, $A_y = 5$ and $A_z = 2$. If we want to reflect that vector with respect to the xy plane, we simply change the sign of the z component, which is perpendicular to the xy plane. Show on your diagram that this procedure gives the desired result. How would we reflect the original vector with respect to the xz plane?

ALICE: Returning to our main argument, we see that the second reflection, the one relative to $|s_0\rangle$, is geometrically the same as the first reflection. The trick is to think about a different set of axes related to $|s_0\rangle$. We express $|s_1\rangle$ as the sum of a part $|s_1\rangle_{\parallel}$ parallel to $|s_0\rangle$ and a part $|s_1\rangle_{\perp}$ perpendicular to $|s_0\rangle$. See Figure 12.4. We reflect the state vector $|s_1\rangle$ relative to the $|s_0\rangle$ direction, by reversing the sign of $|s_1\rangle_{\perp}$, the part of $|s_1\rangle$ perpendicular to $|s_0\rangle$. The result of that process is $|s_2\rangle = |s_1\rangle_{\parallel} - |s_1\rangle_{\perp}$, shown in Figure 12.4.

BOB: It turns out that we can construct a quantum operator that accomplishes both the reflection and the multiplication of the perpendicular part by-1. First, note that the part of $|s_1\rangle$ parallel to $|s_0\rangle$ can be written as

$$|s_1\rangle_{\parallel} = |s_0\rangle \underbrace{\langle s_0|s_1\rangle}_{\substack{\text{projection of}\\ |s_1\rangle\text{along }|s_0\rangle}} .$$

(12.5)

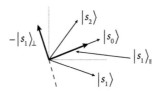

Fig. 12.4 The two-dimensional state space diagram shows $|s_1\rangle$, $|s_1\rangle_{\parallel}$, and $|s_1\rangle_{\perp}$, the components of $|s_1\rangle$ parallel to and perpendicular to $|s_0\rangle$. The dashed line is perpendicular to $|s_0\rangle$. The reflection of $|s_1\rangle$ with respect to $|s_0\rangle$ produces $|s_2\rangle = |s_1\rangle_{\parallel} - |s_1\rangle_{\perp}$.

Remember again that the Dirac bracket $\langle s_0 | s_1 \rangle$ gives us the projection of $|s_1\rangle$ along $|s_0\rangle$. In fact, since both $|s_0\rangle$ and $|s_1\rangle$ are normalized, the bracket is the cosine of the angle between the vectors: $\langle s_0 | s_1 \rangle = \cos\theta$. The part perpendicular to $|s_0\rangle$ is what is left over:

$$|s_1\rangle_{\perp} = |s_1\rangle - |s_1\rangle_{\parallel}. \tag{12.6}$$

The reflection relative to $|s_0\rangle$ changes $|s_1\rangle_{\perp}$ to $-|s_1\rangle_{\perp}$. So, we have

$$|s_2\rangle = |s_1\rangle_{\parallel} - |s_1\rangle_{\perp}. \tag{12.7}$$

We can find a single operator that does all of this by writing out the state vector $|s_2\rangle$:

$$|s_2\rangle = |s_1\rangle_{\parallel} - |s_1\rangle_{\perp} = 2|s_0\rangle \langle s_0 | s_1 \rangle - |s_1\rangle. \tag{12.8}$$

We see that a reflection operator, when it acts on $|s_1\rangle$, does the trick:

$$
\begin{aligned}
R^{(s_0)} &= 2\,|s_0\rangle \langle s_0| - \mathrm{I} \\
&= 2\mathrm{P}_{s_0} - \mathrm{I},
\end{aligned} \tag{12.9}
$$

just what we should have expected based on the second line of Eq. (12.4). Note that we introduced another projection operator, this time involving the state $|s_0\rangle$. When the operator $R^{(s_0)}$ acts on the state $|s_1\rangle$, we obtain the new state vector $|s_2\rangle$ as indicated by Eq. (12.8):

$$R^{(s_0)}|s_1\rangle = |s_2\rangle = 2|s_0\rangle \langle s_0 | s_1 \rangle - |s_1\rangle. \tag{12.10}$$

CARDY: Hmm. I think I see how this works, but it would be helpful to have a specific example to check my thinking about the details.

BOB: Let's see how this plays out for two qubits. We start with the equal-amplitude superposition of basis states $|s_0\rangle = (1/2)\,(|00\rangle + |01\rangle + |10\rangle + |11\rangle)$ and, for our example, choose the target state to be $|w\rangle = |01\rangle$. Our two-dimensional subspace is spanned by $|w\rangle$ and

$$|\psi_{\text{other}}\rangle = \left(1/\sqrt{3}\right)(|00\rangle + |10\rangle + |11\rangle). \tag{12.11}$$

The factor $1/\sqrt{3}$ assures that $|\psi_{\text{other}}\rangle$ is normalized.

The first reflection $R^{(\psi_{\text{other}})}$ applied to $|s_0\rangle$ leaves the amplitudes of the basis states in $|\psi_{\text{other}}\rangle$ the same but changes the amplitude of $|01\rangle$ from ½ to –½. Hence, the state vector $|s_1\rangle$ after the first reflection is

$$|s_1\rangle = \frac{1}{2}(|00\rangle - |01\rangle + |10\rangle + |11\rangle). \tag{12.12}$$

For the next step, we need the Dirac bracket $\langle s_0| \, s_1\rangle$, which in this case gives

$$\langle s_0| \, s_1\rangle = (½)(½) + (½)(-½) + (½)(½) + (½)(½) = ½. \tag{12.13}$$

Try It 12.4

Put $|s_1\rangle$ in column vector form and $\langle s_0|$ in row vector form and calculate $\langle s_0| \, s_1\rangle$. Do you get the result in Eq. (12.13)?

BOB: We now write Eq. (12.8), using the equal amplitudes for $|s_0\rangle$, the amplitudes for $|s_1\rangle$ from Eq. (12.12), and $\langle s_0| \, s_1\rangle = ½$ from Eq. (12.13):

$$|s_2\rangle = (½ - ½)|00\rangle + (½ + ½)|01\rangle + (½ - ½)|10\rangle + (½ - ½)|11\rangle$$
$$= |01\rangle. \tag{12.14}$$

So, if we measure the superposition state $|s_2\rangle$, we will always get the "answer" 01, the label of our target state. This is a rather special case, because we found the desired state after only once through the algorithm. That is unique to two-qubit systems, but it does show starkly the advantage of the quantum algorithm over the classical one. Remember that a two-qubit system has four basis states. If we had four playing cards, one with a purple dot, we might need to flip cards up to three times to find the purple dot card. In the quantum case, we did it in only one "flip" after the initial setup.

Grover proved that the number of iterations needed for the quantum algorithm increases as $\sqrt{N}$, while, as we have seen, the number of classical algorithm steps increases as N.

Amplitude Amplification Approach

A general feature of the Grover algorithm used in many other quantum algorithms is the notion of "amplitude amplification." The process increases (amplifies) the amplitude associated

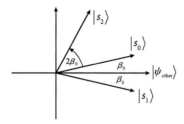

Fig. 12.5 The state space diagram shows the results of two reflections in the Grover algorithm. $|s_0\rangle$ starts at an angle β_0 with respect to $|\psi_{\text{other}}\rangle$. After the two reflection operations, $|s_2\rangle$ forms an angle $3\beta_0$ with respect to $|\psi_{\text{other}}\rangle$.

with the target state $|w\rangle$ while simultaneously decreasing the magnitude of the other states' amplitudes. It's like playing with a pair of dice where one is "loaded" to roll a "1," increasing the odds of rolling snake eyes. Geometrically, this is equivalent to getting the superposition state pointing along $|w\rangle$.

To see where the $\sqrt{N}$ number of steps comes in, note that the average amplitude for the state $|\psi_{\text{other}}\rangle$ is about $1/\sqrt{N}$. For each step, the amplitude for the target state $|w\rangle$ increases by about $1/\sqrt{N}$ (for large N). Thus after $\sqrt{N}$ steps, the amplitude is about equal to 1 for $|w\rangle$ and close to 0 for all the other states.

BOB: I will show you a neat geometric way of seeing why we get the answer in one step for two qubits. See Figure 12.5. After doing the two reflections (and multiplying by -1) the angle between $|s_2\rangle$ and $|\psi_{\text{other}}\rangle$ is $3\beta_0$, where β_0 is the angle between the initial uniform superposition state $|s_0\rangle$ and $|\psi_{\text{other}}\rangle$. For the two-qubit case, the angle β_0 is $\pi/6$. After one application of the algorithm, the angle between $|s_2\rangle$ and $|\psi_{\text{other}}\rangle$ equals $3 \times (\pi/6) = \pi/2$. That tells us that the superposition state points along the target state vector $|w\rangle$ and we are done!

Try It 12.5

Use the column and row vector forms for $|s_0\rangle = (1/2, 1/2, 1/2, 1/2)$ and $|\psi_{\text{other}}\rangle = (1/\sqrt{3}, 0, 1/\sqrt{3}, 1/\sqrt{3})$ to find $\langle s_0| s_1\rangle$, which is $\cos\beta_0$. Show that $\beta_0 = \pi/6$. Note that we need to have the two state vectors normalized to get the correct $\cos\beta_0$.

ALICE: Going back to the general case, we see from Figure 12.6 that each successive application of the reflection adds $2\beta_0$ to the angle between the state $|s_k\rangle$ and $|\psi_{\text{other}}\rangle$, effectively moving the superposition state further from $|\psi_{\text{other}}\rangle$ and closer to $|w\rangle$; so, after k iterations, the angle between $|s_k\rangle$ and $|\psi_{\text{other}}\rangle$ is

$$\beta_k = (2k + 1)\beta_0 \tag{12.15}$$

as shown in Figure 12.6.

ALICE: You can extend that method to get another estimate of how many iterations you need to use to get the state $|s_k\rangle$ to within some specified tolerance of the target state. As you might realize, the more qubits you have in the system, the closer the original equal superposition state $|s_0\rangle$ will lie along $|\psi_{\text{other}}\rangle$. So, more iterations of the algorithm are needed to get $|s_0\rangle$

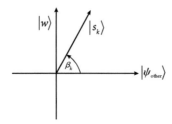

Fig. 12.6 A schematic diagram of the state vector $|s_k\rangle$ and the angle β_k after k iterations of the Grover algorithm.

Table 12.1 Angle results for n qubits. Results to three significant figures.

n	$\cos \beta_0$	β_0 (degrees)
2	0.866	30.0
3	0.935	20.7
4	0.968	14.5
5	0.984	10.2
10	0.999	1.79
50	1.00	1.71×10^{-6}

pointing close to $|w\rangle$. Table 12.1 provides some numerical examples. The values of $\cos \beta_0$ are found from the Dirac bracket $\langle s_0 | s_{\text{other}} \rangle$.

CARDY: I see that the initial angle between the uniform superposition state $|s_0\rangle$ and $|\psi_{\text{other}}\rangle$ decreases rapidly as the number of qubits grows.

ALICE: That's right, Cardy, because with more qubits, most of the basis states are in $|\psi_{\text{other}}\rangle$, with only one, $|w\rangle$, not included; so, subtracting $|w\rangle$ from $|s_0\rangle$ to produce $|\psi_{\text{other}}\rangle$ doesn't change the state vector very much. As the iterations proceed, the angle β_k increases as $\beta_k = (2k + 1)\beta_0$, where k is the number of iterations; so, it may take many iterations before the superposition state is aligned with $|w\rangle$.

Try It 12.6

Challenge: Show that for an n-qubit system (with $N = 2^n$), $\cos \beta_0 = \sqrt{1 - 1/N}$. Hints: For the uniform superposition state, the amplitude for each basis state is $1/\sqrt{N}$. The amplitude for the non-zero terms in $|\psi_{\text{other}}\rangle$ is $1/\sqrt{N-1}$ and there are $N - 1$ non-zero terms. Use that result to check the listings in Table 12.1.

ALICE: We can use our geometric picture to estimate the probability of getting the target state if we carry out a measurement on the superposition state $|s_k\rangle$ after k iterations. The measurement postulate tells us that the probability of getting the target state is the square of the amplitude of the superposition state projected onto $|w\rangle$. So, we have

Probability of observing $w = |\langle w | s_k \rangle|^2$. (12.16)

From Figure 12.6, we see that $\langle w | s_k \rangle = \cos(\pi/2 - \beta_k) = \sin \beta_k$; so, we find

Probability of observing $w = \sin^2 \beta_k$
$$= \sin^2(2k+1)\beta_0 .$$
(12.17)

We found previously that $\cos \beta_0 = \sqrt{(N-1)/N}$ and hence $\sin \beta_0 = 1/\sqrt{N}$. If N is large, then the sine will be small, and we can use the approximation $\sin \beta_0 \approx \beta_0 = 1/\sqrt{N}$.

You can use a calculator or a spreadsheet to verify that $\sin \beta_0 \approx \beta_0$ for small β_0 if the angle is measured in radians. Or look at a graph of $\sin \beta_0$ versus β_0. For small β_0, the graph is almost a straight line with a $45°$ slope. There are more formal proofs, but those two methods should convince you that the result is correct.

CARDY: How did you get $\sin \beta_0 = 1/\sqrt{N}$?

Try It 12.7

Answer Cardy's question. Hint: From trigonometry we have $\cos^2\theta + \sin^2\theta = 1$ for any angle θ.

ALICE: Assembling these results, we see that the probability of observing $|w\rangle$ is approximately $\sin^2\left((2k+1)/\sqrt{N}\right)$. We get the largest probability when the argument of the sine function is equal to $\pi/2$. That means that we want to use k^* iterations, where k^* satisfies

$$(2k^* + 1)/\sqrt{N} \approx \pi/2$$
(12.18)

or

$$k^* \approx \pi\frac{\sqrt{N}}{4} - 1/2$$
$$\approx \pi\sqrt{N}/4.$$
(12.19)

CARDY: What happened to the ½ in the previous equation?

ALICE: Because we are thinking about large N (for example, $N > 100$), ½ will be small compared to the $\sqrt{N}$ term so we can safely ignore it without significantly changing the result. This type of thinking happens a lot in math and science. The important conclusion is that

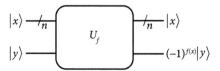

Fig. 12.7 The function gate for the Grover algorithm. The upper-register states are n-qubit states as indicated by the diagonal line with the subscript n.

Eq. (12.19) shows that the number of iterations grows as $\sqrt{N}$ for the quantum algorithm. We argued before that the classical algorithm requires about $N/2$ "flips" on average. This result illustrates the advantage of the quantum algorithm in reducing the number of iterations that are needed to find the target state.

CARDY: That's pretty cool.

BOB: As we have seen with other quantum algorithms, it is customary to formulate the algorithm in terms of a function gate (an "oracle") because those function gates can be built from simpler gates such as CNOTs, X, Y, Z, etc. For the Grover algorithm, the function gate is shown in Figure 12.7. This is similar to the function gate for the Deutsch algorithm, but with the upper register extended to n qubits.

The system state after the function gate has acted is

$$U_f\left(|x\rangle\,|y\rangle\right) = \left(\sum_{x=0}^{2^n-1}(-1)^{f(x)}\,|x\rangle\right) \otimes |y\rangle = \left\{(-1)^{f(x)}\,|x\rangle\right\} \otimes |y\rangle. \tag{12.20}$$

The function gate is unique to the target state we are trying to find. That, in essence, is how we label the target state. In other words, the function gate is constructed so that $f(w) = 1$ while $f(x) = 0$ for all the other xs. You may remember this is the function that was the subject of this chapter's first Try It.

The function gate is often defined in an equivalent version, but one that is more easily implemented in terms of simple quantum gates (see Figure 12.8).

In the alternate version, we assume that the lower-register input state is $|y\rangle = \frac{1}{\sqrt{2}}(|0\rangle - |1\rangle)$. You can then convince yourself that when $f(x) = 1$, the $y \oplus f(x)$ term switches the two right vectors in the input state $|y\rangle$ and that switch is equivalent with this $|y\rangle$ to multiplying the state vector by -1.

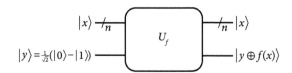

Fig. 12.8 An alternate representation of the Grover algorithm function gate.

Try It. 12.8

Challenge: Work out the details given in the previous paragraph.

ALICE: Don't worry if those details are not yet clear to you, Cardy. It took me a few minutes in a quiet place with a sharp pencil (with an eraser!) and several clean sheets of paper to convince myself of the details.

BOB: Figure 12.9 shows the quantum circuit for the Grover algorithm. Note again that this is just the n-qubit generalization of the Deutsch circuit in Figure 11.3. The gate U_A implements the reflections of the state vectors that we described before. In practice, you would build in k^* repetitions of U_A to get close to the maximum probability of observing the target state when the measurement device works on the upper-register superposition state. Let's go through the circuit for a two-qubit system. The upper-register state on the far left is then $|00\rangle$. That state interacts with a $H^{\otimes 2}$ Hadamard gate while the lower register interacts with a one-qubit Hadamard gate.

CARDY: I get it! I see that for a two-qubit system, we have four basis states. The top Hadamard gate generates an equal-amplitude superposition of the four basis states. I also see that the lower-register Hadamard gate produces the superposition state $|y\rangle = \frac{1}{\sqrt{2}}(|0\rangle - |1\rangle)$ that Alice mentioned before.

ALICE: I'm glad you are recognizing the general structure of the circuit from the quantum algorithms we studied before. Let's continue with the general Grover circuit step by step and see how it aligns to the geometric method.

STEP 1: The upper-register initial state is subject to the appropriate Hadamard gate to produce an equal superposition of the n-qubit basis states. The lower-register single-qubit state becomes $(1/\sqrt{2})(|0\rangle - \sqrt{1})$. At that stage, the system state is

$$|\psi_1\rangle = \underbrace{\{1|x\rangle\}}_{\text{upper register}} \otimes \underbrace{\left(\frac{|0\rangle - |1\rangle}{\sqrt{2}}\right)}_{\text{lower register}}. \tag{12.21}$$

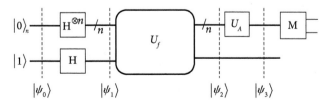

Fig. 12.9 The quantum circuit for the Grover algorithm. The vertical dashed lines indicate where each system state $|\psi_j\rangle$ occurs.

STEP 2: The function gate now acts on this state to produce the system state

$$|\psi_2\rangle = \underbrace{\left\{(-1)^{f(x)}|x\rangle\right\}}_{\text{upper register}} \otimes \underbrace{\left(\frac{|0\rangle - |1\rangle}{\sqrt{2}}\right)}_{\text{lower register}}. \tag{12.22}$$

In words, the function gate has flipped the sign of the amplitude of the target state (the one with $f(x) = 1$), while leaving the other amplitudes unchanged. This is the equivalent of the first reflection of the superposition state in our geometric approach.

STEP 3: We implement the equivalent of the second reflection whose main effect is to increase the superposition state's amplitude of the target state while reducing the other amplitudes in the superposition state. That is what the gate U_A in Figure 12.9 does. As we saw with our geometric method and its translation into quantum operators (Eq. (12.10)), all we need is a gate to implement that operation. Let's assume we have such a gate. Mermin, 2007 (see Chapter 4) shows how to implement the Grover algorithm function gates using simple quantum gates.

Flipping About the Mean

BOB: Some authors call this set of operations "flipping about the mean." The mean m of the amplitudes in the superposition state is their sum divided by the number of basis states $N = 2^n$. So, for each amplitude a, we calculate $\delta = m - a$ and then subtract δ from a to get the new amplitude $a' = 2m - a$. Suppose we have four amplitudes $(1/2, -1/2, 1/2, 1/2)$. The mean is 1/4. The new amplitudes are shown in the following table:

CARDY: I see that Table 12.2 shows that the result of flipping about the mean is to increase the amplitude of the target state (the one that started with the negative amplitude). While decreasing the amplitude of the other states. I also recognize that, as Bob mentioned before, a two-qubit system is special because the Grover algorithm finds the target state after only one iteration of the algorithm: The target state amplitude is 1 while the other amplitudes are 0.

BOB: We can also give a graphical representation of flipping about the mean. In Figure 12.10, we represent the amplitudes of the basis states by the heights of rectangles. The target state $|01\rangle$ is given a negative amplitude, while all the other amplitudes are positive. The lower part of the figure shows the amplitude of the states after flipping about the mean.

Table 12.2 Flipping about the mean for a two-qubit state.

Original a	1/2	−1/2	1/2	1/2
Mean $m = 1/4$				
New $a' = 2m - a$	0	1	0	0

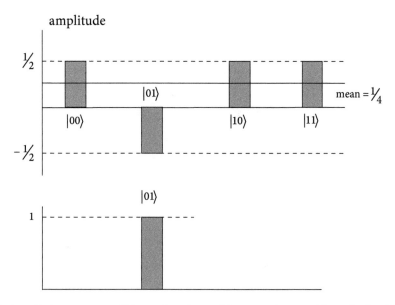

Fig. 12.10 A graphical representation of "flipping about the mean" for a two-qubit system. The amplitudes of the basis states are represented by the heights of the shaded rectangles. The target state $|01\rangle$ starts with a negative amplitude (top part of the figure). The lower part of the figure shows the amplitudes after flipping about the mean. $|01\rangle$ has an amplitude equal to 1. All the other amplitudes are 0.

ALICE: Another graphical representation is often used for displaying state amplitudes. As shown in Figure 12.11, we plot the amplitudes as columns in a rectangular grid. The amplitude is represented by the height of the column. Positive amplitudes go up, while negative amplitudes go down. Figure 12.11 shows the amplitudes for the uniform superposition state of a two-qubit system.

Figures 12.11 to 12.13 take you through the sequence of flipping about the mean for the two-qubit system with target state $|0\rangle\,|1\rangle$. Similar diagrams are widely used in QC to show state amplitudes after gates operate on the states. Comparisons with the output of an actual QC tells you about the gate "fidelity": how faithfully the QC has been able to implement the gate operation. In practice, the commonly used column graphs show what are called the probability density matrix elements for the system. Those elements involve products of various state amplitudes, rather than the amplitudes themselves.

BOB: Figure 12.14 shows a graphical representation of flipping about the mean when the number of states N is large. In that case, the initial mean amplitude is close to $1/\sqrt{N}$. After flipping about the mean, only the state with the negative amplitude has any substantial change in amplitude. If we make that increased amplitude negative and do another flip about the mean, that amplitude becomes even larger, and the other amplitudes decrease by a small amount. Continuing that process, we find that the target state amplitude eventually approaches 1 and the other amplitudes approach 0.

CARDY: The graphical representation was helpful because I can see the huge increase in the amplitude of the desired state.

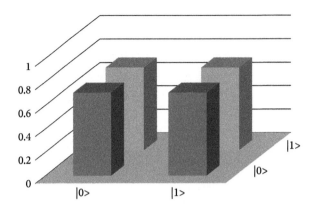

Fig. 12.11 The amplitudes for a uniform superposition state for a two-qubit system. The height of the column represents the amplitude. The state labels are given by the product of the labels along the two "horizontal" directions. $|0\rangle |1\rangle$ is in the front row, far right.

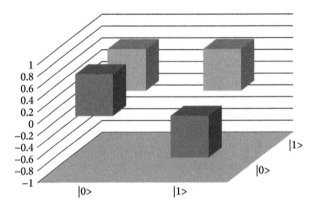

Fig. 12.12 The amplitudes for a two-qubit superposition state in which the target state $|0\rangle |1\rangle$ has been given a negative amplitude.

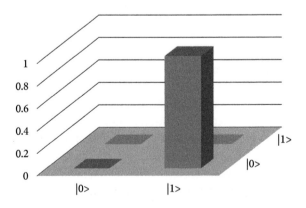

Fig. 12.13 The amplitudes for a two-qubit superposition state after one "flip about the mean." The target state $|0\rangle |1\rangle$ amplitude now equals 1 and all other amplitudes are 0.

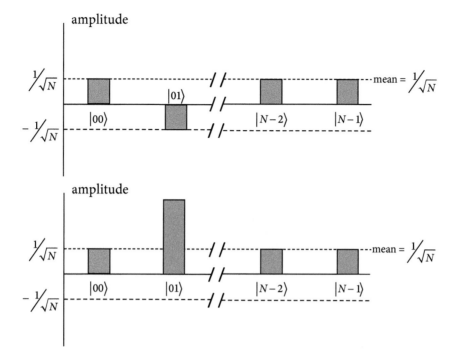

Fig. 12.14 A graphical representation of flipping about the mean for an n-qubit system.

BOB: Let's express the flipping about the mean in terms of an operator like the one in Eq. (12.10) acting on the state $|s_k\rangle$ to produce $|s_{k+1}\rangle$:

$$|s_{k+1}\rangle = R^{(s_{k-1})}|s_k\rangle = 2|s_{k-1}\rangle\langle s_{k-1}|s_k\rangle - |s_k\rangle. \tag{12.23}$$

For a general set of qubit states, the new coefficients a'_j are given by

$$a'_j = 2m - a_j = 2 \times \left(\frac{1}{2^n}\sum_{i=0}^{2^n-1} a_i\right) - a_j = -\left(1 - \frac{1}{2^n}\right)a_j + \frac{1}{2^{n-1}}\sum_{i=0,i\neq j}^{2^n-1} a_i. \tag{12.24}$$

Looking at the first expression in Eq. (12.24), we see that if $a_j \approx m$, it will be almost the same after the operation. But the amplitude of the target state is initially $-m$ and after the operation we get $a'_j \approx 3m$. But remember that these approximations are valid only for large N.

For $n = 2$, we have

$$a'_j = 2m - a_j = \frac{1}{2}\sum_{i=0}^{3} a_i - a_j = -\frac{1}{2}a_j + \frac{1}{2}\sum_{i=0,i\neq j}^{3} a_i. \tag{12.25}$$

Table 12.3 Flipping about the mean for a three-qubit superposition state.

Original a	$1/\sqrt{8}$	$1/\sqrt{8}$	$-1/\sqrt{8}$	$1/\sqrt{8}$	$1/\sqrt{8}$	$1/\sqrt{8}$	$1/\sqrt{8}$	$1/\sqrt{8}$
Mean m	0.265							
$a' = 2m - a$	0.177	0.177	0.884	0.177	0.177	0.177	0.177	0.177

The matrix that carries out this calculation for a two-qubit system is

$$\begin{pmatrix} a'_0 \\ a'_1 \\ a'_2 \\ a'_3 \end{pmatrix} = \frac{1}{2} \begin{pmatrix} -1 & 1 & 1 & 1 \\ 1 & -1 & 1 & 1 \\ 1 & 1 & -1 & 1 \\ 1 & 1 & 1 & -1 \end{pmatrix} \begin{pmatrix} a_0 \\ a_1 \\ a_2 \\ a_3 \end{pmatrix} \quad (12.26)$$

Try It 12.9

Use this matrix with the two-qubit state $\frac{1}{2}\left(|00\rangle - |01\rangle + |10\rangle + |11\rangle\right)$ and show that the results match the results in Table 12.2.

BOB: Now let's see how this works for a three-qubit system.

CARDY: With three qubits, we have eight amplitudes, right?

BOB: That's right. In Table 12.3, we show the results when we choose the target state $|010\rangle$ in the third column.

BOB: Once again we see that the flipping about the mean leads to an increase in the amplitude of the target state and a decrease in the amplitudes of the other states. Figure 12.15 displays a column chart of the Table 12.3 results. Note that the target state amplitude is much larger than the amplitudes of the other states.

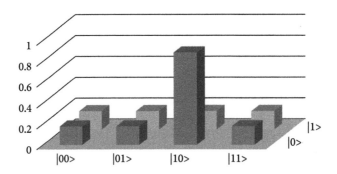

Fig. 12.15 A column chart representation of the Table 12.3 results. The height of a column is the amplitude for that state.

Try It 12.10

Verify the numbers in Table 12.3. Carry out the next step by flipping about the new mean.

ALICE: Let's pause for a moment and summarize what we have learned about the Grover algorithm.

- The Grover algorithm allows us to find a basis state that has been chosen by making its amplitude negative while the amplitudes of all the other basis states are positive.

- We then follow a set of operations on the equal-amplitude superposition state. The result of that set of operations is to increase the amplitude associated with the chosen state while reducing the amplitudes associated with the basis states that are not chosen.

- The operations can be viewed as flipping state vector amplitudes relative to the previous state in the sequence—in effect rotating the state vector closer to the chosen vector direction.

- From those results, we can estimate how many operations are needed to identify the chosen state. We found that the number increases as $\sqrt{N}$, where N is the number of qubit basis states in the system. The corresponding classical algorithm takes about $N/2$ steps. Thus, the Grover algorithm demonstrates the advantage of the quantum algorithm over the classical one.

- Also, we saw how to implement those operators using a Grover algorithm quantum circuit.

CARDY: Before we go on, I need to ask why we need faster searches. Google can find anything in less than a second.

BOB: There are several answers. First, Google gives you only a list of possible search outcomes. In most cases, you need to go through the list and find the ones closest to what you need. The Grover algorithm finds the best match automatically. The second answer may seem a bit dodgy, but it illustrates an important principle: The Grover algorithm was not really meant to be a practical search algorithm. It simply shows that there is a quantum algorithm that is much faster than the corresponding classical one. That expands our horizons about what is possible with quantum computing and why we might want to develop other, more practical quantum algorithms.

12.3 Error Correction

BOB: We now need to look at a general problem in computing and communications: error correction.

CARDY: Why error correction? Aren't computers supposed to follow exact rules and do exactly what we tell them to do?

BOB: Yes, that is what we often assume, but we need to realize that for computers and other forms of digital information processing to work, we need to send information from one place to another. Sometimes the distances are quite short, say within your smartphone, but other times we use fiber optic links under the oceans, or we send the information via radio waves and satellites orbiting Earth. That information is almost always in a sequence of 1s and 0s. But as the signals propagate, the radio waves, for example, might be contaminated with electrical noise from other sources such as power lines and lightning strikes. Fortunately, modern information technology has matured to the point that errors in those messages are infrequent. But when they do occur, if they are not corrected, the received signal might be garbled to the extent that the output of the process is basically nonsense. Error detection and error correction, because of their importance, have evolved into rather sophisticated schemes.

For a classical computer, information is often sent in *clusters* of 1s and 0s, of just seven or eight bits. The most common type of error in the transmission is to have one of the bits change from 0 to 1 or 1 to 0. This is called a "bit flip" error. When we think about error correction for quantum information transmission, the situation becomes even more complex because the state of a qubit, the way we send quantum information, can change relatively easily due to interactions between the qubit and its environment. Not only can the basis states $|0\rangle$ and $|1\rangle$ be interchanged (the equivalent of a bit flip), but the amplitudes of a superposition state can change, thereby corrupting the information carried by the qubit's state. Let's look at a simple example that illustrates the basic features of error correction.

Let's tackle classical error correction first. The first step in error correction is to detect that an error has occurred. One way to do this would be to send the same message twice. We then compare the two messages. If they are the same, we are reasonably confident that no error occurred. That confidence is based on the relatively rare occurrences of errors in the transmission system. But since the errors are rare, we are wasting a lot of communication time by sending the message twice. A better scheme is to append another bit to the sequence that gives us some information about what is in the message. This additional ("ancillary") bit is often called a "parity bit" because it is used to indicate that some number associated with the message is even or odd. In mathematics and computer science, "parity" almost always means even or odd.

ALICE: Let me give you a specific example of the use of a parity bit in classical information transmission. Suppose the message we want to send is seven bits long—0110101, for example. To aid in error detection, we can append a 1 to the right of the sequence if the number of 1s in the sequence is even and a 0 if the number of 1s is odd. For our example, we have four 1s in the message. Since four is an even number, we would append a 1 to the right and send the message as 01101011.

When the message is received, the receiver counts the number of 1s and compares that number to the odd number expected (odd because we include the parity bit). If any of the original 1s flipped to a 0 or any of the original 0s flipped to a 1, there would be a disagreement between the expected number of 1s and the received number.

CARDY: But what if two bits flipped? Then we might not detect the error.

ALICE: That's correct. But remember we assume that the errors occur with only very small probabilities, so the probability of having two errors in the same message is extremely small. It might happen, but it will be rare enough that we don't need to worry about that possibility.

CARDY: Okay, we detect that an error has occurred. How do we find out which bit was affected? I think we need to know that to correct the error.

ALICE: That's where we must be clever. One option is simply to request that the message be resent if an error has been detected. But there are ways we can find the bit in which the error occurred and make the correction directly without having to communicate further with the sender.

One of the simplest schemes is to "pad" the bit that you want to send with other bits, in particular just add two more of the same bit. For example, if you want to send 0 to Bob, you would actually send 000. If a one-bit error occurs during transmission, Bob will receive 100, 010, or 001. Here's the trick to spot the error: designate the bits as b_0, b_1, and b_1. When Bob receives the three bits, he evaluates $b_0 \oplus b_1$ and $b_0 \oplus b_2$. If the results are 0 and 0, then Bob knows there was no one-bit error. In fact, exactly the same process works if you want to send a 1. You launch 111 to Bob, and he carries out the same binary addition without carry. If he gets 0 and 0, then he knows the message is okay in the sense that it does not contain a one-bit error.

Try It 12.11

Check this with the message bit $= 1$.

ALICE: What is more amazing is that if there is a one-bit error, the results of the binary additions let you know which bit was affected. For example, if $b_0 \oplus b_1 = 1$ and $b_0 \oplus b_2 = 0$, then we know that bit b_1 was the affected bit and we should flip it to get the correct bit sequence. This is a variation on the parity test, but here the result of the binary addition tells us whether the two bits are the same or different.

Try It 12.12

(a) Verify that the example is correct. (b) Work out which bit needs to be flipped if Bob gets 01 and 11 from the parity tests. Hint: the answers are in Table 12.4.

Try It 12.13

Verify the results stated in Table 12.4 for the two possible cases: (a) The message sent was 000 but the message received was 010. (b) The message sent was 111 and the message received was 011.

ALICE: The take-home message is that the parity test allows you to find the bit where the error occurred, and the procedure does not depend on knowing the value of that bit.

Table 12.4 The corrective action required to correct a one-bit error.

$b_0 \oplus b_1$	$b_0 \oplus b_2$	Corrective action
0	0	No change
1	1	Flip b_0
1	0	Flip b_1
0	1	Flip b_2

BOB: We have already hinted at why error detection and correction are even more important for QIS and QC than they are for classical systems. The practical problem is that it is difficult to isolate a qubit completely from its environment, particularly if we want to act on the states of the qubit to implement quantum algorithms. Moreover, the qubits we use (spinning electrons or neutrons, superconducting currents, trapped ions, and so on) are systems that have relatively small differences in the energies associated with the qubit quantum states. That means it doesn't take much jostling from interactions with nearby atoms and molecules to cause transitions between states (corresponding to classical bit flips), and even worse, even smaller amounts of jostling lead to changes in the amplitudes and relative phases of the components of a superposition state. We have seen how useful those superposition states are for quantum algorithms. For all those reasons, we must be prepared to detect and correct errors.

CARDY: So, let's see if I understand. Quantum states are more fragile than classical states, particularly the type used in classical computing. Second, there are several different types of quantum errors, while for classical information systems bit flips are the major problem. Finally, quantum errors can build up gradually over time, rather than just a sudden bit flip. So, quantum error correction is both more necessary and more difficult.

Couldn't we just measure the states of the qubits and if they are not what we had expected, we would conclude that an error had occurred and then we simply rebuild the state with the correct amplitudes?

BOB: That does sound like a reasonable way to proceed, but we need to keep in mind that having a qubit interact with a measurement device simply projects the qubit's state onto one of the measurement device's basis states. In quantum systems in a state of superposition, the measurement gives us only limited information about the qubit's state before the measurement. Consider photon polarization. Suppose I have a vertically polarized photon. If I am using a $45°$ measurement basis, the photon will be in a state of superposition. I have a 50:50 probability of measuring either outcome in my basis states. The same will occur if I use a horizontally polarized photon instead of a vertical one. In addition, after the measurement the photon polarization state has changed to be one of the measurement basis states. The original photon state is gone. So, we need to develop a quantum error-detection and -correction procedure that does not involve measurements on those qubits.

The quantum "trick" is like what we did with classical error detection and correction: We combine additional qubits with the original qubit or qubits in such a way that measurements on the additional qubits, usually called *ancillary qubits*, can tell us if an error has occurred and what to do to correct it. Adding ancillary bits is a common and useful strategy in quantum

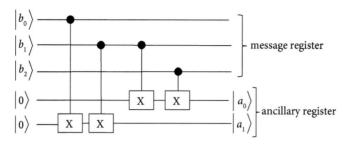

Fig. 12.16 A quantum circuit for detecting bit-flip errors. The message is carried by the top three qubit states called the message register.

algorithms. The only operations allowed on the original qubits are those of quantum gates; no interactions with measurement devices are permitted.

The quantum error-detection and -correction scheme we will examine is analogous to the three-bit parity check procedure we looked at in classical computing. We embed the basis states, which of course form the core of any quantum message we want to transmit, into a three-qubit system using the following rule:

$$|0\rangle \rightarrow |000\rangle \qquad |1\rangle \rightarrow |111\rangle . \tag{12.27}$$

A general state for a message is a linear superposition of those states: $|\psi\rangle = a\,|000\rangle + b\,|111\rangle$.

CARDY: That's an entangled state. Right?

ALICE: Exactly right. As usual, employing entangled states is what makes the quantum algorithm work.

For simplicity's sake, let's look first at the quantum equivalent of one-bit flip errors. For example, the states $|001\rangle$, $|010\rangle$, and $|100\rangle$ are single-bit flip errors associated with $|000\rangle$. Later, we will see how to generalize the procedure for other types of errors. As we mentioned before, we want to detect and correct those errors without having to carry out measurements on these qubits because we know that measurements disturb the state of the system. To do this, we use the states of two ancillary qubits to detect the error and to correct it. First, let's think about the detection part.

ALICE: The quantum circuit shown in Figure 12.16 uses CNOT gates to connect the two ancillary qubit states to the three qubit states in our "message" register. It is not too hard to work out that if all three of the message register bits are the same, both the ancillary states will be $|0\rangle$. That is an indication that we don't have any one-bit flip errors. Table 12.5 lays out the possibilities for the circuit in Figure 12.12.

The crucial point is that the ancillary qubit states allow us both to detect the error and to pinpoint its location amongst the three message qubit states because they flip to $|10\rangle$, $|01\rangle$, or $|11\rangle$ when there is an error that changes any of the message register states.

CARDY: Wow! It seems that for a big quantum computer, you're going to need a lot of qubits to make the error correction work. I think I'll invest in companies that supply qubits. That is obviously going to be a large market.

ALICE: Nice idea, but we are far away from having qubits be a commercial commodity like integrated circuits in classical computing.

Table 12.5 One-bit-flip errors for the message $|000\rangle$ and the corresponding ancillary register states in Figure 12.16.

Message register			Ancillary register	
$|b_0\rangle$	$|b_1\rangle$	$|b_2\rangle$	$|a_0\rangle$	$|a_1\rangle$
$|0\rangle$	$|0\rangle$	$|0\rangle$	$|0\rangle$	$|0\rangle$
$|0\rangle$	$|0\rangle$	$|1\rangle$	$|1\rangle$	$|0\rangle$
$|0\rangle$	$|1\rangle$	$|0\rangle$	$|1\rangle$	$|1\rangle$
$|1\rangle$	$|0\rangle$	$|0\rangle$	$|0\rangle$	$|1\rangle$

Table 12.6 Toffoli gate truth table. $|\bar{a}\rangle$ means $|NOT\ a\rangle$.

$|c_0\rangle$	$|c_1\rangle$	$|s\rangle$
$|0\rangle$	$|0\rangle$	$|a\rangle$
$|0\rangle$	$|1\rangle$	$|a\rangle$
$|1\rangle$	$|0\rangle$	$|a\rangle$
$|1\rangle$	$|1\rangle$	$|\bar{a}\rangle$

Let's get back to the question of what to do now that we have detected an error. How can we carry out the error correction using gates? First, recall that the operator X is equivalent to a NOT operation. For a quantum state, NOT is equivalent to a bit flip. So, to correct bit flip errors, we just need to operate with X on the appropriate state. Ideally, we want this process to be automated so that the errors are corrected without outside intervention.

One way to do this is to have the X operators be controlled by the ancillary states since the results in Table 12.4 link those states to the states that need to have their bits flipped if an error has occurred. With the CNOT gate, the control state determines whether states $|0\rangle$ and $|1\rangle$ are flipped. What we need now is a generalization of that gate with *two* control states. In other words, we need a doubly controlled NOT gate, which is called a Toffoli gate.

BOB: Toffoli gates were invented in 1980 by Tommaso Toffoli, an Italian American electrical and computer engineer. Table 12.6 gives the truth table for the Toffoli gate and Figure 12.17 shows the quantum circuit diagram.

Try It 12.14

Convince yourself that Table 12.6 and the Toffoli gate circuit in Figure 12.17 are consistent with one another.

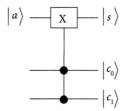

Fig. 12.17 The Toffoli gate circuit diagram. The lower two wires are the control states.

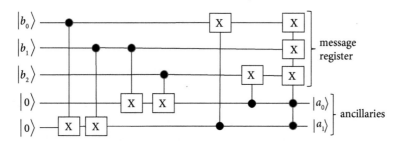

Fig. 12.18 Quantum error-detection and error-correction circuit.

To carry out the error correction for our one-bit flip example, we put Toffoli gates on the three message qubit wires and make the control connections to the two ancillary state wires as shown on the right in Figure 12.18. We have used a condensed diagram rather than including three separate Toffoli gates. The X operations on the three message register qubits are all controlled by the two ancillary qubits.

Try It 12.15

Suppose Alice wants to send $|000\rangle$. Assume that the only errors are $|001\rangle$ and $|010\rangle$. Show that the correction circuit in Figure 12.14 works as desired. Repeat the analysis for the message $|111\rangle$ and the one-bit error $|101\rangle$.

Try It 12.16

Challenge: Construct an argument that shows that the correction circuit also works for errors in superposition states. Assume the message is $a|000\rangle + b|111\rangle$ and the one-bit error is $a|001\rangle + b|110\rangle$. Hint: All the circuit operators are linear operators.

CARDY: You mentioned that there are other kinds of quantum errors besides bit flips. How do we take care of them?

ALICE: The more general quantum error correction schemes are often just generalizations of what we have done. The other common errors are phase errors. For example, a state amplitude a_0 might turn into $-a_0$. That change is called a phase flip. For those kinds of errors, you can use controlled Z gates in place of (or in addition to) the X gates. We will talk about other kinds of phases and phase changes in Chapter 15. It is also helpful to embed the message qubit into systems with more register qubit states (seven is a commonly used number). Furthermore, the error circuits may have more than two ancillary states. The references at the end of this chapter provide more details, but be warned that the circuits and the corresponding formalism can get complicated.

12.4   Quantum Computational Chemistry

BOB: One of the most promising applications of quantum computing for near-term applications is the calculation of the structures and energies of molecules. Why is this important? One of the fundamental principles of chemistry is the link between form and function. For example, how well a drug molecule interacts with a target virus depends on the form (shape) of the molecule and how well it connects to the part of the virus the drug is designed to attack. Catalysts are another good example, where the shape of the catalyst materials determines the effectiveness of their ability to promote other chemical reactions. More generally, the properties of most complex materials depend on the shape of the molecules that make up the material.

The other key principle is that the different shapes are associated with different energies and it takes energy input or output for the molecule to change from one shape to another. Here is a specific example: When your eye "detects" light, a photon is absorbed by a rhodopsin molecule in the retina of your eye. The absorption of the photon's energy changes the rhodopsin from a more-or-less straight-line shape (called "cis") to a bent form (called "trans"). That change in shape leads to an electrical signal sent over your optic nerve from the retina to the brain, telling the brain that light has been detected. Knowing the energies associated with those two rhodopsin shapes is important in predicting the efficiency of light detection.

CARDY: But how do you calculate those energies and structures?

BOB: Scientists can calculate those energies through various methods based on quantum mechanics. The key idea is to think about how the atoms that make up the molecule come together to form the molecule. You may remember from your high school chemistry course that atoms have a central nucleus, made up of protons and neutrons, and that electrons are found in "clouds" centered on the nucleus. There is an attraction between the protons in the nucleus and electrons and a repulsion between the electrons and between the positively charged nuclei. These interactions form part of the basis for molecular bonds and the energy associated with them.

CARDY: It is coming back to me. I do remember the idea of clouds of electrons around the atomic nuclei and something about how those clouds rearrange themselves when the atoms come together to form molecules.

BOB: In that process, some of the electrons that were associated with individual atoms rearrange themselves into the chemical bonds that hold the molecule together. In effect, some of the electrons stay around their original atomic nuclei; others are shared among several of the atoms, forming molecular bonds. The task is to calculate the energy associated with those arrangements of atoms within the molecule. The configuration with the lowest energy is usually the most stable shape of the molecule.

The difficulty is that a typical molecule, particularly those that are of interest in biological settings, have hundreds of atoms and thousands of electrons. Keeping track of all those particles and how their energies change is quite a challenge. Nevertheless, chemists have developed various techniques that are reasonably successful in finding those energies.

But, to return to our drug example, if you want to explore dozens of molecules as potential therapeutic candidates, the computational time required strains even our best classical supercomputers. Quantum computing offers alternative ways of doing these calculations much more quickly, so that less time is needed to explore the various candidate molecules.

In recent years, quantum chemistry algorithms have been developed that combine classical computations with quantum computations. These have been applied so far only to relatively small molecules, but the techniques appear to be quite promising. Along the way, quantum computer scientists realized that to make these calculations work on an actual quantum computer, quantum error corrections need to be applied to account for the uncontrolled interactions of the qubits with other qubits and with stray electric and magnetic fields. Fortunately, those quantum error-correction methods seem to be working as well.

ALICE: Because quantum chemistry algorithms make use of concepts in quantum mechanics that are more advanced than others that we have used for describing QC, we will give you just an overview of how the algorithms work, focusing on those features that quantum computational chemistry algorithms have in common with the other algorithms we have studied.

The key chemical concept is the notion of an orbital, which is a quantum state that describes the electron cloud around a nucleus. In doing these calculations, chemists identify possible orbitals for the electrons. But in the actual molecules, only some of the orbitals will have an electron in the described state; we say those orbitals are *occupied*. If there is no electron there, the orbital is *unoccupied*. In this picture, there are two possible states for the orbital—occupied and unoccupied—so it is not surprising that we can use qubit states to keep track of the orbitals.

Unfortunately, it is hard to calculate the exact orbitals except for the simplest molecules. In practice, we calculate approximate orbitals using many theoretical tools that have been developed over the past several decades. We then assume that the quantum state for the molecule is given by a superposition of these approximate states. Denoting the approximate orbitals by $|\phi_p\rangle$ where p is an index to keep track of the orbitals, we write the state for the electrons in the molecule as

$$|\psi\rangle = a_1 |\phi_1\rangle + a_2 |\phi_2\rangle + \ldots \tag{12.28}$$

As usual, the as are the amplitudes associated with each of the orbitals.

The next step is to calculate the average value of the energy associated with the state $|\psi\rangle$. Bob mentioned the attraction and repulsion of the protons and electrons, but we also need to take the kinetic energy of the electron into account. There are other energy contributions from the magnetism associated with the electron's spin and some similar effects for the nuclei. But for our purposes, we may neglect them because those energies are much smaller than the electrical ones. In the quantum description of molecules, the energy, expressed as an operator (matrix), is called the Hamiltonian.

CARDY: Is that like Alexander Hamilton?

ALICE: A different Hamilton, but equally deserving of a Broadway musical. The quantum energy operator is named after William Rowan Hamilton, an Irish mathematician and physicist, who developed, among many other important mathematical tools, theoretical methods focusing on energy. He lived from 1805 to 1865. Alexander died in 1804. You should listen to the YouTube video listed under Further Reading at the end of the chapter.

CARDY: All that is interesting, but weren't we talking about quantum computational chemistry?

BOB: Yes . . . back to work. The crucial point is that the calculated energy depends on the structure of the molecule through the distances between the nuclei and through the amplitudes in Eq. (12.29). In most approaches to computational chemistry, we vary the amplitudes until we find the lowest value of the energy for that structure and then vary the structure to get the overall minimum in energy. That lowest energy, the "ground state" energy, is the signature of the most stable state of the molecule.

CARDY: Why do they call it the "ground state"?

BOB: The name came from the notion you may have heard about in your high school physics course: When an object (like a ball) interacts with the Earth by gravity, there is an energy associated with the position of the ball relative to the center of the Earth. In everyday life, the closest the ball can get to the center of the Earth is when the ball is on the ground, on the surface of the Earth. That is the situation in which that gravitational energy for the Earth–ball system is smallest. Hence, the name "ground state" for the lowest energy state of any quantum system.

BOB: As I mentioned before, we associate a qubit state with each of the orbitals and then use a quantum circuit, as illustrated in Figures 12.19 and 12.20, to adjust the amplitudes. The gates labeled with Θ_n are sometimes called rotation gates because they contain, among other operations, rotation operators, which we will discuss in Chapter 15. The rotation gates are different from the gates we have seen before because they are adjustable by changing the parameters Θ. The circuit diagram is simplified and does not show the connections that adjust the parameters.

We arrange the measurement devices so that when the qubits interact with the measurement devices, we get a 1 if the orbital is occupied and a 0 for an unoccupied orbital. From

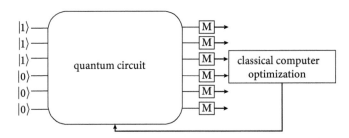

Fig. 12.19 Schematic diagram of the system used for quantum computational chemistry for a six-qubit system.

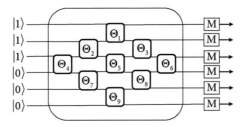

Fig. 12.20 A schematic diagram of the quantum circuit for quantum computational chemistry for a six-qubit system. The boxes with the symbols Θ_n are called rotation gates and the symbols indicate parameters that are used to modify the qubit states using the two-qubit gates. In this diagram, six states are used to describe six molecular orbitals.

that information, we can use a classical computer to calculate the energy and to make decisions about how to vary the quantum circuit parameters shown in Figure 12.20 to lower the energy. When we get to the minimum energy, small variations in the parameters will not lead to significant changes in energy; so, the method is called the variational quantum method. In quantum jargon, we say that we are searching for the energy eigenstates of the system. As we noted before, "eigen" is the German word for "characteristic." So, the implementation of quantum computers uses what is called the variational quantum eigensolver (VQE). To make all this work efficiently, we need to start with good guesses about the approximate orbitals. In practice, classical computers can provide those approximate orbitals to be used in the initial superposition state.

The quantum circuit shown in Figure 12.20 has six qubits and was adequate to predict the energies of the two isomers of diazene H_2N_2, that is, two hydrogen atoms and two nitrogen atoms.

CARDY: I just checked a periodic table of the elements online and found there are seven electrons for each nitrogen atom and one for each hydrogen atom, making for a total of 16 electrons in diazene. How can you make this work with just six qubits?

ALICE: Great question. The "trick" is to recognize that most of the nitrogen atoms' electrons stay close to the nitrogen nucleus. Only a few join the two electrons from the hydrogen in rearranging themselves to make the chemical bonds; so, we only need to use those in our calculation. The other trick is to recognize that for a molecule like diazene, the spin-up electrons are distributed almost exactly like the spin-down electrons; so, we cut the number of required qubits in half. The steps involved in the molecular orbital calculation are:

1. Specify the molecule: structure, charge, etc.

2. Generate approximate orbitals to set an initial state.

3. Generate a quantum circuit that represents the initial molecular orbital as a superposition of the approximate orbitals with adjustable parameters.

4. Measure the output states from the quantum circuit and feed the results to a classical computer to calculate the energy and to find the changes to the parameters needed to go to a lower energy.

5. Continue the process until you reach a minimum in the energy for that structure.

6. Repeat with changes in structures (distances between the nuclei and angles between the chemical bonds).

In step 4, a classical calculation uses the information about how that energy changed when the rotation matrices parameters were varied. That allows us to estimate the parameter changes that will lower the energy. Those new parameters are then used in the quantum algorithm again. The process continues until the minimum in the energy for that configuration of the molecule is found. Then the structure is changed (e.g. in a diatomic molecule, the distance between the atoms is changed) to map out how that energy depends on the internuclear distance. The distance that gives the lowest of the energies is then the equilibrium internuclear distance for that molecule. For more molecules with three or more atoms, you also have angles between the bond directions; so you need to explore those as well to find the overall most stable structure.

BOB: Let's summarize what the quantum computational algorithm has in common with other algorithms:

1. Start with an initial state of the multi-qubit system.
2. Act on that state with various "function boxes" and gates designed to implement the algorithm.
3. Subject the resulting state to a measurement device.
4. Based on the outcome from the measurement, make adjustments that will drive the multi-qubit state closer to the desired target.

All of this sounds great—and it is—but as the number of two-qubit gates increases, the systematic noise in the QC increases. The number of "layers" of gates as shown in Figure 12.20 dictates what is called the "depth" of a quantum circuit. As the depth increases, the level of "noise" in the quantum circuit increases. Here "noise" means that the amplitudes of the qubits change in unpredictable ways due to unwanted interactions with the environment, usually through electric and magnetic fields. Therefore, error-mitigation methods of the types described in the previous section must be deployed. The good news from recent work is that the error-correction methods seem to work well for the quantum circuits used for quantum computational chemistry.

A recent paper (Rubin et al., 2020) shows how these methods can be applied to chains of up to 12 hydrogen atoms and to the diazene molecule H_2N_2, also written as $(NH)_2$. The diazene molecule is particularly interesting for quantum chemistry because it has two shapes (isomers) with just slightly different energies. One isomer is called trans-diazene, which has the two hydrogen atoms on opposite sides of the N–N bond, while cis-diazene has the two hydrogen atoms on the same side. The quantum computational chemistry approach can account for that small energy difference. Although diazene is a relatively simple molecule, the reported results, including the effectiveness of the quantum error-correction methods, are a cause for optimism that enhanced quantum computers, with more qubits, will be able to tackle much more complicated molecules.

If you would like to try quantum computational chemistry yourself, the IBM Qiskit website has a VQE that you can program to calculate the energy of simple diatomic molecules (molecules with two atoms) as a function of the distance between the nuclei of the two atoms.

CHAPTER SUMMARY

- Quantum algorithms are being developed in many fields, outperforming and working alongside classical computing methods.

- Searching for strings of numbers or text is commonplace. We do it all the time in our personal lives, searching for specific words on webpages or values on spreadsheets. The Grover algorithm implements a search among basis states by manipulating a superposition state so that the desired state coefficient increases while the others decrease. Because the probability of measurement increases with the square of the coefficient, this process quickly increases the probability of finding the desired state. Geometrically, this process reflects a pointer state vector across other vectors in state space. This state space is defined with the desired state as one of the axes with the other axis being the sum of all the other states' vectors. First, the uniform superposition vector is reflected across the non-target axis and is then reflected across the initial superposition. This brings the state vector closer to the target state axis, which means that the target state coefficient has increased in the superposition. This process is then repeated until the target state coefficient is close to 1, implying that the probability of getting the target state upon observation is close to 1.

- Correcting errors in transmitted information is another important computational task. Commonly, these errors are found by attaching additional bits to the message and then comparing all the bits. For example, instead of sending a single digit (1), three digits (111), are sent. The three digits are then compared to check if they are different from one another. If they are, a bit-flip error has occurred. Qubit error detection presents a more complex problem. Merely comparing the bits does not work because measurement forces the qubit into the basis state of the measurement device, causing a loss of information about the original state of the qubit. This difficulty is overcome by using ancillary register states as the control states for a Toffoli gate, which acts on the message register states.

- Finally, quantum algorithms have the power to model complex molecules to predict their shapes based on calculating changes in the ground state energy of different configurations. The molecule will assume a shape that minimizes this energy. This is a computationally expensive process to do classically because the shape is affected by the interactions of each individual atomic particle with all the other particles in the molecule. The quantum algorithm uses, of course, superposition, to more efficiently evaluate the changes in the orbital configuration based on changing input parameters of the model. A classical computer calculates the energy change for this model compared to previous models. This iterative method uses the results of the previous energy changes to determine the next configuration to test.

FURTHER READING

L. Grover, "Quantum Mechanics helps in searching for a needle in a haystack." https://arxiv.org/abs/quant-ph/9706033. This paper is quite readable given what you have learned in this chapter.

B. Kain, "Searching a quantum database with Grover's search algorithm." *American Journal of Physics* 89, 618–626 (2021). The paper includes an IBM Qiskit implementation of Grover's algorithm.

N. D. Mermin, *Quantum Computer Science*, Chapter 4 shows how to build the Grover algorithm function circuits from simple qubit gates.

The Wikipedia article on quantum error correction provides a more detailed overview of error-correction methods (https://en.wikipedia.org/wiki/Quantum_error_correction).

A musical song about William Rowan Hamilton, done in the style of the recent musical *Hamilton*, is available at https://www.youtube.com/watch?v=SZXHoWwBcDc.

Rubin et al., "Hartree-Fock on a superconducting qubit quantum computer." *Science* **369** 1084–1089 (2020). This describes quantum chemistry calcuations for chains of hydrogen atoms and for diazene molecules.

The IBM Qiskit website url is https://qiskit.org/. The site has many tutorial materials and allows you to write a quantum computer program to run on either a quantum computer simulation or on one of the actual IBM quantum computers.

Cambridge Quantum Computing provides quantum software, a quantum development platform (t|ket⟩™), enterprise applications in the area of quantum chemistry (EUMEN), quantum machine learning (QML), and quantum augmented cybersecurity (IronBridge™). https://cambridgequantum.com/about/.

Mathematics is the queen of the sciences and number theory is the queen of mathematics.

Attributed to Carl Friedrich Gauss

13.1 RSA Encryption and Factoring

ALICE: The Shor factoring algorithm is perhaps the most famous of the "classic" quantum computing algorithms. The problem Shor wanted to solve is how to factor large numbers. This is an interesting problem because factoring large numbers allows you to break many encryption codes. Bob and I want to alert you, Cardy, (and our dedicated readers) that this chapter is rather challenging. It uses results from number theory and some more advanced math called the Fourier transform. We have worked hard to explain all the steps without the advanced math, but there are many steps. We provide examples and graphs to help build intuition about the results, but we know from our own experience it takes time to get your head wrapped around these ideas.

Once we get to the quantum part of what Shor did, we will see familiar tools from other quantum algorithms: superposition states, quantum function gates, and measurement devices. Nevertheless, expect to spend more time on this chapter than you did on the others. We hope that the results of your efforts will be satisfying in providing more insight into how encryption works and how quantum algorithms are engineered to take advantage of the nature of quantum states. But given the length of the arguments, I suggest that you feel free to try out the algorithms without going through all the details of the justifications for the steps. Once you see what the algorithm does, you can come back later to look at the detailed explanations.

BOB: This chapter does two things: First, we introduce the RSA encryption algorithm, which is a classical cryptographic procedure; breaking the RSA encryption involves factoring large numbers, which is what the Shor algorithm was designed to do. We also explain the Shor algorithm. Both the RSA encryption algorithm and the Shor algorithm require a modest amount of number theory. In the second part of the chapter, we explain the number theory results because many algorithms, both classical and quantum, make use of them. In fact, the number theory used here shows how we reduce the factoring problem to the problem of finding the period of a periodic function. The Shor algorithm provides a slick way of using superposition states to find the period.

Why should we worry about factoring numbers? Isn't that something you do in middle school math and the SAT exam and never use again? It turns out that factoring is not just an intellectual or mathematical curiosity. The difficulty of finding the factors of large numbers

Quantum Computing: From Alice to Bob. Alice Flarend and Bob Hilborn, Oxford University Press.
© Alice Flarend and Robert C. Hilborn (2022). DOI: 10.1093/oso/9780192857972.003.0013

lies behind some widely used classical encryption procedures, designed to make it difficult for anyone to intercept and decode digital messages. That is of obvious interest to financial, governmental, and medical institutions. If an efficient factoring scheme could be developed, those encryption procedures could be broken into.

CARDY: All I remember about factors are simple things like $3 \times 5 = 15$; so, 3 and 5 are factors of 15. How does that lead to encryption?

ALICE: The basic idea behind encryption is that you use an encryption "key" to code letters and numbers just the way the dots and dashes in Morse code stand for letters of the alphabet or, at a more sophisticated level, what the Germans did during the Second World War with the famous Enigma code implemented by the Enigma machine, which changed the encryption key daily. If you can find the encryption key, you can decode the encrypted messages. Of course, for digital computers, the key will be a sequence of 0s and 1s. For example, you might have a key which contains $8 \text{ bits} \times 26 \text{ letters} = 208$ bits in which the successive 8-bit sequences each represent a letter in the English alphabet. We saw an example of an encryption key in Chapter 8's treatment of the BB84 encryption method.

The RSA encryption system—named after Riverst, Shamir, and Adleman, who developed the algorithm in 1977—takes a different tack. The RSA algorithm, an example of public key distribution, makes the encryption numbers public but keeps other information needed to decode the message secret. RSA is an asymmetric encryption algorithm because you use one number to encrypt a message and a different number to decrypt (decode) the encrypted message. One of the encryption numbers is a large integer that is a product of two large prime numbers. To decode the encrypted message, you need to know those prime numbers. In other words, you need to factor that large number. It turns out that finding those prime number factors is a difficult computational task and essentially impossible because of time constraints for a classical computer if the number has several hundred digits. The Shor algorithm uses quantum methods to do that factoring. So far, quantum computers have been able to factor only small numbers (15 and 21) but there is no fundamental barrier to scaling the process to large numbers. We just need quantum computers with more qubits.

CARDY: Remind me what a prime number is.

ALICE: A prime number is one that is evenly divisible (no remainder) only by 1 and by itself. The smallest prime numbers are 2, 3, 5, 7, 11, 13, and so on. Note that besides 2, all prime numbers are odd numbers. The fundamental theorem of arithmetic states that all positive integers (also called natural numbers) except for the number 1 can be expressed as products of prime numbers, and furthermore that product (except for the order of the prime numbers within the product) is unique. If the product contains only 1 and the number itself, then that number is a prime number. For example, the number 91 can be written as $7 \times 13 = 91$. Since 7 and 13 are both prime numbers, they are the only prime numbers whose product equals 91.

Before stating the RSA encryption algorithm, which itself is somewhat complicated, we will explore a few aspects of number theory used in the algorithm. As we mentioned before, number theory is the mathematics of the integers 0, 1, 2, 3 .... For most of our work, we will need only positive integers. It sounds simple, but there are several results which are far from obvious. But, with some persistence and attention to detail, we can work our way through the steps.

13.2 Number Theory Warm-Up

ALICE: An important number theory tool for RSA encryption, and many other classical algorithms, is the mod function, which I like to call the remainder function. Most mathematicians call it the modulus function or the modulo operation. Let's explore it before looking at the RSA algorithm.

We start with some simple arithmetic. Remember we are dealing only with positive integers. For any two positive integers x and y (assume $x < y$), we can always find an integer Q that is the largest integer such that $Qx < y$. Then we introduce the remainder $R = y - Qx$ when you divide y by Qx. In other words, we have $y = Qx + R$, a linear relationship between x and y. In the language of everyday arithmetic, Q is the quotient, x is the divisor, and y is the dividend—the number to be divided.

Try It 13.1

For $x = 5$ and $y = 22$, show that $Q = 4$ and $R = 2$. Check that $y = Q\,x + R$.

The mod (or modulus) function gives you the remainder part without needing to specify the quotient. That will be handy in dealing with large numbers. The mod function for x and y is written as

$$R = y \ \mathrm{mod} \ x, \tag{13.1}$$

where R is the remainder after dividing y by $Q\,x$. Mathematicians like to say "R equals y, modulo x."

BOB: Actually, we use the mod function every day when we talk about clock time: Three hours after ten o'clock (13 o'clock) is 1 o'clock because we use mod 12 for clock time. 1 is the remainder: $1 = 13 \ \mathrm{mod} \ 12$.

CARDY: Cool! I can't wait to tell my friends that they are mod!

BOB: Be careful! You don't want your friends to feel like leftovers from a division problem.

ALICE: For a fixed value of x, the remainders go through a cycle from 0 (for numbers evenly divisible by x) to $x - 1$ as y increases. That cyclic nature will play a role in what follows.

Fortunately, most computer languages and spreadsheets have a mod function to do the calculation for you. Often, they take the form $R = \mathrm{MOD}[y, x]$, where y is the dividend, x is the divisor, and R the remainder. But if you are stuck on a desert island and need to calculate by hand, the arithmetic is straightforward: Simply divide y by x. Then R is the integer remainder. Of course, for large numbers the division can be rather tedious.

Try It 13.2

For $x = 11$, calculate $R = y \ \mathrm{mod} \ x$ for $y = 0$ through 24. Is the result cyclic?

ALICE: We will need one more variation on the mod operation. If you think about it for a minute, you will realize that for a given divisor x, there are infinitely many dividends (y) that give the same remainder, each with its own value of Q. We will see that the RSA encryption algorithm makes use of that fact.

Try It 13.3

Show that $3 = 15 \bmod 12$ and that $3 = 27 \bmod 12$; so, 15 and 27 have the same remainder mod 12. Note that $15+12 = 27$.

ALICE: If $y \bmod x = u \bmod x$, that is, y and u have the same remainder when divided by x, then the difference between y and u is evenly divisible by x. (We normally use $|u - y|$ for the difference to restrict ourselves to positive integers.)

Let's do a numerical example. 17 and 32 have the same remainder for the divisor 5. In fact, $2 = 17 \bmod 5$ because $17/(3 \times 5)$ has the remainder 2 and so does $32/(6 \times 5)$. So, we write $32 \bmod 5 = 17 \bmod 5$ (the remainders are equal), and we see that the difference $32 - 17 = 15$ is evenly divisible by 5.

Try It 13.4

Find the following remainders: $23 \bmod 2$, $121 \bmod 2$, $15 \bmod 3$, $16 \bmod 3$, $19 \bmod 5$, and $24 \bmod 5$. Are any of the remainders the same for the given divisors? If so, check if the difference between the dividends is evenly divisible by the divisor.

BOB: Before we move on, it might be helpful to look at a graph that shows some of the features we have been talking about. Figure 13.1 displays a plot of the remainder R as a function of the dividend y with the divisor fixed at $x = 7$. Note that the plot is periodic in y. The values of y for which R is the same are said to be congruent (mod 7).

Try It 13.5

The difference between successive numbers that are congruent (mod 7) is called the "repetition period" (or just "period") of that congruence. What is the period of congruence for the results shown in Figure 13.1?

CARDY: What does periodic mean here? When I think of periodic, I imagine something that repeats itself in time: periodic appointments and the like.

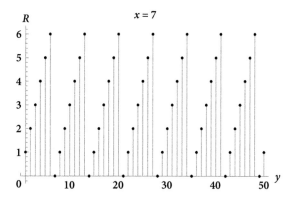

Fig. 13.1 A plot of $R = y \bmod x$ with $x = 7$.

ALICE: Here periodic is similar but we don't have to interpret the numbers as times or anything specific. In more formal terms, a function $f(x)$ is periodic with period r if $f(x) = f(x + r)$ for all x.

Now let's look at functions with exponents. In particular, let's use $y = a^b$ and consider $a^b \bmod N$. We will find that there are integers r such that

$$a^b \bmod N = a^{b+r} \bmod N. \tag{13.2}$$

The value of r that works in Eq. (13.2) depends on a and N. Of course, if r works, so does $2r$, $3r$, and so on. We say that $a^b \bmod N$ is a periodic function of the exponent b with period r. Eq. (13.2) is an example of "modular exponentiation," which plays a major role in RSA encryption and the Shor factoring algorithm. In section 13.8, we will show you how to solve that kind of equation.

CARDY: I thought that R was a remainder. So, what is r?

ALICE: To match the notation commonly used in QIS and QC, we will use lower case r to indicate the repetition period of a periodic function and keep upper-case R for remainders. So, here r means "repetition period."

Figure 13.2 displays a plot of $a^b \bmod 77$ with $a = 5$, as a function of the exponent b. Note that the results are periodic with respect to b, though that periodicity might not be obvious. To convince yourself that there is some periodicity there, look at the sequence of filled circles with b between 8 and 20 and then between 38 and 50.

Try It 13.6

Describe the results displayed in Figure 13.2. Are the results periodic with respect to b? If so, what is the period r?

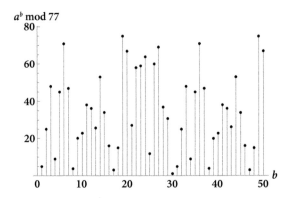

Fig. 13.2 A plot of a^b mod 77 with $a = 5$. The results are periodic as a function of the exponent b. See the main text for a hint about the periodicity.

13.3 The RSA Algorithm

ALICE: With that bit of number theory under our belts, we are ready to tackle the RSA algorithm. Here is how it goes. Bob wants to send me an encrypted message. The RSA algorithm has the following steps.

1. I create a large number N by multiplying together two fairly large prime numbers p and q; so $N = pq$.

2. I also create an encryption key K with $K < N$ and check that K has no common factors (other than 1) with $(p - 1)(q - 1)$. I send the integers K and N to Bob. I may use an unencrypted channel to do so.

3. Let's assume that Bob expresses his message as a number m. Using the values of K and N I have sent him, Bob calculates another number c such that $c = m^K \bmod N$ and sends the number c to me.

4. In the meantime, I have calculated an integer d such that

$$1 = Kd \bmod (p - 1)(q - 1), \tag{13.3}$$

which means that $(p-1)(q-1)$ divides $Kd-1$ evenly, that is, with no remainder. I need the two prime numbers p and q to calculate d.

5. To decode the message from Bob, I solve $m = c^d \bmod N$.

CARDY: I have lots of questions. First, why do we limit ourselves to just two prime factors? More would seem to be better.

ALICE: We use just two prime factors because two are sufficient for our purposes. Also, remember the fundamental theorem of arithmetic: Once we have chosen p and q, we know that they are the only two prime factors of N.

CARDY: My next question is why is the product $(p - 1)(q - 1)$ used in steps 2 and 4? And how do you check that there are no common factors when you choose K?

ALICE: The answer to the first question is a bit complicated. We will show you the details in section 13.8. But now, let's read Eq. (13.3) to make sure we understand what it is telling us. It says that if we divide Kd by $(p-1)(q-1)$ we get the remainder 1, which, as we saw before is equivalent to $Kd = Q \times (p-1)(q-1) + 1$ for some integer Q. In other words, $Kd - 1$ is evenly divisible by the product $(p-1)(q-1)$.

For your second question, finding common factors is straightforwardly done with what is called Euclid's method or Euclid's algorithm. Many computer programming languages and spreadsheets have functions that allow you to find the "greatest common divisor," the largest integer that divides both factors evenly. If that greatest common divisor is 1, then the two numbers have no common factors. We will explain how to find the greatest common divisor in section 13.9.

Let's get back to the algorithm. In step 3 Bob sends me the number c. Since I am well versed in the RSA method, I know that I can find the message string m from c by using

$$m = c^d \bmod N. \tag{13.4}$$

That is, the message (viewed as a binary number) is the remainder when c^d is divided by N. So, we are left with another modular exponentiation problem.

CARDY: Oh, my! A lot went by there. Is this going to be important for message security?

ALICE: I certainly sympathize with you, Cardy. That is how I felt the first time I saw this algorithm; I was totally baffled. There is a lot of number theory behind these calculations, and it takes some patience and concentration to work through the arguments. The main point is that unless someone knows the secret number d, which I calculated from p and q, the secret prime factors, there is no way to decipher Bob's message m from the number c he sends to me.

Let's try this with a numerical example. Suppose I choose two prime numbers $p = 59$ and $q = 47$, for which $N = pq = 2773$. Next, I pick a number $K < N$ randomly and check to see if K divides $(p-1)(q-1) = 2668$ evenly or not. If not, I am ready to go to the next step. If it does divide that product evenly, I pick another K and try again. Note that if K divides the product evenly, then $K + 1$ will not. Let's suppose I pick $K = 17$, which does not divide 2668 evenly; so, I am good to go on. Finally, I need to compute an integer d such that $1 = Kd \bmod 2668$, from which I find $d = 157$.

CARDY: I know I sound like a broken record, but how do you solve for d?

ALICE: There are several ways to do this. The easiest is to use Euclid's algorithm (section 13.9). Also, many classical computer languages implement the process in a modular inverse (or similarly named) function.

BOB: Let's continue with the example. Next, I carry out the encryption calculation for my message m. Let's suppose that my message (in decimal digits) is 73. So, I calculate the integer c given by

$$c = m^K \bmod N = 73^{17} \bmod 2773 = 928. \tag{13.5}$$

Then I send $c = 928$ to Alice over an unencrypted channel. Note that I don't know (or need to know) the numbers p, q, and d, which Alice has and keeps secret.

Once Alice receives $c = 928$, she uses $d = 157$ to solve for m from $m = c^d \mod N$ and finds $a = 928^{157} \mod 2773 = 73$. Alice has decoded my message! Note that if an eavesdropper doesn't know d (which Alice has kept secret) or the factors p and q, there is no way to find m.

CARDY: The process seems to work but when I tried to evaluate some of the exponentiations, my spreadsheet and my hand calculator both gave up because the results were too large.

BOB: That's where modular exponentiation comes in. As Alice said, we will go through the details of how to handle modular exponentiation in section 13.8.

13.4   The Shor Factoring Algorithm

BOB: We are now ready to look at the Shor factoring algorithm and how it allows us to break the RSA encryption. The key feature is finding the two prime factors of N. That would seem to be a relatively straightforward thing to do. We just go through the list of prime numbers less than N and find those that divide N exactly. Once we have found two factors that are prime numbers, we are done; we could use that information to break the RSA encryption. But, if you think about how many numbers you need to check for large values of N, you should be able to convince yourself that the task is hopeless even with today's supercomputers to help you out. For example, if N has 300 digits, we have about $\sqrt{10^{300}} = 10^{150}$ prime numbers to try (this is just a rough estimate based on the fact that one of the factors of N will be less than $\sqrt{N}$). Let's say that your computer can check one number in 10^{-8} sec. It would then take about 10^{142} seconds to complete the factoring. Of course, you might get lucky and find a factor early on, but on average, it's going to take a long time.

CARDY: That does sound like a long time! But I don't have a good comparison.

BOB: In round numbers, the age of the universe is 10 billion years. There are about 10^7 seconds in one year; so, the age of the universe in seconds is about 10^{10} years times 10^7 seconds per year $= 10^{17}$ seconds, way smaller than the 10^{142} seconds needed to complete the factoring of the 300-digit number.

CARDY: It certainly looks like we need a better way of factoring large numbers.

BOB: We certainly do. As we have seen several times before, the strategy is to reduce one problem to another, hopefully simpler and less time-consuming, problem. The basic idea behind the Shor algorithm is to use some number theory ideas that we met up with in the RSA encryption procedures to reduce the factoring problem to one of finding the period of a periodic function. The Shor algorithm also provides a quantum method to find that period for large numbers much more quickly than can be done on a classical computer. Once you have the period, it takes just a few relatively simple calculations to find the prime factors.

Let's go through the steps in the Shor algorithm and then do a few examples to see how they work. We will then tackle the quantum method that speeds up the period-finding part. In sections 13.8 and 13.9, we provide details about why the steps work.

Just as a reminder, unless otherwise specified, all the numbers we will deal with are positive integers. Here are the steps in the Shor factoring algorithm to find two prime factors of some number N. Note that they are close to the reverse of the RSA algorithm's steps.

1. Choose a random integer $a < N$. Check to see if a divides N evenly (no remainder). If it does, we have found a factor. That is unlikely; so, we proceed with the following steps.

2. Find the smallest integer r such that $1 = a^r \mod N$. In other words, we are looking for a number r such that $a^r - 1$ is evenly divisible by N.

3. If r turns out to be an even number, proceed with step 4. Otherwise, return to step 1 and choose another random number a.

4. Calculate $s = a^{r/2} \mod N$. Note that we need r to be an even number to make this work.

5. Calculate the prime factors p and q from

$$p = \gcd(s - 1, N)$$
$$q = \gcd(s + 1, N), \tag{13.6}$$

where $\gcd(s - 1, N)$ means greatest common divisor—the largest number that divides both $s - 1$ and N exactly.

6. Check that the results satisfy $N = pq$.

Try It 13.7

Construct a flow chart for the Shor factoring algorithm. Then use the flow chart to work through the examples that follow.

Let's try out the algorithm with two examples. Here is one you could do in your head: $N = 15$. The two prime factors are 3 and 5; so, it will be easy to check that the algorithm gives the right answer. But let's follow the steps anyway.

1. Randomly choose $a = 2$. Obviously, a does not divide 15 evenly.

2. Find the smallest r such that $1 = 2^r \mod 15$. We recognize that the smallest r that works is $r = 4$ because $2^4 = 16$ and $16/15 = 1$, remainder 1. We also check that $a^r - 1$ is evenly divisible by 15, which obviously works in this case.

3. r is even; so, we proceed to step 4.

4. Calculate $s = a^{r/2} \mod N = 2^2 \mod 15 = 4$.

5. Find $p = \gcd(s - 1, 15) = \gcd(3, 15) = 3$ and $q = \gcd(s + 1, 15) = \gcd(5, 15) = 5$.

6. Check that p and q are factors of 15 by calculating $N = pq$.
   So, we have successfully found the two prime factors of 15, namely, 3 and 5.

> **Try It 13.8**
>
> Use $N = 15$ and $a = 4$. Do the four steps work? If not, what is the problem? How can you get around that problem?

BOB: Now, let's try a more difficult example: $N = 77$.

CARDY: Hey, that isn't so hard. I think that 7 and 11 work.

BOB: You are right, Cardy. But some of the intervening math using our algorithm is a bit more challenging. You will need a calculator or spreadsheet to calculate some of the results.

1. Randomly choose $a = 4$, which does not divide 77 evenly. So, it is not a factor.

2. We find that $1 = 4^{15} \bmod 77$; so, the repetition period $r = 15$.

3. But r is not even; so, we need to try a different value of a. With $a = 5$, we find $1 = 5^{30} \bmod 77$. Since the repetition period $r=30$ is even, we are ready to go on.

4. We then find $s = a^{r/2} \bmod 77 = 34$.

5. Find $p = \gcd(s - 1, 77) = \gcd(33, 77) = 11$ and $q = \gcd(s + 1, 77) = \gcd(35, 77) = 7$.

6. Since $7 \times 11 = 77$, the prime factors of 77 are 11 and 7, just as Cardy said.

CARDY: Cool! And we saw that the period of 30 in the data shown in Figure 13.2. Well, I see that the algorithm works, but I don't' see why it works. Why does finding r that satisfies $1 = a^r \bmod N$ get us the factors?

13.5 Connecting Periodic Functions and Factoring

ALICE: To answer Cardy's question, we will look at the general connection between finding the period of a function and factoring a number. At least for me, if someone had told me there is such a connection, I would have been highly skeptical. The concepts seem like they have no relationship. But number theory has many surprises of that sort. If you are willing to accept that connection, feel free to skip ahead to section 13.6.

To see how the connection comes about, we again assume $N = pq$ and follow the algorithm's steps to find p and q. The first step seems totally unmotivated: We choose an integer $a < N$ and check to see that a is not a factor of N; that is, N is not evenly divisible by a. If it is divisible by a, we have found a factor and we are essentially finished. So, we assume that a is not a factor of N.

CARDY: Does a have to be a prime number?

ALICE: Good question. It is not necessary for a to be a prime number. It just can't divide N evenly. Mathematicians like to say that a and N must be *coprime*.

Once we have chosen a, the next step in connecting the periodic function to factoring is to find the smallest even integer r for which

$$a^b \bmod N = a^{b+r} \bmod N. \tag{13.7}$$

This is where the periodicity comes in. Note that in Eq. (13.7), you could replace r with $2r$, $3r, \ldots$. They would all work. That is why we insist on finding the *smallest* integer r that works. We also require r to be positive; so, we can eliminate $r = 0$. (For $r = 0$, Eq. (13.7) is valid, but not very informative.) The function $f(b) = a^b \bmod N = f(b + r) = f(b + 2r) \ldots$ is periodic with repetition period r. Of course, the integer r depends on both a and N, but not on b. We saw that periodicity in Figure 13.2. Another example of the periodicity of a modular exponential function is shown in Figure 13.3.

Try It 13.9

Find the period of the results displayed in Figure 13.3. Hint: Look at the plot where b is in the range near [4,8] and search for the range of b where that pattern appears again.

ALICE: Figure 13.3 shows why finding the period of such a function is not easy. If you try to find the period by picking a result for a particular value of b and searching for the next appearance of that result, you have many numbers to sort through. And that is for a relatively small value of N, 121 in this case. Imagine what the graph would look like for an N with 300 digits.

Now let's get back to the algorithm. If we choose $b = 0$ in Eq. (13.7), we get

$$1 = a^r \bmod N, \tag{13.8}$$

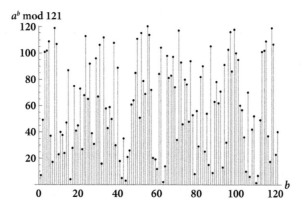

Fig. 13.3 A plot of $a^b \bmod 121$ with $a = 7$ plotted as a function of the exponent b.

which means that there is some integer k such that

$$a^r - 1 = kN. \tag{13.9}$$

Since we are requiring that r be even, we may write Eq. (13.9) as

$$a^r - 1 = \left(a^{r/2} - 1\right)\left(a^{r/2} + 1\right) = kN. \tag{13.10}$$

CARDY: Does that mean that $\left(a^{r/2} - 1\right)$ and $\left(a^{r/2} + 1\right)$ are factors of kN? If they are, it would seem that they are factors of N also. Is that right?

BOB: The first part of what you said is right. But the second part is not right unless $k = 1$. Let me give you a simple example with $N = 5$. We know that 5 and 7 are factors of 35. Now let $kN = 35$ with $k = 7$ and $N = 5$. We see that 7 is not a factor of $N = 5$ even though it is a factor of $35 = kN = 7 \times 5$.

CARDY: Got it!

ALICE: We now invoke some number theory arguments. Since r is the smallest even integer for which $a^r - 1$ is evenly divided by N, we conclude that the integers $\left(a^{r/2} - 1\right)$ and $\left(a^{r/2} + 1\right)$ are *not* evenly divided by N. That means that N must have a factor (excluding 1) in common with $\left(a^{r/2} - 1\right)$ and $\left(a^{r/2} + 1\right)$.

CARDY: Is that last statement obviously true?

ALICE: The proof of the statement is in section 13.8. If we accept that statement for the moment, we can find those common factors by finding the greatest common divisor of $\left(a^{r/2} - 1\right)$ and N and the greatest common divisor of $\left(a^{r/2} + 1\right)$ and N.

CARDY: How do I find a greatest common divisor? Couldn't that take a long time?

ALICE: As I mentioned before, there is a simple algorithm called Euclid's algorithm to do that. The details of the algorithm are in section 13.9. Also, many spreadsheets and computer programming packages include greatest common divisor as a function, often denoted by $\gcd[x, y]$ or $GCD[x, y]$.

CARDY: Why do we use $s = a^{r/2} \bmod N$?

ALICE: For the Shor algorithm, the number $a^{r/2}$ is often very large. To work with smaller numbers, we can use the $s = a^{r/2} \bmod N$ from the Shor step 4 in the gcd functions. Why does that work? First, note that $\gcd(a^{r/2} - 1, N)$ will always be less than N. That suggests using a remainder (also less than N) to replace $a^{r/2}$; s is the appropriate remainder.

We finish off the connection between factoring and period-finding by noting that if we use

$$p = \gcd(s - 1, N) \text{ and } q = \gcd(s + 1, N),$$

we have found two factors p and q of N. We can check to see if they are indeed factors of N simply by multiplying them, and we can also check that they are prime numbers by using the greatest common divisor algorithm.

Let's work through a period-finding and factoring example. Suppose we want to find the prime factors of 91. First, we use $R = a^b \bmod N$ with $a = 8$ and to find the period r. Table 13.1 gives the numerical results with b in the range [0,10].

Table 13.1 The exponent b and the reminder $R = 8^b$ mod 91.

b	0	1	2	3	4	5	6	7	8	9	10
$R = 8^b$ mod 91	1	8	65	57	1	8	65	57	1	8	64

ALICE: We see that the repetition period for 8^b mod 91 is $r = 4$. Next, we calculate $(a^{4/2} - 1) = 63$ and $(a^{4/2} + 1) = 65$. For the final step, we calculate gcd$[63, 91] = 7$ and gcd$[65, 91] = 13$. The factoring result is that $7 \times 13 = 91$.

13.6 Period Finding and Fourier Analysis

CARDY: Okay, I am willing to accept all of that, but where do quantum algorithms come in?

BOB: Excellent question. It is important to note that all the steps, except for finding the period r, can be carried out effectively with classical algorithms. So, how do we figure out the period using a quantum algorithm? For relatively small numbers N, we can use the equivalent of Figure 13.3, but it should be obvious that the results soon become complicated enough that it is hard to see the period by eye. In Figure 13.3 the periodicity is hard to spot because the period is long compared to the range of b plotted.

ALICE: A standard way to find periodicities in science, engineering, and technology is a mathematical tool called Fourier analysis. The basic idea is to compare the signal you want to analyze to standard periodic functions: usually sines and cosines. You adjust the repetition period of the sines and cosines and find those periodicities that give you the best matches with your data. There are many efficient implementations of Fourier analysis in almost every modern scientific computing package. What Shor realized is that there are quantum algorithms that in principle will do the Fourier analysis in a much shorter time compared to the time required by a classical computer, particularly when it comes to large numbers with numerically large periodicities.

CARDY: Is that Charles Fourier who promoted utopian socialism around 1800? I read about that in my sociology class.

ALICE: No, this Fourier was Jean-Baptiste Joseph Fourier, who lived about the same time, but worked for Napoleon, did lots of wonderful mathematics, wrote books about Egypt, and recognized what we now call the greenhouse effect—how certain gases in the atmosphere can lead to climate warming. Another surprising tidbit: The mathematics behind Fourier analysis was mainly developed by others: Clairaut, Lagrange, Euler, Gauss, and others. What Fourier recognized was that the methods would work for any function, an assertion now known as Fourier's theorem.

CARDY: The history of math is certainly a bit weird. I have another question. We used sines and cosines previously the way I learned about them in trig: properties of angles in right triangles and the relationships among adjacent sides, opposite sides, and the hypotenuse. My question is how did we get from the ratio of the sides of a triangle to something that is periodic?

ALICE: Good question, Cardy. In trig, we usually are worried about things like $\sin \theta$ and $\cos \theta$ where θ lies between $0°$ and $360°$ or, equivalently, between 0 and 2π radians. But it is perfectly fine to think of θ as a mathematical variable that can take on any numerical value. If you do that, you find that sine and cosine repeat whenever the variable goes from θ to $\theta + 2\pi$ if we use radian units. That periodic property is useful in both finding the period of other functions and then representing those functions as sums of sines and cosines. Those are the basic ideas behind Fourier analysis, which is often used both in classical signal processing and in quantum algorithms.

Try It 13.10

Use a calculator or spreadsheet to show that

$$\cos\left(45°\right) = \cos\left(45° + 360°\right) = \cos\left(45° + 720°\right)$$

if your device uses degrees, or if your device uses radians

$$\cos\left(0.2\right) = \cos\left(0.2 + 2\pi\right) = \cos\left(0.2 + 4\pi\right).$$

ALICE: Let's first look at the traditional Fourier analysis (sometimes called the Fourier transform) method and how it allows us to find the period of a "signal." We plot a pure cosine function and another more complex periodic function with a period different from the pure cosine (Figure 13.4).

You can see that because the cosine's period does not match the period of the other function, their plots do not align very well. If you multiplied the two curves together and averaged the results you would get a relatively small number because the positive parts of the product would tend to be cancelled by the negative parts of the product. However, if you adjusted the period of the pure cosine, you could find a value for which the two curves match as shown in Figure 13.5. Then the average product would be reasonably large.

Now let's apply that same idea to modular exponentiation. In Figure 13.6, we plot a^b mod 77 with b running from 1 to 200. We subtracted the average value from each of the terms; so, the function "oscillates" centered around 0 on the vertical axis, just the way cosines and sines do. We have also plotted a cosine function whose period is 30 (to match the period

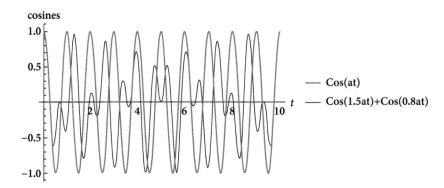

Fig. 13.4 A plot of a pure cosine function (gray curve) and another periodic function (dark curve).

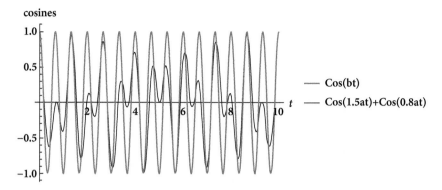

Fig. 13.5 The frequency of the pure cosine (gray curve) has been adjusted so there is a good match between the pure cosine and the other periodic function.

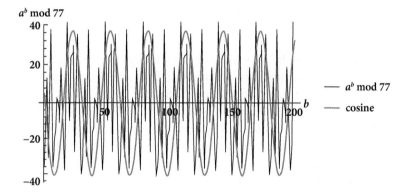

Fig. 13.6 A plot of $a^b \bmod 77$ with $a = 5$ and with b ranging from 1 to 200 along with a cosine function with a period of 30 (in units of b). The two periodicities match reasonably well.

for this case). We have moved the cosine function to the left a bit to match the "peaks and valleys" of the $a^b \bmod 77$ function.

If we calculate the product of the cosine times the $a^b \bmod 77$ "signal" and add up the results over the entire signal, we find the sum of the products is large if the cosine is a good match to the signal and a smaller number if it is not. When the sum of the products is large, we know the signal has a periodicity nearly the same as that of the cosine and that the peaks and valleys of the cosine are aligned with those of the signal. Why does this lead to a large number? If there is a good match, when the cosine is positive, we have a positive signal (on average) and when the cosine is negative, we have a negative signal (on average); so overall the product contributions are positive. If the cosine period is not a good match for the signal, there will be more or less equal amounts of a positive product and a negative product giving an overall contribution near zero. Figure 13.7 shows a plot of that situation.

ALICE: For the situations shown in Figures 13.6 and 13.7, the sum of the products is about 840 for the first case with a period of 30 (in units of b). In the second case the period of the

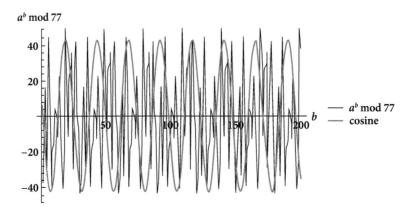

Fig. 13.7 A plot of a^b mod 77 with $a = 5$ and with b ranging from 1 to 200 and a cosine function with a period of 25 (in units of b). The two periodicities do not match very well.

cosine function is 25 and the sum of the products is about 3. So, we conclude that the signal has a period of about 30.

The full Fourier analysis method automatically runs through a range of periodicities to find the ones that best match the signals. Most often, both cosines and sines are used together. That method avoids having to do any left–right shifting to find the best match.

For the quantum Fourier algorithm, we introduce what is called the *discrete* Fourier transform. This method was devised to work with discrete (distinct, unconnected) sampled signals that arise in all sorts of digital signal processing, including audio and image processing used on your electronic devices. They are called "discrete" because the data are discrete (separated) samples of an independent variable, rather than continuous.

Let's focus on an audio signal that varies in time. The extension to spatial samples (pixels) for images is straightforward. In the discrete time case, a signal $f(t)$ is recorded at discrete times $t = \{t_0, t_1, \ldots, t_{N-1}\}$ for a total of N samples. If the samples are taken at equal time intervals Δt starting with $t_0 = 0$, we can write the sequence of times as $t = \{j\Delta t\}$ with $j = 0, 1, 2, \ldots N - 1$. DVD audio files use a sampling rate of 48,000 samples per second; so, the time Δt between samples j and $j+1$ is $1/(48000 \text{ samples/sec}) = 0.000021$ seconds. Audio CDs use 44,100 samples per second.

As we mentioned previously, the basic idea of Fourier analysis is to multiply the signal by cosines and sines and add up the results over the length of the signal. To keep things as simple as possible, we will use just cosines. The cosine functions we use are

$$U_F(j, k) = \cos(2\pi j k/N). \tag{13.11}$$

In Eq. (13.11), j and k are integers that run from 0 to $N - 1$. When $j\,k/N$ is an integer, the cosine will equal 1, its largest possible value. To simplify notation and allow for other generalizations, we have introduced a function $U_F(j, k)$ in Eq. (13.11). In more general situations U_F might be some combination of cosines and sines.

Figure 13.8 shows a plot of $U_F(j, k) = \cos(2\pi j k/N)$ as a function of the index j. Note that the function has a maximum whenever $j\,k/N$ is an integer (including 0), as promised.

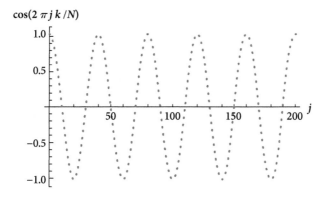

Fig. 13.8 A discrete-sample plot of the discrete function $U_F(j, k) = \cos(2\pi j\, k/N)$ with $N = 200$ and $k = 5$ as a function of the index j.

Try It 13.11

Verify from Figure 13.8 that the plot has a period of 40. Calculate k/N for that plot. Is that what you expect for a period of 40?

ALICE: Using this mathematical machinery, we define the discrete Fourier cosine transform of $f(t)$ as

$$\tilde{f}_k = \sum_{j=0}^{N-1} f(j\Delta t)\, U_F(j, k), \qquad (13.12)$$

where $\tilde{f}_k$ is the so-called Fourier amplitude, which is labeled by the index k. Eq. (13.12) is just the sum of products of the signal at a particular time and the cosine at the same time. Here the time is indicated by the index j.

Once we have found $\tilde{f}_k$ for all values of the index k, we may write the original function as a sum of the U_Fs:

$$f(t_j) = \sum_{k=0}^{N-1} \tilde{f}_k\, U_F(j, k). \qquad (13.13)$$

An aside: In writing Eqs. (13.12) and (13.13), we have omitted "normalization" factors that are not important for our general argument. The important point is that $\tilde{f}_k$ for specific integers k will be large if $f(t)$ and U_F "line up" as j runs from 0 to $N-1$. Those values of k then tell us about the periodicity of the signal $f(t)$.

The Fourier method has much the same flavor as superposition states in quantum mechanics. Fourier's theorem says that any periodic function can be written as a sum (a superposition)

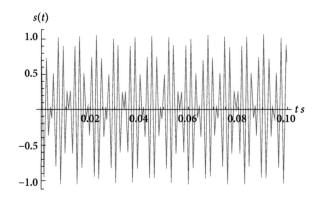

Fig. 13.9 A plot of a periodic function corresponding to a sum of two cosines.

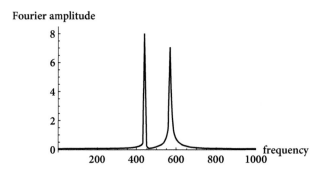

Fig. 13.10 A plot of the discrete Fourier "spectrum" of the signal in Figure 13.9. The peaks occur at the two frequencies 440 Hz and 572 Hz used to generate the signal in Figure 13.9.

of cosine and sine functions with appropriate amplitudes. So, we can view Eq. (13.13) as expressing the signal $f(t)$ as a superposition of "basis states" $U_F(j, k)$. In fact, though we won't go through the details, those basis states are orthogonal and normalized in the sense that the product of two of them, summed over the range of the indices, gives 0 if the indices are different (different basis states) and 1 if the indices are the same (and if we have included the appropriate normalization factors).

BOB: All of this may seem rather abstract; so, let's show how this works in a case that might be familiar from introductory physics. Suppose we record an audio signal that is a mix of 440 Hz (middle A on the musical scale) and 572 Hz. Figure 13.9 shows the combined audio signal. Figure 13.10 shows what is called the discrete Fourier spectrum of the signal, which shows peaks at 440 and 572 Hz as expected. We can think of the original audio signal as being a superposition of a 440 Hz signal and a 572 Hz signal. The period T of the 440 Hz signal is $1/(440 \text{ Hz}) = 0.0023$ s. The 572 Hz signal has a period of 0.0017 s.

13.7   Quantum Fourier Analysis

ALICE: As we mentioned before, the heart of the Shor algorithm is a quantum formulation of Fourier analysis that is much faster than the classical Fourier transform in finding the period of a periodic function. The quantum method is similar to the discrete Fourier transform for a classical signal.

We want to show how the quantum Fourier analysis allows us to find the smallest period r of the function

$$f(x) = a^x \bmod N, \tag{13.14}$$

with $f(x) = f(x+r)$, excluding $r = 0$ as mentioned previously. We have shown in section 13.5 that if we know the period, we can find the factors of N. We have used x as the exponent to match what other authors have used and to link to the way we have previously labeled multi-qubit basis states. As usual, a, x, r, and N are integers.

As we proceed through the algorithm, you should pay attention to where we use super-position states and measurements, which we have seen to be key features of other quantum algorithms.

The first ingredient we need is the quantum implementation of the function $f(x) = a^x \bmod N$ as a quantum function gate U_f. We won't worry about the details of that im-plementation, but we note that it is usually based on binary modular exponentiation (see section 13.8). As we have seen in other quantum algorithms, the function gate has two "registers" of input states: an upper register and a lower register, as shown in Figure 13.11.

As usual, we start off with computational basis states. Both the upper and lower register states are n-qubit states with n chosen so that $2^n = N$, the number we want to factor. The upper-register states are then acted on by Hadamard gates to produce a superposition state consisting of equal contributions of all the basis states labeled by the binary numbers x, where x is in the range $[0, N - 1]$. We have seen this procedure in several of the other quantum algorithms. After the Hadamard gate acts on the upper-register states, the system state is

$$|\psi_1\rangle = H^{\otimes n}|\psi_0\rangle = \underbrace{\frac{1}{\sqrt{N}}\sum_{x=0}^{2^n-1}|x\rangle}_{\text{upper register}} \underbrace{|0\rangle_n}_{\text{lower register}}. \tag{13.15}$$

In analogy with some of the other quantum algorithms, the modular exponentiation function gate U_f (not to be confused with the Fourier function U_F) then acts on the state in Eq. (13.15)

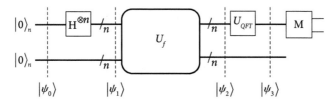

Fig. 13.11 A schematic of the Shor period-finding circuit with a function gate U_f, upper- and lower-register inputs and outputs, a gate U_{QFT} representing the quantum Fourier transform, and n measurement devices M for the upper register. Only one measurement device is shown.

to produce the state

$$|\psi_2\rangle = U_f |\psi_1\rangle = \frac{1}{\sqrt{N}} \sum_{x=0}^{2^n-1} \underbrace{|x\rangle}_{\text{upper register}} \underbrace{|f(x)\rangle}_{\text{lower register}}. \tag{13.16}$$

This state is a superposition of the products of all the basis states and the states labeled by the results of the periodic function in Eq. (13.14) evaluated at x. After we process this state with the operations described below, we will let the upper-register states interact with measurement devices. The measurements outcomes will give us information on the desired period.

Next, we apply the quantum Fourier transform gate to the upper-register states. The quantum Fourier transform gate's action is defined by its action on each of the basis states $|x\rangle$:

$$U_{QFT} |x\rangle = \frac{1}{\sqrt{N}} \sum_{y=0}^{2^n-1} U_F(x,y) |y\rangle. \tag{13.17}$$

In Eq. (13.17), the U_{QFT} function acts on a particular basis state $|x\rangle$ to produce a superposition of all the basis states, labeled by y in the sum. Each basis state in the sum is multiplied by $U_F(x,y)$, itself a periodic function with the indices j and k replaced by the integers x and y. In the full-blown quantum Fourier transform, U_{QFT} is a combination of sines and cosines called the "complex exponential" function, which we shall talk about in Chapter 15. For now, all we need to know is that U_F is periodic in x and y.

The result of the quantum Fourier transform gate acting on $|\psi_2\rangle$ is

$$|\psi_3\rangle = U_{QFT} |\psi_2\rangle = \frac{1}{N} \sum_x \sum_y \underbrace{U_F(x,y) |y\rangle}_{\text{upper register}} \underbrace{|f(x)\rangle}_{\text{lower register}}. \tag{13.18}$$

Let's read off what is in the state. The $U_F(x,y) |y\rangle$ is the representation of applying the quantum Fourier transform operator to the basis states $|x\rangle$ in the upper register. The $|f(x)\rangle$ parts are unchanged because nothing has operated on the lower-register states in going from $|\psi_2\rangle$ to $|\psi_3\rangle$.

Finally, we let the upper-register states interact with the n measurement devices. The result will be some value of y. We shall see shortly that the most probable result of the upper-register y measurement will be $y = N/r$. Once we have that value of y, we can find the period r from $r = N/y$.

We now want to figure out the probability of getting a particular measurement value y. As with many other quantum algorithms, we don't always get precisely the answer we want. But we want to arrange the algorithm so the probability of getting the right answer is close to 1. As promised, that result gives us information about the period r of the function $a^x \bmod N$, and we know from the analysis of the Shor factoring algorithm that will allow us to factor the number N.

To find the probability, we follow the usual generalized Born rule for multi-qubit states: We find the appropriate amplitudes, sum them, and then square the result to get the probability. The critical observation is that we have several amplitudes that contribute because the function $f(x)$ is periodic in x. That is, we have $f(x_0 + qr) = f(x_0)$, where q is any integer such that $x = x_0 + qr < N - 1$. Then, the probability of getting the result y from the upper-register measurement is

$$\frac{1}{N} \left| \sum_{x:\, f(x)\,=f(x_0)} U_F(x, y) \right|^2 = \frac{1}{q_{max} N} \left| \sum_{q=0}^{q_{max}} U_F(x_0 + q\,r, y) \right|^2, \tag{13.19}$$

where q_{max} is the largest integer less than $2^n/r$, approximately the number of periods of length r in the range $[0, 2^n - 1]$. The quantum efficiency in the algorithm occurs because we need to evaluate only a relatively small number of terms compared to the number that would be required with the classical discrete Fourier transform.

Try It 13.12

Find n for $2^n = N = 1000$.

BOB: This is pretty abstract; so, let's use the explicit cosine form of U_F:

$$U_F(x_0 + q\,r, y) = \cos[2\pi(x_0 + q\,r)y/N]$$
$$= \cos[(2\pi x_0 y/N) + (2\pi\, q\, r\, y/N)] . \tag{13.20}$$

In the second line of Eq. (13.20), we see that the term $2\pi x_0 y/N$ is a constant that offsets the results when we sum over the integer q. It corresponds to shifting the cosine left or right but does not involve the periodicity. So, for our purposes we just focus on $\cos[2\pi\, q\, r\, y/N]$. Recall that this is just like the function plotted in Figure 13.8. Note that $\cos(2\pi k)$ equals 1 if k is an integer. The term has its maximum value 1 when $r\, y/N$ is an integer and it will be close to 1 when $r\, y/N$ is close to being an integer. That tells us that the most probable result of the upper-register y measurement will be $y = N/r$ from which we find the desired period $r = N/y$. So, we see that the quantum Fourier transform method is likely to give us the desired period, but there is some probability that it won't, much like the probabilistic Simon algorithm. Of course, once you have a value for r, you should check that it is indeed a period of $a^x \mod N$.

CARDY: But how do I do that?

BOB: Just make use of the classical modular exponentiation algorithm (section 13.8) to show that $1 = a^r \mod N$, which is equivalent to showing that $a^r - 1$ is evenly divisible by N. You should also check that r is the smallest integer period by showing that $r/2$ does not work.

Let's get back to an important question. How likely are we to get the "correct" result? It turns out that the sum in Eq. (13.19) in the general quantum Fourier transform method can

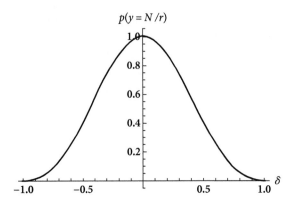

Fig. 13.12 A plot of the (un-normalized) probability given in Eq. (13.21) as a function of δ with $N = 10^7$ and $r = 30{,}000$.

be expressed as a geometric series. The result is that the probability given by

$$p(\delta) = \frac{1}{q_{max} N} \frac{\sin^2(\pi \delta \, q_{max} \, r/N)}{\sin^2(\pi \delta \, r/N)}, \tag{13.21}$$

where $\delta = y - N/r$ indicates how close the integer y is to N/r. In Chapter 15, once we have introduced complex numbers and the complex exponential function, we will see how Eq. (13.19) leads to Eq. (13.21).

CARDY: What is a geometric series?

BOB: Oops! That is a bit of math we should explain. A geometric series is a sum of terms, each of which is the previous term multiplied by a common factor R. In plain English, that means that R in the successive terms in the sum has exponents that range from 0 to $n - 1$. For example, in the series

$$a + aR + aR^2 + \ldots aR^{n-1} = a\frac{1 - R^n}{1 - R}, \tag{13.22}$$

R is the ratio of successive terms in the series and the expression on the right is the result of adding the series.

So, we see from Figure 13.12 that the probability is sharply peaked when $\delta = 0$, that is when $y = N/r$, and the probability remains high even for $y = N/r \pm 0.5$. That means the quantum Fourier transform is highly likely to give us the "correct" answer, though there is some (smaller) probability that it will miss the target. In such a case, we ought to repeat the analysis several times to assure we get the right answer.

13.8   Modular Exponentiation

ALICE: We promised you that we would go through some of the mathematical details of the RSA algorithm. If you are not interested in the details, feel free to skip to the next section.

Modular exponentiation means that we want to find c, given general integers a, b, and N, where

$$c = a^b \bmod N. \tag{13.23}$$

In other words, c is the remainder of dividing a^b by N. We could use a brute force method to find c: calculate a^b and then divide by N to find the remainder c. The difficulty is that a, b, and N might be large numbers, which need to be stored as exact binary sequences, and a^b is extremely large and must be stored exactly. Also, carrying out the division by N will be time-consuming. However, once we have chosen a and N, there is a simple algorithm to find c that avoids the large number issues:

1. Set $c = 1$, $b' = 0$.
2. Increase b' by 1.
3. Set $c = (a \cdot c)^{b'} \bmod N$.
4. If $b' < b$, go to step 2. Otherwise, c is the solution to $c = a^b \bmod N$.

Note that b' serves as an index to keep track of the loops through the algorithm. Here is a flow chart for the algorithm:

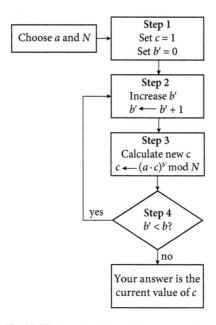

Fig. 13.13 Flow chart for modular exponentiation.

Try It 13.13

Use the algorithm to find 4^6 mod 17. That is easy to evaluate directly by hand, but it is a good exercise to see how the algorithm operates.

ALICE: You will notice that as we go through the loop in the Figure 13.13 flow chart, we build up b factors of a in step 3, but the mod operation keeps the numbers relatively small (less than N) at each step, although there may be many steps. In the straightforward version of the algorithm, there are b steps.

There is another "trick" that may drastically reduce the number of operations needed to perform modular exponentiation, while keeping the small number method described above. It is a combination of the previous method and a more general principle called *exponentiation by squaring* (also known as *binary exponentiation*).

First, we convert the exponent b to binary form; that is, b can be written as a string of 1s and 0s:

$$b = \sum_{j=0}^{n-1} k_j 2^j, \tag{13.24}$$

where n is the number of bits required to represent b, and the coefficients k_j are either 0 or 1.

An example might be helpful: Suppose $b = 17$ (decimal). In binary form $b = 10001$ and $n = 5$. Hence, we may write

$$b = 10001 = 1 \times 2^4 + 0 \times 2^3 + 0 \times 2^2 + 0 \times 2^1 + 1 \times 2^0$$

from which we read off

$$k_0 = 1, \ k_1 = 0, \ k_2 = 0, \ k_3 = 0, \ k_4 = 1 \ .$$

Try It 13.14

Do the same analysis for $b = 15$.

ALICE: Let's continue with our binary exponentiation. Using Eq. (13.24) and employing the sum rule for exponents $x^y x^z = x^{y+z}$, we can write a^b as

$$a^b = a^{\sum k_j 2^j} = a^{k_0 2^0} a^{k_1 2^1} \ldots a^{k_{n-1} 2^{n-1}} \tag{13.25}$$

Why does this help? As we go through the algorithm loop, any time we have a $k_j = 0$, we can skip that multiplication step because $a^0 = 1$. That reduces the number of steps considerably in most cases.

Let's do an example of binary exponentiation. We choose the exponent $b = 11$ (decimal) $= 1011$ (binary). The algorithm uses a process similar to the previous modular exponentiation except that we have

$$a^b = a^{11_{10}} = a^{1011_2} = a^{1 \cdot 2^3} a^{0 \cdot 2^2} a^{1 \cdot 2^1} a^{1 \cdot 2^0}.$$

Thus, we need to do only three multiplications (instead of 11) because whenever there is a 0 in the exponent, we may skip that multiplication since it is equivalent to multiplying by 1. This process is called binary exponentiation because we are using powers of 2. There are efficient implementations of this method in common programming languages such as Python and Mathematica.

To wrap up, we point out that all the RSA encryption steps can be carried out efficiently using classical computers. Quantum computation comes in when we try to beat the encryption by finding the prime factors p and q to decipher the message.

13.9   Greatest Common Divisor

ALICE: In this section, we will explain how to find the greatest common divisor of two integers a and b, a task that we used to factor N back in section 13.3. Of course, you could always use "brute force" methods and first find all the divisors of the first integer—looking for integers that divide a with no remainder. Then do the same for b. Then look at those divisors to find the divisors that are common to both and select from that subset the one that is the greatest. This works fine for relatively small numbers. But it takes a long time for larger numbers.

A more efficient way makes use of what is called Euclid's algorithm (or the Euclidean algorithm), which was written up more than 2,000 years ago in Euclid's famous book *Elements*. Many authors claim that this algorithm is the oldest known algorithm, but there is no historical evidence that Euclid himself invented it. What is important is that he made it readily available to subsequent generations of mathematicians. As we mentioned before, greatest-common-divisor functions are available in most programming languages and spreadsheets.

Here is how the algorithm works:

1. Let a be the larger of the two integers, a and b. Then find the remainder r_1 from dividing a by b. If the remainder is 0, then b is the greatest common divisor of a and b. If it is not, go to step 2.
2. Divide b by r_1 to find the remainder r_2. If r_2 is not equal to 0, proceed to step 3.
3. Divide r_1 by r_2 to find a new remainder r_3.
4. Continue these steps until the remainder is 0. The next-to-last remainder (the one that is non-zero) is the largest common divisor of a and b.

CARDY: Seems pretty easy. Would you show me an example with actual numbers?

BOB: Here is one. What is the greatest common divisor of 350 and 245? The steps are

$$350/245 = 1, \ r_1 = 105$$
$$245/105 = 2, \ r_2 = 35$$
$$105/35 = 3, \ r_3 = 0 \ . \tag{13.26}$$

This example took only three divisions. The last non-zero remainder is 35; so, the greatest common divisor (gcd) of 350 and 245 is 35.

Try It 13.15

Draw a flow-chart diagram like the one for modular exponentiation. Use it to work out the examples in the following Try It.

Try It 13.16

Use the Euclidean algorithm to find the gcd of 77 and 33 and the gcd of 77 and 35, the numbers we used in our example of the factoring algorithm.

CARDY: Thanks! The algorithm seems to work, but I don't quite see why it works.

ALICE: Cardy, you are going to turn into a mathematician because you always want to delve into the details of "why." That is a good habit to cultivate even if you are not going to be a professional mathematician. It turns out that the proof of the algorithm is not difficult. You can easily find several of them with video explanations on the internet.

BOB: Now let's tackle the proof of the statement about two numbers having a common factor that we made use of in section 13.4.

What we want to prove is that if N does not divide $a^{r/2} \pm 1$ evenly, then they have factors in common with N (excluding 1 as a factor), a statement we used in section 13.4.

Try It 13.17

As an example of Bob's statement about factors in common, use $N = 77$ and $a^{r/2} = 34$. Find the factor that N has in common with 33 and the factor that N has in common with 35.

The proof involves the strategy called "proof by contradiction": we assume that our desired conclusion is *not* true and then show that the assumption leads to a contradiction. We then infer that our assumption must be false, and hence that the desired conclusion is true. So, we start with the assumption that $a^{r/2} - 1$ and N do *not* have a factor in common. That means that the greatest common divisor is 1:

$$\gcd\left(a^{r/2} - 1, N\right) = 1. \tag{13.27}$$

A general property of gcds is

$$z = \gcd(x, y) \Rightarrow mx + ny = z, \tag{13.28}$$

where m and n are integers, one of which may be negative.

CARDY: How do we prove *that*?

ALICE: Cardy, I'm glad you asked. Eq. (13.28) is known as Bézout's identity. The proof of that identity is not hard but it would take us deeper into number theory. You can find several versions of the identity's proof online. You could also try a few numerical examples to convince yourself that it works.

Try It 13.18

The greatest common divisor of 12 and 42 is 6. Show that those numbers satisfy the Bézout's identity with $m = 11$ and $n = -3$ and also with $m = 4$ and $n = -1$.

ALICE: To simplify notation, let us introduce $c = a^{r/2} - 1$ and $d = a^{r/2} + 1$. Then for our case, Eqs. (13.27) and (13.28) tell us that

$$mc + nN = 1. \tag{13.29}$$

We now multiply both sides of Eq. (13.29) by d:

$$mcd + ndN = d. \tag{13.30}$$

Recall from Eq. (13.10) that $cd = kN$, which when used in Eq. (13.30) gives us

$$mkN + ndN = d. \tag{13.31}$$

Dividing both sides of Eq. (13.31) by N yields

$$mk + nd = d/N. \tag{13.32}$$

The last equality in Eq. (13.32) tells us that d is evenly divisible by N; that is, the result of the division is an integer. But that result contradicts the statement that d is not divisible evenly by N. Thus, our assumption that c and N have no common factor (other than 1) is wrong. To say it another way, c and d must have common factors with N, which we can find through the gcd process.

CARDY: This chapter has given me a real math workout! My brain is exhausted. But I think I have a pretty good grasp of the general principles.

ALICE: That is the main point, Cardy. The details are important if you are going to implement the Shor factoring algorithm on your quantum computer, but the real take-home message is the type of thinking that goes into constructing the algorithm: framing the question so it is amenable to QC methods and making clever use of superposition states, quantum function gates, and strategically placed measurements to arrive at your result.

 CHAPTER SUMMARY

- Encryption employs a key K that is typically used with other numbers to encode and decode the message. The RSA protocol for encryption involves factoring a product N of two large prime numbers p and q. It also uses the value of d which satisfies:

$$1 = Kd \bmod (p-1)(q-1)$$

- The prime factors, p and q, and the number d are kept secret. K and N may be shared publicly, and the sender uses them to encrypt the message. For large N, classical computers are unable to do the calculations fast enough to determine p, q, and d in a practical amount of time.

- The Shor algorithm is a famous quantum algorithm that can find the encryption key faster by making use of the periodic nature of the remainder or mod function. In effect, the Shor algorithm allows us to break the RSA encryption and other encryption methods based on factoring large numbers, the basis for a large portion of our internet security. The algorithm does this by making use of quantum Fourier transforms which approximate periodic functions as a series of cosine and sine functions. The periods of the cosine and sine functions give information about the period of the original function.

- In terms of quantum gates and a quantum circuit, the Shor algorithm has an upper register and a lower one with the upper register put into a state of superposition using Hadamard gates. The system state is acted on by a function gate U_f with $f(x) = a^x \bmod N$. The function is periodic $f(x) = f(x+r)$, where r is the period of the function. Next the upper register is operated on by a quantum Fourier transform gate. Finally, we send both registers to measurement devices from which we find $y = N/r$ and solve for $r = N/y$. Once we find the period r, we can find the prime factors p and q by straightforward classical computations.

 FURTHER READING

"Fundamental theorem of arithmetic," Wikipedia provides nice explanations of the theorem, why it is important, and how it is proved. There are many excellent YouTube videos on modular arithmetic, including congruence, greatest common divisors, and modular exponentiation.

M. Suhail Zubairy, *Quantum Mechanics for Beginners* (Oxford University Press, Oxford, 2020). Chapters 13 and 16 have good introductions to quantum secure communications and the Shor algorithm.

14 Fundamental Quantum Issues

I will argue here that the measurement problem is not a real problem.

Nancy Cartwright

14.1 Introduction

ALICE: In the previous chapters, we have introduced several different types of quantum algorithms and quantum information processing schemes. All those procedures rely heavily on quantum superposition states, particularly entangled states. In this chapter, we will explore some of the conceptual issues raised by entanglement. Although understanding these issues doesn't directly impact our ability to use quantum algorithms, they do highlight the differences between classical computational resources and quantum computing resources.

We will also look at how other fundamental aspects of quantum physics are connected to QIS. Although these topics are not directly related to QC, we hope you are intrigued enough about these fundamental issues to read on.

CARDY: All that sounds cool. I have come to realize that the basic concepts are important even though I may not use them every day.

ALICE: That's highly commendable and an important professional skill, no matter what you do. Understanding the "why" gives you the tools to find new applications.

As we dig into those concepts, we ought to be a bit modest about our current state of knowledge of the quantum world and say that, as far as we know, that is the way the world works. It would be highly presumptuous to claim that we have arrived at the final truth. The history of science and technology is littered with now-defunct claims of achieving the final truth. On the other hand, no one has any idea of what an alternative to quantum mechanics would look like or how an alternative could reproduce the successes of quantum mechanics while offering a different picture of the nature of reality.

14.2 Bell's Theorem and Quantum Weirdness

ALICE: The first issue we will look at is the nature of reality. That sounds really profound, but it is an issue that underpins the consequences of entangled superposition states—the main ingredient in many QIS procedures. Thanks to the theoretical work of the late physicist John Bell and the experimental work of dozens of scientists around the world, we now have experimental

Republished with permission of Oxford University Press - Books (US&UK), from How the laws of physics lie, Nancy Cartwright, 1983; permission conveyed through Copyright Clearance Center, Inc.

Quantum Computing: From Alice to Bob. Alice Flarend and Bob Hilborn, Oxford University Press.
© Alice Flarend and Robert C. Hilborn (2022). DOI: 10.1093/oso/9780192857972.003.0014

Fig. 14.1 The participants in the Foundations of Quantum Mechanics Conference. Amherst College, Amherst, MA June 10–15, 1990. John Bell is in the first row, far right. Author Bob is in the middle of the second row from the top. Photograph used by permission from Amherst College. Frank Ward, photographer.

proof that nature really does work in what seems to be the crazy way described by entangled states, and that any efforts to sneak around the strictures of quantum mechanics are doomed to fail.

BOB: In 1990, I was a participant in the Foundations of Quantum Mechanics Conference held at Amherst College, Massachusetts (see Figure 14.1). Among the participants were John Bell, David Mermin, Anton Zeilinger, Danny Greenberger, and many other notables in the development of QIS and the interpretations of the foundations of quantum mechanics. In the 1960s, John Bell had introduced what became known as Bell's theorem, which provided a way to test the quantum predictions of correlations among measurement outcomes against the predictions of classical physics. In 1990, only a few experiments testing Bell's theorem were available and each of them had several loopholes that prevented them from being definitive tests of the theorem. Much of the discussion at the conference focused on what Bell's theorem and those experiments were actually telling us about the nature of reality. Sadly, John Bell passed away about four months later, but the conversation has continued.

David Mermin had developed a simple model that displayed, with a minimum of mathematics, the contradictions between what seems like a straightforward, commonsense prediction of the results of a set of measurements and the quantum predictions for those outcomes. In essence, Mermin's model is a specific instance of the general results embodied in Bell's theorem. We will use a slight variation on Mermin's model to illustrate Bell's theorem.

Here is the scenario: Imagine that we have put together a device that emits two correlated qubits. Those two qubits later encounter two measurement devices as shown in Figure 14.2. The devices are identical and each has three settings for the measurement basis states. For now, we

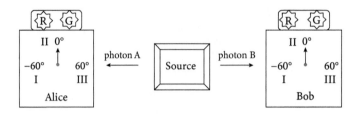

Fig. 14.2 A sketch of the apparatus used to illustrate Bell's theorem. The source emits two photons, one of which travels to the right and one travels to the left. They interact with the measurement devices and either a green light or a red light indicates the measurement outcome. Each measurement device has three different measurement bases available. The bases are labeled I, II, and III or $-60°, 0°$, and $60°$.

will call the settings I, II, and III. Later we will translate them to linear polarization direction settings for photons, which is why they have degree values associated with them.

Here's how the measurement devices work. Each device has two lights, one red and one green, that indicate the measurement results. When a qubit interacts with one of the measurement devices, things are arranged such that only one light goes on—the red one or the green one—but never both, just as we have seen for polarizing beam splitters and photons. When the two qubits interact with the two measurement devices (one on the left and one on the right) set to the *same* settings—both I, both II, or both III—we find that both red lights go on or both green lights go on but in an apparently random pattern. On average, you get 50% red lights and 50% green.

The question is what determines whether the red lights or the green lights go on. A classical model does not allow for random results. There must be a definite cause. It seems obvious that the qubits must carry some instructions or have some property that determines how they interact with the measurement devices and which light goes on. For example, you might imagine that the qubits have an electrical charge: plus charge or minus charge. When the entity interacts with the measurement device, which we might assume is sensitive to the sign of the electrical charge, we could arrange that plus charge makes the green light go on while minus charge leads to the red light going on.

Since there are three settings for each device in the Bell experiment, it seems reasonable to think that the qubits have three properties (or instructions) built in to indicate which light should go on when they interact with the measurement devices. We will represent those instructions as a set of boxes $\boxed{R}\boxed{R}\boxed{G}$ with one box for each of the possible device settings. Reading from left to right, we have the I setting result (R in this example), the II setting result (R again), and on the right, the III setting result (G). These sets of instructions are sometimes called "hidden variables" because we have not been able to see them with available experiments. Thinking classically, it seems something like hidden variables must be there determining the measurement outcomes.

As part of the classical model, we assume that there are no other communications channels between the right and left devices. The devices respond solely to the interaction with their local qubit. That kind of model is called a "local realistic," or better, a "local deterministic" model: The results are determined (with no randomness) by those local instructions. The only reason the red and green lights seem to go on randomly is that with successive trials the source sends

Table 14.1 The results of running the Figure 14.2 device. The top row indicates the possible qubit instruction sets. The first column on the left indicates the detectors' settings; I, II, or III. The other boxes display A when both measurement results "agree" (both devices show a red light or both show a green light) and D when the results "disagree" (one shows red, the other shows green). The last row is to be filled in Try It 14.1.

	$\boxed{R}\boxed{R}\boxed{R}$	$\boxed{G}\boxed{G}\boxed{G}$	$\boxed{R}\boxed{R}\boxed{G}$	$\boxed{R}\boxed{G}\boxed{R}$	$\boxed{G}\boxed{R}\boxed{R}$	$\boxed{R}\boxed{G}\boxed{G}$	$\boxed{G}\boxed{R}\boxed{G}$	$\boxed{G}\boxed{G}\boxed{R}$
I, I	A	A	A	A	A	A	A	A
II, II	A	A	A	A	A	A	A	A
III, III	A	A	A	A	A	A	A	A
I, II	A	A	A	D	D	D	D	A
I, III	A	A	D	A	D	D	A	D
II, I	A	A	A	D	D	D	D	A
II, III	A	A	D	D	A	A	D	D
III, I	A	A	D	A	D	D	A	D
III, II								

out qubits with different instruction sets and we don't know in advance, nor can we control, what the instruction set is.

CARDY: Sounds good. So, from what you said it would seem that to account for the results when the settings are the same on the right and on the left—both green or both red—the entities would each have to have the hidden variables $\boxed{R}\boxed{R}\boxed{R}$ or $\boxed{G}\boxed{G}\boxed{G}$ with each occurring 50% of the time. Is that right?

ALICE: That's precisely it. The R or G in the box means that there is some qubit property that leads the lit light to be red or green depending on the property and on the measurement device setting.

Now let's look at what happens when the settings are different. There are six possible combinations of different device settings. See Table 14.1. If we list my device's setting on the left, Bob's device's setting on the right, the six possibilities for different measurement device settings are (I, II), (I, III), (II, I), (II, III), (III, I), or (III, II). To see what our local deterministic model predicts, let's look at what happens for the remaining six possible property sets:

$$\boxed{R}\boxed{R}\boxed{G} \ \ \boxed{R}\boxed{G}\boxed{R} \ \ \boxed{G}\boxed{R}\boxed{R} \ \ \boxed{R}\boxed{G}\boxed{G} \ \ \boxed{G}\boxed{R}\boxed{G} \ \ \boxed{G}\boxed{G}\boxed{R}. \tag{14.1}$$

Since there are nine different combinations of measurement device settings and a total of eight sets of instructions, we fill in a nine-by-eight table with the lit light results predicted by our model. Remember that we are assuming that each pair of qubits is produced with the same set of three properties by the source to account for having the same lights lit when the settings are the same. The entries in the table labeled A mean the measurement results agree (we both see red or we both see green). The ones labeled D mean that I see red and Bob sees green or vice versa.

Cardy, can you see a pattern? What are the first four entries of the last row? We will leave the other entries as a Try It.

CARDY: I think I see the pattern. Are the results for the first four A, A, D, and D?

Try It 14.1

Fill the final four results boxes in the bottom row of Table 14.1 and explain your reasoning. **Challenge**: fill in all Table 14.1 boxes with R/R, G/G, R/G, or G/R to indicate which lights are lit in each case.

ALICE: You got it, Cardy. Now we are ready to analyze the frequency of those results. We ask a very simple question: What fraction of all the results listed in the table gives rise to the same light being lit in both devices? The lights could be both red or both green.

CARDY: I counted the number of A results and got 48. Since we have nine rows and eight columns for a total of 72 possible results, we get the same lights lit $48/72 = 2/3$ of the time.

ALICE: That's right, Cardy. We assume that we have constructed the source device to produce the instruction sets with equal probability and we constructed the measurement devices to use I, II, or III settings with equal probability. Then, if we look at the results of a long sequence of source launchings and measurement device detections, our model predicts that we should see two green lights or two red lights lit 2/3 of the time. Of course, that means we should see red/green or green/red 1/3 of the time. We could also check our prediction that each detector displays 50% red and 50% green lights on average.

Try It 14.2

Verify from the listings in Table 14.1 that our model predicts that each detector should, on average, display 50% red and 50% green lights lit.

ALICE: It turns out we don't need to make any assumption about how often the source makes the various instruction sets. Think about it in the following way. Let's look at the specific types of columns in Table 14.1: ones where the instructions are the same for all detector settings and ones where they are different. In columns one and two, we will always get the same color lights lit in the two detectors regardless of their settings. So, if the source produces only $\boxed{R}\boxed{R}\boxed{R}$ or only $\boxed{G}\boxed{G}\boxed{G}$ or only some mixture of $\boxed{R}\boxed{R}\boxed{R}$ and $\boxed{G}\boxed{G}\boxed{G}$, we will have the same colors lit in every trial. That is, the probability of getting lights of the same color is 1. That is the maximum value a probability can be. Now count the number of times we get the same colors for Table 14.1 columns three to eight with different qubit instruction sets. Do you find that five out of the nine detector settings give lights of the same color? For those columns, the probability of getting two lights of the same color is 5/9, much smaller than that for the first two columns. Look at each of the columns individually and note that they all indicate a probability of 5/9 for getting the same colors. If the source produces the instruction sets corresponding to any one of the third through eighth columns or even a mixture of those columns, we will get the same color lights with a probability of 5/9.

Now let's put the result for the two types of columns together. This may take some thinking. For any mixture of all those instruction sets or all the columns, we will always have a probability of the same color lights greater than or equal to 5/9. For our equal-probability model, where all instruction sets are equally probable, we got a probability of 2/3, larger than 5/9. The overall results of our local deterministic model could be specified as an inequality: The probability of getting both lights red or both green is greater than or equal to 5/9.

What John Bell was able to show (in what is now called Bell's theorem) was that any theory that purports that qubits (or any other quantum entity) have a "deterministic" and "local" set of properties independent of measurements gives predictions that satisfy a similar inequality. Bell's theorem gives a quantitative prediction we can test by doing an experiment.

ALICE: It is now time to see what quantum mechanics predicts for the results of these measurements. We will use linearly polarized photons as the qubits emitted by the source. We have had some experience calculating the results of quantum measurements on such photons and almost all of the experimental tests of Bell's theorem have been carried out with photons using setups that are similar to the one described in Figure 14.2.

We set the source to prepare two linearly polarized photons in the entangled superposition state

$$|\psi\rangle = \frac{1}{\sqrt{2}} |hlp\rangle |hlp\rangle + \frac{1}{\sqrt{2}} |vlp\rangle |vlp\rangle . \tag{14.2}$$

You recall that if this is the state of the two-qubit system, then it is not legitimate to say that one of the qubits has vlp and the other has hlp. To use the language we employed to build our local deterministic model, we see that quantum mechanics does not assign a definite property (or instruction set) to each photon. The state, as usual, tells us only the probabilities of observing hlp or vlp when the two qubits interact with linear polarization measurement devices.

In the photon version, the three measurement device settings correspond to the horizontal linear polarization direction set at $-60°$, $0°$, and $60°$, which are labeled on our measurement devices under I, II, and III. This amounts to having a choice of three different measurement basis states. If the green light is lit, the measurement device has detected hlp relative to the measurement setting axes, and if red, vlp relative to those axes. In fact, we had worked out the details of these basis states in Chapter 9. But let's review the results of choosing the various measurement basis states because the results are fundamental to understanding the quantum results.

If both Bob and I choose the same polarization measurement basis states, that is, we choose the same linear polarization directions for measurements, then we will always get the same measurement results: 50% of the time we will both observe "vertical" polarization (red light lit) and 50% of the time we will both observe "horizontal" polarization (green light lit) relative to that measurement direction. We don't get red or green every time, but Bob and I will always have the *same* result: red or green.

What happens if we each choose different measurement directions? Then we will not always get the identical lights lit. That is, we will sometimes get the same lights lit and sometimes different lights lit. In fact, we calculated the probabilities for those results in Chapter 9. We found that for different measurement basis settings, we both get red or both get green only 25% of the time. And averaging over all possible measurement device settings, we get the same color only 50% of the time; that is, according to quantum mechanics, the probability of our getting the same lights lit is 1/2.

Let's compare the predictions of the two models. We showed previously that our local, deterministic model predicts that we should get the same color with a probability exceeding 5/9, with the exact value depending on the selection of the various possible instruction sets. The quantum model, on the other hand, predicts that the probability of getting the same color exceeds 1/4, with the lower limit occurring when the measurement device settings are different. If we average over all the measurement device settings, the probability for getting the same lights lit is 1/2. In other words, we have distinctly different predictions for the local, deterministic model and for quantum mechanics.

BOB: Cardy, if we carry out actual experiments with the Bell's theorem set up as described here, we find that the probability that Alice and I will observe the same color is 1/2. Which model do the experimental results support—the quantum model or the local, deterministic model?

CARDY: According to what you showed me, the experiments seem to support the quantum model. But the local deterministic model seems so reasonable and matches my commonsense ideas about measurements. The obvious conclusion is that the experiments are wrong or flawed in some fundamental way.

BOB: Your reaction is quite understandable. There are in fact some subtle aspects of the experiments that we need to think about carefully. But after several decades of work, all the questionable aspects have been dealt with and the precision of the experiments has been dramatically improved. There is no longer any reasonable doubt about the validity of the experimental results: The predictions of quantum mechanics for this setup are right and those of a local, deterministic model are wrong.

CARDY: But our local, deterministic model makes sense to me. In fact, from what I know about science, those assumptions make a lot of sense. It seems that it would be impossible to do science without assumptions like those. If quantum mechanics makes different predictions, then maybe there is something missing from quantum mechanics, and we ought to fix it.

BOB: Many scientists, including Einstein, thought there must be something wrong with quantum mechanics, particularly because of its probabilistic nature. As Einstein put it, "God does not play dice." He thought that quantum mechanics was a useful theory, but that eventually someone would figure out a better theory—better in the sense of getting the right predictions but doing away with the probability part.

Let's now unpack the terms "deterministic" and "local." By "deterministic" we mean the commonsense notion that an object like a photon or an electron always has a definite value of its properties such as linear polarization direction or spin orientation and that those properties determine (no probability involved) how the object interacts with measurement devices (or with anything else for that matter). We may not know what the property value is until we measure it, but it is, in some sense, really there.

As we have seen with entangled states such as the two-qubit spin-1/2 state

$$|\psi\rangle = \frac{1}{\sqrt{2}} \left(|\uparrow\uparrow\rangle + |\downarrow\downarrow\rangle \right), \tag{14.3}$$

when I observe the orientation of the spin of one of the qubits, I might get ↑. Then when Alice measures the orientation of the other qubit's spin, she will get ↑. Such states are often called Bell states because of their use in illustrating Bell's theorem.

The "realistic" or "deterministic" account is that the two qubits both had those spin orientations before the measurements; all we did was to observe what is already there. Those qubit properties determine how the qubit interacts with the measurement device and which light gets lit. The lights appear to be red or green randomly only because we don't know which set of properties is launched each time the source emits the qubits. I prefer to use the term deterministic account because "real," "reality," and "realism" are philosophically loaded terms; it would take us a whole other book to dig into the meaning of those words.

Now let's talk about "local." Local means that the outcome of the measurements depends only on the specific measurement device, its settings, and the properties of the entity the device interacts with. The results do not depend on what is going on at the other measurement device, at the source, at Simone's Café, or on some exo-planet on the other side of the galaxy. There are two parts to that requirement. One is practical: We have to be sure we exclude effects such as the vibrations of the table on which the equipment is set up, "hidden" signals that might travel through the electrical power lines, and so on. We might also worry about whether choosing the measurement device settings affects the source or somehow affects the measurement devices themselves from a distance. We can even arrange to choose the device settings so they go into place *after* the source has emitted the entities. That rules out the possibility that the measurement device settings influence what the source sends out.

All those kinds of "hidden signal" effects can be ruled out by having the detectors far enough apart so that no signal can travel from one to the other during the time interval between when Alice's measurement device carries out its measurement and when my measurement device does its measurement. It is easy to make sure there is not enough time for a hidden signal to travel from one detector to another since all signals have an upper speed limit: $\approx 3 \times 10^8 \text{m/s}$. This is the speed of light and the speed of gravitational waves and, as far as we know, it is nature's fundamental speed limit. So, if we make sure that the detectors are far enough apart that a signal can't get from one to the other between two measurements, then we can rule out that kind of influence.

Having done all of this, the actual measurements agree with the quantum predictions, not the hidden signals' deterministic prediction. The bottom line of this argument is that having a local, deterministic model, although it may conform both to our intuition about how systems interact (the deterministic part) and to our knowledge about the limits of sending signals from one part of the world to another, does not conform to the way the world works.

Bell's theorem generalizes this argument for different measurement setups and different questions about the correlations among the measurement results. In all cases for which experiments have been carried out, the experimental results agree with the predictions of quantum mechanics. The conclusion is that these experiments rule out any reasonable local, deterministic alternative to quantum mechanics.

CARDY: You slipped in a qualifier, "reasonable."

ALICE: Yes, you could always come up with some contrived theory that gives a local, deterministic account of the experiments. For example, you might propose that there is a mysterious matrix that surrounds us all and the entire universe, and the effects of the matrix "cause" what we call quantum probabilities. But we have no evidence that such a matrix exists, and we don't like to invent special things that explain just one phenomenon and don't have any other effects.

BOB: Another conclusion from these experiments is that the strange non-local correlations between measurements of entangled states are a fundamental part of the way nature works. Einstein called it "Spooky action at a distance." (Einstein was really good at catchy aphorisms!) We saw examples of that in Chapter 9, where the result of my observation on one qubit affected the probabilities for what Alice would observe on another qubit, as long as the two-qubit state was entangled. Those effects may be "spooky" when looked at from a deterministic viewpoint, but they seem to be an integral part of the quantum world.

I struggled to understand those effects for many years because they seem so counter-intuitive. I have finally accepted them by remembering they only show up for conditional probabilities, for which, I must admit, my intuition is almost always wrong, even for classical probabilities.

Let me explain. When we say that my measurement probabilities depend on Alice's observation results, we are really saying that *if* Alice observes her qubit state and gets spin-up (for example), *then* my measurement probabilities will be different from those when Alice observes spin-down. So, my probabilities are conditioned (that's the *if–then* part) upon Alice's observations. Of course, all the measurement probabilities are conditioned upon the state in which the two qubits are prepared. The muddle occurs when we try to give a mechanistic ("deterministic") account of those correlations. In some sense, we just need to "go with the flow" and accept these strange results as part of the way the world works.

We can turn the arguments around and ask: If quantum states have these strange effects, why don't we see them with everyday objects? Part of the answer has to do with the fact that everyday objects (which range from biological cells up to the entire universe) are built out of many, many qubits. That sounds straightforward but providing a detailed account of how that explains local determinism for everyday objects has turned out to be rather difficult. In particular, it has been challenging to specify a border between quantum land and classical, Newtonian land.

You might be interested in looking at the results of the experiment, mentioned in Chapter 9, carried out by a team of Chinese scientists (Yin et al., 2017). They used a source of entangled photons emitted by a laser system in an Earth-orbiting satellite. The photons were linearly polarized and emitted in pairs in an entangled two-photon state. One of the photons was directed to a receiving station on Earth and the other went to another receiving station 1,200 km away. The receivers each had measurement devices that could select different linear polarization measurement directions. The large distances between the two receivers and the satellite ruled out secret signals that might explain the observed correlations. The results agreed with the predictions of quantum mechanics and ruled out local, deterministic models (à la Bell's theorem) to a high degree of precision

CARDY: How high a degree of precision?

BOB: Cardy, you may have heard of standard deviations as a quantitative measure of uncertainties. If the standard deviation, a measure of the uncertainty in the experimental results, is small enough, we can clearly decide between the two sets of predictions and see how well they match the experimental results. In most areas of science having the difference between the predictions of one of the theories and the data larger than five times the standard deviation (sometimes called a "5-sigma" result) is viewed as providing very strong evidence for one set of predictions over the other. In the Chinese experiment, the differences were almost 300-sigma. That means there is no practical way the differences between the local, deterministic predictions and the experimental results could have happened "by accident."

CARDY: I guess we are stuck with quantum mechanics.

ALICE: Let's put a positive spin on this: All the machinery of entangled states seems to be a valid description of nature; therefore, we can be confident that the QIS and QC methods that we build from entangled states and other quantum theoretical components will provide a solid underpinning for our work.

BOB: I would like to sell bumper stickers and T-shirts with phrases that capture the importance of Bell's theorem. How about "Revel in the randomness" and "Think globally, act non-locally"?

CARDY: Sounds good to me! I'll take three large T-shirts as soon as you have them ready.

14.3 The Measurement Problem

ALICE: In Chapter 5 where we introduced quantum measurements, we mentioned what is called the "measurement problem." The difficulty is trying to find a theoretical description, ideally within the framework of standard quantum mechanics, for what happens when a qubit (or any other quantum entity) interacts with a measurement device.

As we have seen, the probability interpretation of quantum mechanics (the Born rule) comes into play with measurements. Although we haven't said much about how quantum states evolve in time, we have implicitly assumed that if I produce a qubit in a superposition state and launch the qubit to Bob, the state stays the same unless some interaction with other qubits intervenes. If those interactions do occur, we describe the changes in a state via matrices that represent reversible processes. The problem is that measurements, as normally construed, are irreversible. The measurement postulate asserts that the appropriate state for the qubit after a measurement has been made is the basis state associated with the observed value of that property—spin-up or spin-down, for example. But the postulate does not describe a mechanism by which that happens. How do we account for the irreversibility? That is one aspect of the measurement problem. The measurement problem continues to generate spirited debates among scientists and philosophers.

CARDY: Do you mean that when philosophers get tired of talking about Kant, Hegel, and Heidegger, they worry about quantum mechanics?

ALICE: Well, certainly not all philosophers do that. But those who specialize in the philosophy of science find quantum mechanics to be a fertile field in which to plant the seeds of their ideas about reality and the nature of science. Bob will explain more about the measurement problem.

BOB: At a practical level, we know how to make a measurement device work. For linearly polarized photons, for example, we use a polarizing beam splitter to separate the qubits depending on the measurement outcome and then we use a photodetector to indicate in which state the photon was found. The detection of the photon is an irreversible event. We can then store the result by turning on a light that goes with the appropriate channel or in a number that gets stored in a classical computer. So, the basic issue is how to reconcile the irreversibility of the measurement process with the normally reversible quantum processes among qubits.

The general strategy is to invoke arguments based on the number of qubits (or other quantum entities) that make up the measurement device. If the number of qubits is large enough, the interactions between the original qubit and the measurement qubits get spread out over so many processes that, "for all practical purposes" (as John Bell wrote), the overall measurement process is irreversible even though each contributing individual process is reversible. We use that same kind of argument in describing thermal energy ("heat") flow from a hot object to a cool object. We know that for all practical purposes, we never see thermal energy flow spontaneously from cool regions to hot regions. Individual particles in the cool region may gain or lose kinetic energy, but if more of them gain and fewer lose kinetic energy, the region's temperature goes up. Since energy is conserved, the other region's average energy must go down and consequently its temperature goes down. If we do want to move thermal energy from a cool region (the inside of your house during the summer) to a warmer region (the air outside your house in summer), we must use a gadget such as an air conditioner to move that energy in the desired direction.

For quantum mechanics, the difficulty has been in finding a good theoretical criterion to decide when a process is irreversible enough to count as a measurement. A measurement device is essentially a classical machine; so, we can say that the measurement problem is a problem in defining the boundary between the quantum world and the classical world.

Absent an agreed-upon solution to the measurement problem, we use our practical knowledge of what works for actual measurements, hold our noses, and proceed to build all sorts of interesting quantum gadgets, including quantum computers. In practical terms, we know how to make measurements and to check those measurement results with the theoretical predictions of quantum mechanics. That combination has worked extraordinarily well so far, which to me indicates that we are just not asking the right questions about measurements.

Some authors have decided that ordinary quantum mechanics is perfectly fine, but that we need to add additional ideas to account for what happens in a measurement. How to describe that reduction from many possible outcomes to one outcome is another form of the measurement problem in quantum mechanics You may have heard of parallel universes, consistent histories, quantum Bayesians, and so on, which are the signatures of various attempts to figure out that connection. The Further Reading section has citations to books and articles where you can find further information about these issues.

14.4   Measurement and Decoherence

ALICE: The measurement issues spill over into another aspect of quantum states—an aspect that is of major concern for QC. This problem is called *decoherence*. We design our measurement devices to interact with our qubits at specified times and locations. However, the qubits are not completely isolated from their surroundings—their environments. Most often the environment has electromagnetic fields (thermal radiation, light, electric fields) and molecular interactions, for example, that affect our qubits. Those interactions mean that we need to take the quantum states of the environment into account when describing the behavior of our qubits. In essence, the overall system (qubits plus the environment, which is of course nothing more than a vast collection of qubits itself) is described by an entangled state. We distinguish this situation from a measurement interaction because often (almost always) we have no control over the interaction of our qubits with the environment. As we discussed in the previous section, we build a measurement device so that only certain of its states are correlated with our qubit's states.

To take the environment into account, we could write the state of qubit plus environment as

$$\underbrace{|A_0\rangle}_{\text{qubit state}} \underbrace{\{c_a|e_a\rangle + c_b|e_b\rangle + ...\}}_{\text{state of the environment}}, \tag{14.4}$$

where $|e_a\rangle$ is a state of the environment and c_a is the corresponding state amplitude. Unfortunately, we don't know anything about those amplitudes, and in most cases they are changing with time. That means that the relative phases in any qubit superposition state we prepare become washed out (they suffer decoherence) due to interactions with the environment. This is bad for quantum computing! One of the tricks of the trade in building QCs is minimizing the effects of the environment on the computational qubits. Of course, we can't eliminate such effects completely and that is one reason a lot of effort goes into developing quantum correction routines to detect and correct the effects due to unintended interactions with the environment.

CARDY: What does a "state of the environment" mean? It seems to me that it would be almost impossible to figure out that kind of state because the environment has so much stuff in it.

BOB: You are right, Cardy. In writing Eq. (14.4), we are just schematically showing what the state of the system might look like, not that we could actually figure out those states and do calculations with them.

ALICE: Decoherence is an example of the automotive makers' old motto: "If you have a lemon, sell lemonade." There are ways of using these interactions to enhance the sensitivity of a qubit to its environment in a controlled way; so, the qubit acts as an enhanced sensor for the environmental conditions. For example, we might be able to use a qubit as an enhanced sensor to monitor magnetic fields or electromagnetic waves. In these situations, we are focusing on just one aspect of the environment, and in that case Eq. (14.4) makes good sense.

BOB: As an example, let's look at an experiment (Lachance-Quiron et al., 2020) that shows that a qubit can be used to detect the smallest possible magnetic excitation in a magnetic crystal. The excitation is called a "magnon" (in analogy with photon for electromagnetic radiation). The crystal was yttrium iron garnet, often used in microwave, optical, and data-storage devices.

The experiment set up an interaction between the magnon and the qubit via a microwave radiation device similar to the ones used in microwave ovens. That interaction leads to a shift in the energies associated with the qubit states depending on the number of magnons present in the crystal. (Under these experimental conditions, the qubit has two energy states, $|E_0\rangle$ and $|E_1\rangle$.) The changes in the energy states then lead to changes in the results of measuring the qubit state. For example, if no magnon is present, you are more likely to measure the energy state E_1, while observing E_0 is correlated with having one or more magnons present. It turns out that we can measure those qubit energy changes with sufficient precision that we end up with enhanced sensitivity to the presence of the magnons compared to traditional magnetic field measurement methods. Enhanced quantum sensing might be the most practical use of QIS even if QCs never reach their potential.

14.5   The Relationship between Classical and Quantum Computing

ALICE: The discussion in the previous sections leads to the interesting question about the relationship between classical computing and quantum computing. The indistinct boundary between quantum states and quantum measurements carries over to that relationship. In Chapter 2, we said that classical computing states—the high voltages and low voltages in the typical modern computer—are persistent states: We can look at them to determine what gates should be doing, and we can connect them with wires to many gates at the same time without disturbing the state. In fact, with most computers today, you can turn off the computer and when you turn it back on, it remembers what you had been doing just before you turned it off. The classical computing states are nothing if not persistent.

BOB: There is nothing magical about that persistence, at least from an engineering point of view. It has been known for well over a century that those states can be made persistent by building in feedback mechanisms that keep the circuitry in the desired state until you are ready to have it switch states. In principle, this could be mechanical feedback like an on–off electrical switch in your home. The lamp on my desk has pull-chain switches to turn the light bulbs on and off. If I pull on the chain, I hear a click indicating that some mechanical process has turned the switch from the on position to the off position or vice versa. In electronic computers, engineers build in "flip-flops" and other switch-like mechanisms that play the same role for the high-voltage and low-voltage states of the circuits.

So, once again the notions of irreversibility and persistence come into play. The classical computer switches can be quite small with millions of them built into a single integrated circuit "chip." Such switches have even been built from single molecules that switch from one state to another by changing their shapes. The switching becomes irreversible in the sense that the molecule needs a "nudge" to go from one shape to the other.

From these perspectives, a classical computer is just a quantum computer that has been designed to have only two persistent states, which we traditionally call 0 and 1. I don't think that anyone is going to build a quantum computer and then add feedback so that it behaves like a classical computer. Today's classical computers are highly refined systems that do their jobs quickly and reliably, except when they fail a few hours before one of my important deadlines! But as a matter of principle, a classical computer is just a quantum device built from billions of qubits, organized to provide persistent computational states.

14.6   Entanglement and Quantum Measurements

ALICE: Entangled states also show up, perhaps surprisingly, in some descriptions of quantum measurements. John von Neumann (1903–57), one of the pioneers of quantum theory, proposed that measurements in quantum mechanics should be understood as creating entangled states between the quantum system state (a qubit's state in our language) and the measurement system (usually built out of a lot of qubits). At first sight, that formulation seems to contradict Bob's mantra "measurement crushes entanglement." Let's explore these ideas.

A measurement device is engineered, based on our knowledge of physics, so that it has a state $|M_0\rangle$ correlated with the Alice's basis state $|A_0\rangle$, for example, and another state $|M_1\rangle$ correlated with Alice's basis state $|A_1\rangle$. That means that if we observe the measurement device in the state $|M_1\rangle$ after the measurement interaction, then we know the incoming state had a state component along $|A_1\rangle$. If you understand those last two sentences, you have come a long way in developing your understanding of QIS!

In von Neumann's picture, measurement systems are set up so that the output of a measurement device is expressed as

$$\frac{1}{\sqrt{2}} \left\{ |A_0\rangle \, |M_0\rangle + |A_1\rangle \, |M_1\rangle \right\}. \tag{14.5}$$

Often authors will say, pointing to an expression like Eq. (14.5), that the measurement interaction leaves the measurement device in a superposition of the two states $|M_0\rangle$ and $|M_1\rangle$. Schrödinger's infamous cat, viewed as a measurement device for radioactive decay, is, in this view, both dead and alive. Based on our discussion of entangled states in previous chapters, that kind of statement is just plain wrong. We have seen that if a system of qubits is described by an entangled state, we cannot say anything about the states of the individual qubits. In fact, in some sense, the individual qubits don't "have" quantum states if the overall state is an entangled one.

The key question is then: How does the combined object-measurement device system "collapse" to a specific measurement outcome? Von Neumann's model does not address that issue.

If we want a piece of equipment to serve as a measurement device, its state after the measurement interaction must be persistent so we can record the result without changing the measurement device state. For example, in the situation described in Eq. (14.5), if we obtain the result $|M_0\rangle$ for the measurement apparatus, we know from the design of the apparatus that the qubit state must have had a component along $|A_0\rangle$. So, the measurement problem is then the problem of figuring out under what conditions a measurement device state is persistent and not significantly disturbed by itself being measured. We know how to build such devices, and they work reliably. But how do we provide a theory that tells us, for example, how many qubits the measurement device must have to end up in a persistent state.

BOB: In the examples we showed earlier, after the measurement interaction the red light or the green light should be on and both Alice and I (and even Cardy) should be able to look at that state without disturbing it. As we have seen, such persistence does not hold for generic quantum states. Measuring the measurement device (if it is described by an ordinary quantum state vector) would in general put the measurement device into a new state: one of the basis states of the device that is measuring the measurement device. What breaks this chain of measuring the previously measured measurement device? Eq. (14.5) certainly does not solve the measurement problem.

In practice, we think we know how to carry out a measurement on a quantum system using magnets, polarizers and light detectors, lasers, and so on. Once again, we see that the measurement problem is how to describe those results formally within the framework of quantum mechanics itself without bringing in notions such as "macroscopic," "large scale," and so on, which don't seem to be part of quantum mechanics.

14.7　Entanglement and Correlations of Measurement Outcomes

ALICE: In Chapter 9, we saw that entangled quantum states have unusual properties. First, none of the qubits whose system state is entangled has its own state. Second, measurements on one part of the multi-qubit entangled system affect the state we assign to the other parts. At a fundamental level, these issues are all tied to the concept of correlation. For an entangled state, outcomes of measurements carried out on different parts of the system are correlated.

Correlations are quantitative relationships between (or among) sets of data—an issue that is important in statistical analysis. However, for our purposes, we will need only a few simple ideas about correlations. In this section, we will show how correlations among measurement outcomes are related to entanglement. Those results will give us yet another way of determining if a quantum state is entangled. They also tell us that if we want to develop a quantum algorithm that includes correlations, we need to build in entangled states.

CARDY: I recall correlations from my statistics course. It has something to do with if x increases when y increases, then x and y are correlated.

ALICE: That's the basic idea. And indeed, correlations occur in classical systems as well as in quantum systems. So, let's start by thinking about classical correlations. Imagine that we have two cubes: one red and one green. If I randomly send one of the cubes to Bob and he observes that his cube is green, then he knows with certainty that my cube is red and vice versa. In this case there is a definite correlation between the colors that we observe. We might call that *deterministic* correlation because the individual observations are perfectly correlated.

There is another, and very important, type of classical correlation: correlations among stochastic (random) variables. Stochastic means that the results of repeated observations fluctuate from one measurement to the next. In some cases, the variables themselves, such as people's heights, vary among the population being studied. In other cases, for example in measuring outdoor temperatures, the temperature itself might be fluctuating due to changing weather conditions. In such cases, we can have a situation in which the individual observations do not exhibit obvious correlations, but the averages of the observations do. These correlations are the bread and butter of statistical analysis. For example, clinical studies of vaccine effectiveness look for correlations between the vaccine dose and (hopefully) a smaller probability of catching a disease.

BOB: However, statistical correlations can be misleading. In a 2012 paper in the *New England Journal of Medicine*, the author showed that there is a statistical correlation between the chocolate consumption per capita in a country and the number of its scientists who have won the Nobel Prize. This relationship does not hold for all countries, but on average the correlation seems to be there. The author concluded that eating chocolate increases your cognitive capabilities.

CARDY: I am going to run out and grab a few chocolate bars to help me prepare for my finance exam tomorrow.

BOB: Before you invest too much money in chocolate, you need to remember that "correlation is not causation." Perhaps chocolate consumption has nothing directly to do with cognitive abilities but is just a reflection of the wealth of a country, which also leads to more investment in scientific research, which leads to more Nobel Prizes. Also, in this case the author mixed up large group data (the country's chocolate consumption as a whole) and small group data (the Nobel Prize winners). The author failed to check how much chocolate the Nobel Prize winners themselves actually consumed. This cautionary tale reminds us that we must be careful in drawing conclusions from statistical correlations

ALICE: The correlations we need to describe quantum entanglement are relationships among *average* results of measurements. In statistics and probability theory, the average is also called the mean, the expected value, or the expectation value, though I could never figure out who is expecting what. We will stick with "average."

BOB: The history of "expectation value" is kind of interesting. In the 1600s, mathematicians became interested in predicting the outcome of different kinds of games of chance, basically various forms of gambling. They wanted to estimate how much they might expect to win or lose from the games with different rules and different odds. The word "expectation" latter became associated with those calculations, and by the early 20th century the letter E was used to express expected or expectation values.

ALICE: Nice to know. Let's review the basic concept of average. Suppose we have carried out N spin orientation measurements. Each measurement result will be $+1$ (spin-up) or -1 (spin-down). The appropriate quantum operator is the Pauli Z operator. The average of those measurements is given by

$$\langle Z \rangle = (1 + 1 - 1 - 1 - 1 + 1... + 1)/N, \tag{14.6}$$

where we used the angle bracket expression $\langle Z \rangle$ to indicate the average value of the z component measurements.

Try It 14.3

Thirteen students take an exam. Their grades on the exam are 88, 90, 95, 90, 88, 40, 40, 55, 35, 92, 60, 64, and 59. What is the average grade for the class? Note that the average is not equal to any of the actual exam grades.

ALICE: How do averages give us a quantitative way of expressing correlations among measurements? There are many, but we will focus on just one. Suppose we have two stochastic variables F and G—ones that have significant variation across a set of measurements. We will make use of three averages: $\langle F \rangle$, $\langle G \rangle$, and $\langle FG \rangle$. The degree of correlation between F and G is often expressed by the so-called covariance:

$$\text{Covariance}(F, G) = \langle FG \rangle - \langle F \rangle \langle G \rangle. \tag{14.7}$$

If the covariance is equal to zero (or very close to zero compared to the averages), we say that the two variables are uncorrelated.

Try It 14.4

The following table gives the results of 10 sets of measurements of two stochastic variables F and G. Calculate the covariance of the results given in the table. Are the variables correlated or not? Explain your reasoning.

F	6.9	8.0	8.5	4.7	4.6	8.1	0.8	3.7	5.0	5.4
G	2.2	9.1	5.9	5.7	1.5	3.7	8.8	1.5	2.1	4.0

As we have seen, quantum measurements involve probability and randomness; so, we might guess that covariance can be used to express correlations among quantum measurements.

We'll use a two-qubit system to demonstrate explicitly quantum measurement correlations. We'll see that those correlations can often tell us whether the system state is entangled or not. In a sense, this method is a generalization of the entanglement criteria we introduced in Chapter 9.

We focus on a two-qubit system in which each of the qubits is a spin-1/2 object. In Chapter 7, we noted that for those qubits there are two possible states for the spin component along a spatial direction. We will follow common practice and use the direction indicated by a z axis. If we measure a qubit's spin along the z axis, we get either "spin-up" with projection $+1$ (in suitable units) or "spin-down," with projection -1.

If we measure the z projection for a set of identically prepared qubits, each described by a superposition state

$$|\psi_1\rangle = a_\uparrow |\uparrow\rangle + a_\downarrow |\downarrow\rangle, \tag{14.8}$$

we get a random series of $+1$s and -1s. Since we are now expert quantum mechanicians, we know that the probability of getting $+1$ is $a_\uparrow^2$ and the probability for getting -1 is $a_\downarrow^2$, assuming as usual that $a_\uparrow^2 + a_\downarrow^2 = 1$.

BOB: For many situations, it is helpful to write the average in a slightly different way by making use of the probability (relative frequency) of each measurement outcome. For our spin-1/2 qubit, suppose we get spin-up $n_\uparrow$ times and spin-down $n_\downarrow$ with $n_\uparrow + n_\downarrow = N$. That means the measurement outcome is $+1$ for $n_\uparrow/N$ fraction of the measurements and -1 for $n_\downarrow/N$ fraction of the measurements. In other words, the relative frequency of getting $+1$ is $P_\uparrow = n_\uparrow/N$ and the relative frequency for -1 is $P_\downarrow = n_\downarrow/N$. If N is fairly large, we expect that those relative frequencies will be close to the probabilities we obtain from the superposition state amplitudes $P_\uparrow \approx a_\uparrow^2$ and $P_\downarrow \approx a_\downarrow^2$. So, we can write the average in terms of the probabilities for each measurement outcome and the outcome values ($+1$ and -1):

$$
\begin{aligned}
\langle Z \rangle &= (\text{probability of getting spin - up}) \times (+1 \text{ for spin - up}) \\
&\quad + (\text{probability of getting spin - down}) \times (-1 \text{ for spin - down}) \\
&= (+1)\, P_\uparrow + (-1)\, P_\downarrow = (+1)\, a_\uparrow^2 + (-1)\, a_\downarrow^2.
\end{aligned}
\tag{14.9}
$$

The advantage of writing the average this way is that it is easy to generalize: You add up the products of the probabilities for each of the measurement basis states and the value associated with Z in that state. It is important to note that the average will be different for different quantum states, which will have different amplitudes.

CARDY: So, if the probabilities for spin-up and spin-down are equal, it seems that the average $\langle Z \rangle = 0$. Is that right?

ALICE: Yes, that's right. We are now ready to look at the two-qubit system. Let's label the qubits with our usual A and B. The quantities we want to calculate are the average of the Z components of the spin $\langle Z_A \rangle$ and $\langle Z_B \rangle$ for each of the two qubits. The other quantity we need is $\langle Z_A Z_B \rangle$. This means that we measure both qubits' z component of spin. It turns out that in general for an entangled two-qubit state $\langle Z_A Z_B \rangle \neq \langle Z_A \rangle \langle Z_B \rangle$ and for a simple product (non-entangled) state $\langle Z_A Z_B \rangle = \langle Z_A \rangle \langle Z_B \rangle$. The difference between $\langle Z_A Z_B \rangle$ and $\langle Z_A \rangle \langle Z_B \rangle$, the covariance, is a measure of the correlation in the qubits' system state:

$$
\text{Covariance}\, (Z_A, Z_B) = \langle Z_A Z_B \rangle - \langle Z_A \rangle \langle Z_B \rangle.
\tag{14.10}
$$

As a specific example, suppose we have prepared the two qubits in the Bell state

$$
|\psi_2\rangle = |\text{Bell}_{00}\rangle_{AB} = \frac{1}{\sqrt{2}} \left(|\uparrow_A\rangle\, |\uparrow_B\rangle + |\downarrow_A\rangle\, |\downarrow_B\rangle \right).
\tag{14.11}
$$

Since the spin-up and spin-down states are the measurement basis states for the z component of the spin, we know that

$$
Z_A\, |\uparrow_A\rangle = +1\, |\uparrow_A\rangle \text{ and } Z_A\, |\downarrow_A\rangle = -1\, |\downarrow_A\rangle.
\tag{14.12}
$$

When we apply Eq. (14.12) to the parts of the superposition state in Eq. (14.11), we get

$$
\begin{aligned}
Z_A\, |\uparrow_A\rangle\, |\uparrow_B\rangle &= +1\, |\uparrow_A\rangle\, |\uparrow_B\rangle \\
Z_A\, |\downarrow_A\rangle\, |\downarrow_B\rangle &= -1\, |\downarrow_A\rangle\, |\downarrow_B\rangle,
\end{aligned}
\tag{14.13}
$$

with analogous results for measurements on B.

Using those results to calculate the average value of the z component measurements, we find

$$\langle Z_A \rangle = \underbrace{\frac{1}{2}}_{\text{probability for } \uparrow} \underbrace{(+1)}_{\text{result for } \uparrow} + \underbrace{\frac{1}{2}}_{\text{probability for } \downarrow} \underbrace{(-1)}_{\text{result for } \downarrow} = 0, \tag{14.14}$$

just as Cardy predicted. Note that the average value is not equal to either of the individual measurement results. Similarly, we get $\langle Z_B \rangle = 0$. We immediately conclude that $\langle Z_A \rangle \langle Z_B \rangle = 0$.

Now we need $\langle Z_A Z_B \rangle$. Let's write this out for each of the parts of Eq. (14.11):

$$Z_A Z_B \left| \uparrow_A \right\rangle \left| \uparrow_B \right\rangle = (Z_A \left| \uparrow_A \right\rangle)(Z_B \left| \uparrow_B \right\rangle) = (+1)(+1) \left| \uparrow_A \right\rangle \left| \uparrow_B \right\rangle$$

$$Z_A Z_B \left| \downarrow_A \right\rangle \left| \downarrow_B \right\rangle = (Z_A \left| \downarrow_A \right\rangle)(Z_B \left| \downarrow_B \right\rangle) = (-1)(-1) \left| \downarrow_A \right\rangle \left| \downarrow_B \right\rangle . \tag{14.15}$$

From Eqs. (14.15) and (14.9), we find

$$\langle Z_A Z_B \rangle = 1. \tag{14.16}$$

The covariance between these spin measurements is

$$\text{Covariance}\,(Z_A, Z_B) = \langle Z_A Z_B \rangle - \langle Z_A \rangle \langle Z_B \rangle$$
$$= 1 - 0 = 1. \tag{14.17}$$

We conclude that the z component measurements for the two parts of the system described by the state in Eq. (14.11) are correlated and hence the state is an entangled state. In fact, this result tells us that the z spin measurements for the Bell state in Eq. (14.11) are perfectly correlated.

CARDY: Oh, I see it! Whenever we get spin-up for qubit A, we also get spin-up for qubit B, and similarly for spin-down. That means the measurements are completely correlated.

Try It 14.5

Suppose the two-qubit state is the product state $\left| \psi_3 \right\rangle = \left| \uparrow_A \right\rangle \left| \uparrow_B \right\rangle$. What is the value of the covariance of the Z_A and Z_B measurements? What about the state $\left| \psi_4 \right\rangle = \left| \uparrow_A \right\rangle \left| \downarrow_B \right\rangle$? Explain the results.

Try It 14.6

Show that the covariance for z spin measurements for the state $|\psi_3\rangle = |\text{Bell}_{11}\rangle = \frac{1}{\sqrt{2}}(|\uparrow_A\rangle|\downarrow_B\rangle - |\downarrow_A\rangle|\uparrow_B\rangle)$ equals -1. We say that for this state, the spin measurements are perfectly "anti-correlated." Whenever you get spin-up for qubit A, you get spin-down for qubit B and vice versa.

ALICE: Let's sum up what we have learned: For a product (non-entangled) state, the co-variance of two different measured variables equals 0. If the covariance is not 0, the state is an entangled one. However, it turns out that for a given entangled two-qubit state, you can always find two operators whose covariance is 0. That means that getting a covariance equal to 0 does not guarantee having a non-entangled state. If you get a 0 covariance for one pair of operators, calculating the covariance of another pair of operators will usually tell you if the state is entangled or not.

Measurements of correlations can also be used to test the "fidelity" of the implementation of quantum gates in real-life quantum communications devices and quantum computers. If the actual gates can maintain the correlations, the gates are said to have "high fidelity." That means that the errors introduced by the gates are small.

14.8 Another Look at Averages in Quantum Mechanics

ALICE: There is a nice way of writing averages of measurements in quantum mechanics using the Dirac bracket notation. This method often shows up in discussions of QC. The basic idea is that the average can be written, as we saw in the previous section, as a sum of products of probabilities and measurement outcome values. We also know that the probabilities can be written as the squares of the amplitudes associated with a particular state, and that the sum of those squares shows up when we multiply a right (column) state vector by the corresponding left (row) state vector. For example, for a two-qubit system with state amplitudes c, d, e, and f, we have

$$\begin{pmatrix} c & d & e & f \end{pmatrix} \begin{pmatrix} c \\ d \\ e \\ f \end{pmatrix} = c^2 + d^2 + e^2 + f^2. \tag{14.18}$$

The final ingredient is the fact that every measurable property M can be represented by an operator and that the appropriate measurement basis states are the states that satisfy

$$M|\alpha\rangle = \lambda_\alpha |\alpha\rangle$$
$$M|\beta\rangle = \lambda_\beta |\beta\rangle, \tag{14.19}$$

where λ_α and λ_β are the numerical values associated with the operator M. States that satisfy these conditions are called eigenstates or eigenvectors associated with the operator M. So, λ_α and λ_β are called the eigenvalues for M.

How does this tie into average values? For a single qubit, the quantum state vector can always be written as a sum of M's eigenvectors:

$$|\psi\rangle = a_\alpha |\alpha\rangle + a_\beta |\beta\rangle . \tag{14.20}$$

When the operator M acts on this state, we get

$$M|\psi\rangle = a_\alpha M|\alpha\rangle + a_\beta M|\beta\rangle = a_\alpha \lambda_\alpha |\alpha\rangle + a_\beta \lambda_\beta |\beta\rangle . \tag{14.21}$$

In matrix form, Eq. (14.21) becomes

$$(M)\begin{pmatrix} a_\alpha \\ a_\beta \end{pmatrix} = \begin{pmatrix} \lambda_\alpha a_\alpha \\ \lambda_\beta a_\beta \end{pmatrix} . \tag{14.22}$$

The next step might not be obvious, but it turns out to be just what we need to calculate the average of a sequence of measurements of M. We multiply Eq. (14.21) on the left by the left vector associated with the state $|\psi\rangle$:

$$\langle\psi| M|\psi\rangle \Rightarrow \begin{pmatrix} a_\alpha & a_\beta \end{pmatrix}\begin{pmatrix} a_\alpha \lambda_\alpha \\ a_\beta \lambda_\beta \end{pmatrix} = a_\alpha^2 \lambda_\alpha + a_\beta^2 \lambda_\beta. \tag{14.23}$$

Surprisingly, the result is just the average of the M measurements, a generalization of the average value expression in Eq. (14.9). The Dirac bracket is a reminder that the left side of Eq. (14.23) is indeed an average of the measurements of the property represented by the operator M given that the qubit has been prepared in the state $|\psi\rangle$. Some authors like to call the bracket $\langle\psi| M|\psi\rangle$ a "sandwich": The operator M is sandwiched between two slices of ψ bread.

CARDY: Does the sandwich come with quantum fries? Can I have mine with pickles and tomatoes?

ALICE: Indeed, you can put any combination of operators in the bracket. Evaluating the bracket will give you the average of the measurement outcomes for that combination of operators when the system is prepared in the state $|\psi\rangle$.

14.9   Where is *h*?

CARDY: I showed my notes to some friends who are taking a formal quantum mechanics course this semester. They thought the material is pretty cool. But they had a question I couldn't answer. The question was: "Where is h?"

ALICE: Those are very observant friends! Yes, if you study traditional quantum mechanics, there is a fundamental parameter h called Planck's constant, after Max Planck (1858–1947), another of the founders of quantum mechanics. h has the numerical value $6.62607004 \times 10^{-34}$

Joule-second. It shows up whenever you need to know the actual energy or size of a qubit or any other quantum system. For example, the energy E of a photon associated with light with a frequency f (number of oscillations per second) is given by $E = h\,f$. Similarly, the size of an atomic orbital also involves h.

Those energies and sizes are quite important for actually building a QC. But, as we have seen, the quantum algorithms (the logical structures) do not involve h. An analogy for classical computers is that there is no mention of the charge of the electron in any of the classical gates. Their logical structures are abstractions and are independent of the physical system used to implement them. That means that for much of QIS and QC, you don't need to know the actual physical details of the qubits and how they are manipulated, just as for classical computing you don't need to know about the properties of electrons in semiconductors. That's why you don't see h in many descriptions of quantum algorithms.

 CHAPTER SUMMARY

- Bell's theorem provides theoretical predictions that can be used to test local, deterministic (hidden variable) models against the corresponding quantum models of correlations among measurement outcomes for multi-qubit systems. Experiments rule out the local, deterministic models.

- The notion of measurement remains, in principle, somewhat problematic at the fundamental level. However, the "measurement problem" does not seem to provide any constraints in building actual quantum computers.

- The interaction of qubits with their environment leads to decoherence of the qubits' quantum states and thus requires quantum error correction in QIS and QCs.

- Structured decoherence can lead to enhanced quantum sensing devices, which in the short term may be more important than QCs.

- Correlations among measurement outcomes in multi-qubit states can be used to test for entanglement.

- Measurements of correlations can also be used to test the "fidelity" of the implementation of quantum gates in real-life quantum communications devices and quantum computers.

 FURTHER READING

Experiment demonstrating entanglement distribution

J. Yin, Y. Cao, Y.-H. Li, S.-K. Liao, L. Zhang, J.-G. Ren, W.-Q. Cai, W.-Y. Liu, B. Li, H. Dai, G.-B. Li, Q.-M. Lu, Y.-H. Gong, Y. Xu, S.-L. Li, F.-Z. Li, Y.-Y. Yin, Z.-Q. Jiang, M. Li, J.-J. Jia, G. Ren, D. He, Y.-L. Zhou, X.-X. Zhang, N. Wang, X. Chang, Z.-C. Zhu, N.-L. Liu, Y.-A. Chen, C.-Y. Lu, R. Shu, C.-Z. Peng, J.-Y. Wang, and J.-W. Pan, "Satellite-based entanglement distribution over 1200 kilometers." *Science* **356** 1140–1144 (2017).

Using a qubit as an enhanced quantum sensor

D. Lachance-Quirion, S. P. Wolski, Y. Tabuchi, S. Kono, K. Usami, and Y. Nakamura, "Entanglement-based single-shot detection of a single magnon with a superconducting qubit." *Science.* **367** 425–428 (2020).

Bell's theorem and related issues

Daniel F. Styer, *The Strange World of Quantum Mechanics* (Cambridge University Press, Cambridge, 2000) has a nice discussion of experimental tests of Bell's theorem (pp. 41–47) without mathematics but with all the physics.

George Greenstein and Arthur Zajonc, *The Quantum Challenge: Modern Research on the Foundations of Quantum Mechanics*, 2nd edn. (Jones and Barlett, Sudbury, MA 2006). A comprehensive and comprehensible treatment of recent experiments on Bell's theorem, delayed-choice measurements, and so on.

Tanya Bub and Jeffrey Bub, *Totally Random: Why Nobody Understands Quantum Mechanics (A Serious Comic on Entanglement)* (Princeton University Press, Princeton 2018). A visual introduction to the mysteries of quantum mechanics. Definitely not your grandmother's textbook on quantum mechanics. Bell's theorem (inequality) is described on pages 36–41. More information is available at http://www.totallyrandom.info.

Chris Bernhardt, *Quantum Computing for Everyone* (MIT Press, Cambridge, MA, 2019). Bell's inequality, pages 79–84.

Robert Ross, "Computer simulation of Mermin's quantum device." *Am. J. Phys.* **88**, 483–489 (2020). The article shows how to use a classical computer to simulate the quantum predictions for the Bell's theorem device described in this chapter.

Correlations

F. H. Messerli, "Chocolate consumption, cognitive function, and Nobel laureates." *New England Journal of Medicine* **367**, 1562–1564 (2012).

The conclusions about chocolate and Nobel Prizes have been critiqued many times. See, for example, Vladica Velickovic, "What Everyone Should Know about Statistical Correlation." *American Scientist* **103**, 26 (2015). DOI: 10.1511/2015.112.26

David Mermin, "What is quantum mechanics trying to tell us." *American Journal of Physics* **66**, 753 (1988). The author argues that quantum mechanics can be freed of "the tyranny of measurements" by focusing on correlations.

Jed Brody and Gavin Guzman, "Calculating spin correlations with a quantum computer." *American Journal of Physics* **89**, 35 (2021). Discusses correlations in quantum measurements and how to calculate them using a quantum computer accessible online.

15 Complexifying Quantum States

For every complex problem there is an answer that is clear, simple, and wrong.

H. L. Mencken *Prejudices* (1920)

15.1 Complex Numbers and Variables

ALICE: In this chapter, we will introduce one aspect of quantum mechanics that we have purposely avoided up to now: the use of complex numbers and variables. Standard quantum mechanics starts off by postulating that state vector amplitudes are complex numbers. Those complex numbers are necessary to be able to describe the full range of quantum phenomena. However, as we have seen, a large part of QIS and QC can be understood without complex numbers.

CARDY: All I remember about complex numbers is that they have something to do with the square root of -1. I could never get my mind wrapped around that idea. How can you have a number be the square root of a negative number? Plus times plus gives plus. Negative times negative also gives plus. So, what's up?

BOB: Let's back up a bit and see why complex numbers occur in the first place. One place to start is with everyone's favorite mathematical expression: the quadratic formula, which arises from the solution of a quadratic equation (one involving x^2):

$$ax^2 + bx + c = 0, \tag{15.1}$$

whose solutions are

$$x = \frac{-b \pm \sqrt{b^2 - 4ac}}{2a}. \tag{15.2}$$

There are two solutions: one with the $+$ sign and the other with the $-$ sign in front of the square root. The need for some new math occurs when the combination of terms under the square root sign is negative: that is, when $b^2 - 4ac < 0$. This easily happens. For example, if $b = 8$, $a = 2$, and $c = 16$ in Eq. (15.1), we have $b^2 - 4ac = 64 - 128 = -64$. To handle that situation, we can write $\sqrt{-64} = \sqrt{-1}\sqrt{64} = i8$. We introduced the symbol $i = \sqrt{-1}$ with the property that $i \times i = -1$. This is obviously a mathematical entity different from the standard real numbers. Eq. (15.2) gives the solutions

$$x = -2 \pm 2i. \tag{15.3}$$

Quantum Computing:From Alice to Bob. Alice Flarend and Bob Hilborn, Oxford University Press.
© Alice Flarend and Robert C. Hilborn (2022). DOI: 10.1093/oso/9780192857972.003.0015

This is the typical form of what is called a complex number: There is a part with i in it and a part without i. For historical reasons, the part without i is called the "real part" and the part with i is called the "imaginary part." But don't be misled: The imaginary part is just as important as the real part. The combination, usually written with a plus (or minus) sign between the two parts, is called a complex number. If we use general mathematical symbols, we call it a complex variable such as

$$z = u + iv. \tag{15.4}$$

Another important concept associated with complex numbers is the *complex conjugate*. If we have a complex variable $z = u + iv$, its complex conjugate (z^*) is defined as $z^* = u - iv$, assuming that u and v are real numbers. (Note that the superscript * is commonly used to denote the complex conjugate.) To write the complex conjugate of a complex number, we simply change the sign of the imaginary term. This may see almost trivial, but it has far-reaching implications.

For example, the product of a complex number with its complex conjugate gives us the square of the "length" ("magnitude") of the complex number:

$$|z|^2 = zz^* = z^*z = (u + iv)(u - iv) = u^2 + v^2 . \tag{15.5}$$

The term $|z|^2$ is also called the absolute-value-squared of the complex number z. We will shortly see a graphical representation for the complex number z that illustrates that $u^2 + v^2$ does make sense as the square of the length of z.

CARDY: I follow what you are saying, but how and why do complex variables and complex conjugates show up in quantum mechanics?

ALICE: The basic "how" answer is that the state vector amplitudes that we have been using throughout our explorations of QIS and QC are generally complex variables. For example, in a generic quantum state such as

$$|\psi\rangle = a_0 |0\rangle + a_1 |1\rangle \tag{15.6}$$

a_0 or a_1, or sometimes both, need to be complex variables with real and imaginary parts for quantum mechanics to work properly (that is, to describe reality). Probabilities are calculated by multiplying the complex amplitudes by their complex conjugates. This is the generalization of the squared-amplitude rule (Born rule) we used previously. In other words, for the state in Eq. (15.6), the probability of observing $|0\rangle$ or $|1\rangle$ when the state interacts with a measuring device is given by

$$P(0) = a_0^* a_0 = |a_0|^2 \quad P(1) = a_1^* a_1 = |a_1|^2 . \tag{15.7}$$

Let's look at the mathematics of a general complex variable $z = u + iv$. We saw in Eq. (15.5) that $|z|^2 = u^2 + v^2$, which looks a lot like the Pythagorean theorem. In fact, it suggests that a complex number $z = u + iv$ can be represented by a two-dimensional geometric diagram as shown in Figure 15.1:

Fig. 15.1 A complex-plane diagram. Re indicates the real part axis and Im indicates the imaginary part axis.

Figure 15.1 shows a "complex plane" diagram in which a complex variable z is viewed as a superposition of a real part u and an imaginary part v: $z = u + iv$. By convention, the vertical axis is called the imaginary axis (labeled Im) and the horizontal axis is called the real axis (labeled Re).

CARDY: That diagram looks a bit like one of our state-space diagrams. Is there a connection?

BOB: That's a good observation, Cardy, but there are no direct connections. However, this kind of representation is useful whenever we have two parts of a mathematical object, and we want those two parts to be orthogonal.

As we mentioned, you can think of a complex variable as a superposition of a real part and an imaginary part, just like we thought about quantum states as superpositions of basis states in state space. You can also define rotations in the complex plane, that take one complex variable into another, just like we had rotations in state space.

ALICE: Complex amplitudes are also used in other features of quantum state vectors. For example, the left vector corresponding to the state in Eq. (15.6) is written as

$$\langle \psi | = a_0^* \langle 0 | + a_1^* \langle 1 | \tag{15.8}$$

in which the state coefficients are replaced by their complex conjugates.

This carries over to column vector and row vector representations:

$$| \psi \rangle \Rightarrow \begin{pmatrix} a_0 \\ a_1 \end{pmatrix} \qquad \langle \psi | \Rightarrow \begin{pmatrix} a_0^* & a_1^* \end{pmatrix}. \tag{15.9}$$

Try It 15.1

Show that using the column and row vectors in Eq. (15.9) leads to $\langle \psi | \psi \rangle = |a_0|^2 + |a_1|^2$.

CARDY: I see how that works out, but why do we need it? To me, it just looks like an unnecessary complication.

ALICE: That's a good question. Let's use a spin-1/2 system of the type we described in Chapter 7 to see why complex numbers and variables are not just useful, but essential for quantum mechanics.

In Chapter 8, we showed that the states for spin aligned along the x axis (in real space) can be written in terms of the z axis (spin-up and spin-down) basis states:

$$|\rightarrow\rangle = \frac{1}{\sqrt{2}}\left(|\uparrow\rangle + |\downarrow\rangle\right)$$

$$|\leftarrow\rangle = \frac{1}{\sqrt{2}}\left(|\uparrow\rangle - |\downarrow\rangle\right). \qquad (15.10)$$

Let's change notation a bit by adding some subscripts and using up and down arrows to indicate spin orientation (up or down) along the chosen axis. That is, we replace $|\rightarrow\rangle$ with $|\uparrow\rangle_x$. This notation will help us in describing states along a third direction, namely the y direction in real space. Using that new notation, we write Eq. (15.10) as

$$|\uparrow\rangle_x = \frac{1}{\sqrt{2}}\left(|\uparrow\rangle_z + |\downarrow\rangle_z\right)$$

$$|\downarrow\rangle_x = \frac{1}{\sqrt{2}}\left(|\uparrow\rangle_z - |\downarrow\rangle_z\right). \qquad (15.11)$$

CARDY: Ah. We are writing the spin-along-x states as superpositions of the basis states with spin along z. I would think it would be easy to write the states for spin along y. Maybe just swap $+$ and $-$ in Eq. (15.11).

ALICE: Well, let's see what happens if we do that. Here is what Cardy's suggested states look like:

$$|\uparrow\rangle_y \xrightarrow{?} \frac{1}{\sqrt{2}}\left(|\uparrow\rangle_z - |\downarrow\rangle_z\right)$$

$$|\downarrow\rangle_y \xrightarrow{?} \frac{1}{\sqrt{2}}\left(|\uparrow\rangle_z + |\downarrow\rangle_z\right). \qquad (15.12)$$

How do we check if this works? There are two mathematical requirements for basis states: (1) each of the states must be normalized, and (2) the two states must be orthogonal. Both of those requirements are satisfied by Cardy's suggested states.

> **Try It 15.2**
>
> Show that the states in Eq. (15.12) are normalized and orthogonal. Hints: For normalization, the sum of the squares of the amplitudes should be 1. For orthogonality, $_y\langle\uparrow|\downarrow\rangle_y$ should be 0. Use the row vector and column vector forms.

ALICE: But, in addition to the mathematical requirements, there is a physics condition to be satisfied. Remember that if we prepare a spin-1/2 qubit in the state $|\uparrow\rangle_x$ but measure the spin orientation along z, the probability of getting spin-up along z is 0.5 and the probability for spin-down is 0.5. The same result should be true if we prepare the qubit state to be $|\uparrow\rangle_y$ and then measure along x. Quantum mechanical democracy means that x, y, and z should all be treated equally. Let's see if that works with Cardy's proposed y states. There are several ways to do this, but it is relatively easy to use the row and column vectors for the amplitudes in the z basis.

$$|\uparrow\rangle_x \Rightarrow \frac{1}{\sqrt{2}}\begin{pmatrix}1\\1\end{pmatrix}_z \quad |\downarrow\rangle_x \Rightarrow \frac{1}{\sqrt{2}}\begin{pmatrix}1\\-1\end{pmatrix}_z$$

$$|\uparrow\rangle_y \Rightarrow ? \frac{1}{\sqrt{2}}\begin{pmatrix}1\\-1\end{pmatrix}_z \quad |\downarrow\rangle_y \Rightarrow ? \frac{1}{\sqrt{2}}\begin{pmatrix}1\\1\end{pmatrix}_z. \tag{15.13}$$

For a specific example, let's calculate the probability amplitude $_x\langle\downarrow|\downarrow\rangle_y$, which by our previous reasoning ought to have its square equal to 0.5. Here is what we find for Cardy's suggested states:

$$_x\langle\downarrow|\downarrow\rangle_y \Rightarrow \frac{1}{2}\begin{pmatrix}1 & -1\end{pmatrix}\begin{pmatrix}1\\1\end{pmatrix} = \frac{1-1}{2} = 0. \tag{15.14}$$

CARDY: Oops! Something is definitely wrong here. Perhaps it is just this specific amplitude.

Try It 15.3

Calculate the other amplitudes with x states and Cardy's y states. Are the results what you expect?

ALICE: Yes, we have a problem. It turns out, though we won't go through the details, that no combination of purely real number amplitudes in the y states will satisfy all the requirements. So, what do we do? Perhaps there is a fundamental flaw in quantum mechanics, and we need an entirely new theory. That would be a mess! And we would not have written the 14 chapters building up this formalism. We can "fix" the problem by allowing the amplitudes to be complex numbers. For the y spin-1/2 states written in terms of the z basis state, it turns out that only two of the amplitudes need to be complex.

We will work through the details, making some reasonable assumptions to simplify the argument. If you are interested in exploring more of the details, see the Further Reading references listed at the end of the chapter.

First, we write the y spin states with general amplitudes in terms of the z basis states:

$$|\uparrow\rangle_y = \frac{1}{\sqrt{2}}\left(a|\uparrow\rangle_z + b|\downarrow\rangle_z\right)$$

$$|\downarrow\rangle_y = \frac{1}{\sqrt{2}}\left(c|\uparrow\rangle_z + d|\downarrow\rangle_z\right). \tag{15.15}$$

Note that we included the ubiquitous $1/\sqrt{2}$ needed for normalization with two-state systems.

CARDY: OK, that is reasonable, but I feel that we are overworking $\sqrt{2}$.

BOB: Don't worry, Cardy. The $\sqrt{2}$ is resilient and doesn't think much about being used over and over again. It's not very rational in that respect.

ALICE: Ouch! If you two continue anthropomorphizing $\sqrt{2}$, it will need its own emoji. Getting back to showing why we need complex numbers, we will see that only b and d need to be complex numbers to satisfy all our requirements. Consequently, we can use $a = 1 = c$ in Eq. (15.15) to speed up our argument. Normalization requires that

$$\frac{1}{2} + \frac{b^*b}{2} = 1 \quad \text{and} \quad \frac{1}{2} + \frac{d^*d}{2} = 1. \tag{15.16}$$

So, we need to make sure that $b^*b = 1 = d^*d$. We also require that the two y basis states be orthogonal:

$$_y\langle\uparrow|\downarrow\rangle_y \Rightarrow \begin{pmatrix} 1 & b^* \end{pmatrix} \begin{pmatrix} 1 \\ d \end{pmatrix} = 1 + b^*d = 0. \tag{15.17}$$

This adds another requirement: $b^*d = -1$. Finally, let's look at the problematic amplitude

$$_x\langle\downarrow|\downarrow\rangle_y \Rightarrow \frac{1}{2}\begin{pmatrix} 1 & -1 \end{pmatrix}\begin{pmatrix} 1 \\ d \end{pmatrix} = \frac{1-d}{2}$$

Since we want the resulting probability (absolute value squared of the amplitude) to be equal to 0.5, we require that

$$\frac{1}{4}(1-d^*)(1-d) = \frac{1}{2}. \tag{15.19}$$

It is easy to check that $d = \pm i$ satisfies Eq. (15.19). Using $d = i$, we get

$$\frac{1}{4}(1-d^*)(1-d) = \frac{1}{4}(1-d^*-d+d^*d) = \frac{1}{4}(1-i+i+1) = \frac{1}{2}. \tag{15.20}$$

CARDY: I am still a bit shaky with complex numbers. If $d = i$, am I correct in concluding that $d^* = -i$ and then $d^*d = (-i) \times i = 1$ and since $b^*d = -1$, then $b = -i$?

ALICE: Right on target! In this case the two amplitudes b and d have only imaginary parts; they are called "pure imaginary."

Try It 15.4

Challenge: Separate d into real and imaginary parts, $d = f + ig$, and use that form in Eq. (15.20) along with $d^*d = 1$ to prove that $d = \pm i$ is the correct solution with $f = 0$ and $g = 1$.

ALICE: Let's put all of this together in both the state right-vector form and in the column vector form using the conventional choice $d = i$:

$$|\uparrow\rangle_y = \frac{1}{\sqrt{2}} (|\uparrow\rangle_z - i|\downarrow\rangle_z) \Rightarrow \frac{1}{\sqrt{2}} \begin{pmatrix} 1 \\ -i \end{pmatrix}_z$$

$$|\downarrow\rangle_y = \frac{1}{\sqrt{2}} (|\uparrow\rangle_z + i|\downarrow\rangle_z) \Rightarrow \frac{1}{\sqrt{2}} \begin{pmatrix} 1 \\ i \end{pmatrix}_z. \tag{15.21}$$

Using these new coefficients, we should check that we can get a sensible answer when we calculate the expression in Eq. (15.14)

$$_x\langle\downarrow|\downarrow\rangle_y \Rightarrow \frac{1}{2} \begin{pmatrix} 1 & -1 \end{pmatrix} \begin{pmatrix} 1 \\ i \end{pmatrix} = \frac{1-i}{2}, \tag{15.22}$$

and the probability is given by the absolute-value-squared $(1/4)(1+i)(1-i) = 1/2$, which is what we expect.

Try It 15.5

Show that the other combinations of probabilities work out for spin-up and spin-down in the x and y states and in the z and y states.

ALICE: Now that we know that two of the amplitudes need to be purely imaginary and the rest real, we should update the Y gate introduced in Chapter 6. The "correct" form is

$$Y = \sigma_y \Rightarrow \begin{pmatrix} 0 & -i \\ i & 0 \end{pmatrix} = i \begin{pmatrix} 0 & -1 \\ 1 & 0 \end{pmatrix}. \tag{15.23}$$

Since many amplitudes in quantum mechanics turn out to be complex numbers, most people assume that they all are and write row vectors with the complex conjugate amplitudes and probabilities as the absolute-value-squared of the amplitudes even if the coefficients turn out to be real numbers. If it turns out that an amplitude, c for example, is a real number, then we have $c^* = c$; so, assuming that the amplitude is complex does no harm.

Although we used a two-state spin system to demonstrate the need for complex amplitudes in quantum mechanics, the same results show up whenever you have three or more "incompatible" properties. For example, we might have three two-state operators U, V, and W. Incompatible means that the operators do not commute: $UV \neq VU$, $UW \neq WU$, and $VW \neq WV$. The corresponding pairs of basis states are $|u_0\rangle$, $|u_1\rangle$, $|v_0\rangle$, $|v_1\rangle$, and $|w_0\rangle$, $|w_1\rangle$. (The three states don't need to be associated with directions in the physical, three-dimensional world.) If measurements using one basis set give no information about measurements with respect to the other two basis sets, then the properties (and the corresponding operators) are said to be maximally incompatible.

If the properties are maximally incompatible, at least some of the state amplitudes need to be complex. Sometimes the term "non-commuting properties" is used. For example, for the spin-1/2 system, we know that $\sigma_x \sigma_z = -\sigma_z \sigma_x$. In other words, the operators for spin projections along the three mutually perpendicular x, y, and z directions do not commute. The order in which the operators act on states matters.

15.2 Complex Exponentials

ALICE: There is one more complex variable situation we want to discuss because it occurs so frequently in quantum mechanics and hence in QIS and QC. This is the complex exponential function built with the fundamental exponential number $e = 2.71828\dots$. In particular, the famous Euler formula links the exponential function to sines and cosines: a truly amazing and extremely useful relationship. The Euler formula is

$$e^{i\phi} = \cos\phi + i\sin\phi \tag{15.24}$$

for any number ϕ. If we use a real number ϕ as the argument, the cosine part of Eq. (15.24) is the real part of $e^{i\phi}$, while the sine term is the imaginary part.

I always found the Euler formula mind-boggling. It combines algebra (sums and exponentials) with trigonometry (sines and cosines) and complex numbers (i). That's a lot of math in one package!

You can also turn the formula around to write cosine and sine in terms of the complex exponentials:

$$\cos\phi = \frac{1}{2}\left(e^{i\phi} + e^{-i\phi}\right) \qquad \sin\phi = \frac{1}{2i}\left(e^{i\phi} - e^{-i\phi}\right). \tag{15.25}$$

Not only is the Euler formula a useful way to do trigonometry, it also leads to a useful geometric interpretation for complex variables. Multiplying a generic complex variable, for example our old friend $z = u + iv$, by $e^{i\theta}$ (with θ real) is geometrically equivalent to rotating the vector representing z in the complex plane by the angle θ without changing its length:

$$\begin{aligned} e^{i\theta}z &= (\cos\theta + i\sin\theta)(u + iv) \\ &= (u\cos\theta - v\sin\theta) + i(u\sin\theta + v\cos\theta) = z'. \end{aligned} \tag{15.26}$$

Try It 15.6

Challenge: Show that Figure 15.2 is in agreement with Eq. (15.26).

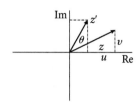

Fig. 15.2 Multiplying z by $e^{i\theta}$ rotates z counterclockwise in the complex plane by the angle θ, to produce another complex number, z'.

ALICE: We can express any complex number z as the product of an amplitude $|z|$ and a complex exponential phase. Again using $z = u + iv$, we have

$$z = u + iv = |z|\{\cos\phi + i\sin\phi\} = |z|\,e^{i\phi}. \tag{15.27}$$

The magnitude of z and the angle ϕ are expressed as

$$|z| = \sqrt{u^2 + v^2}$$

$$\text{with } \cos\phi = u/|z| \qquad \sin\phi = v/|z| \text{ and } \phi = \arctan(v/u). \tag{15.28}$$

There is also a nice geometric way of representing $e^{i\phi}$ as an arrow rotated by ϕ counterclockwise from the real axis. So, e^{i0} is represented by an arrow $\rightarrow$, representing the absence of an imaginary part; $e^{i\pi/4}$ is represented by $\nearrow$ and $e^{i\pi/2}$ is represented by $\uparrow$. The geometric representation makes it easy to see that $e^{i\pi/4} + e^{i5\pi/4} = 0$ because the two arrows $\nearrow + \swarrow = 0$ add to zero. We don't need to put in the numerical values to get that result.

Finally, another useful feature is that the complex exponential multiplied by its complex conjugate gives 1:

$$\left(e^{i\phi}\right)^* \left(e^{i\phi}\right) = \left(e^{-i\phi}\right)\left(e^{i\phi}\right) = e^{i(\phi-\phi)} = e^{i0} = 1. \tag{15.29}$$

15.3   Quantum Fourier Transform

ALICE: The complex exponential function shows up in the quantum Fourier transform method we introduced in Chapter 13 for the Shor factoring algorithm. We had used the operator $U_F(j, k)$ as a cosine function, but in actual practice it is simpler and more general to use the complex exponential form

$$U_F(j, k) = e^{i2\pi j\, k/N}. \tag{15.30}$$

The complex exponential incorporates both sine and cosine contributions and therefore automatically takes care of time shifts (phase shifts) between the signal being analyzed and the sines and cosines taken individually. The complex exponential is also used in operators that express how quantum states change in time. But we won't pursue those ideas in this book.

CARDY: You promised me that when we got to complex exponentials, you would show me how to derive the probability that the Shor algorithm will get the correct period of a function.

ALICE: Thanks for reminding me, Cardy. Let's go through the details. We start by using complex exponentials to express the sum of amplitudes shown in the last term in Eq. (13.19), repeated here

$$\frac{1}{q_{max}N}\left|\sum_{q=0}^{q_{max}}U_F\left(x_0+qr,y\right)\right|^2 . \tag{15.31}$$

Using Eq. (15.30), we have

$$\sum_{q=0}^{q_{max}}e^{i2\pi(x_0+qr)y/N}=e^{i2\pi x_0 y/N}\sum_{q=0}^{q_{max}}e^{i2\pi qry/N}, \tag{15.32}$$

where we temporarily dropped the normalization factors. In the last term of Eq. (15.32), we pulled the part of the exponential that does not depend on q out of the sum. When we calculate the absolute value squared to get the probability, that factor will drop out.

Next, we recognize that the sum in Eq. (15.32) is a geometric series with the common factor $e^{i2\pi ry/N}$. Using the result for the geometric series sum in Eq. (13.22), we find that the sum is

$$s=\sum_{q=0}^{q_{max}}e^{i2\pi qry/N}=\frac{1-e^{i2\pi ryq_{max}/N}}{1-e^{i2\pi ry/N}} . \tag{15.33}$$

To get the probability, we need the absolute value squared $s^*s=|s|^2$:

$$\begin{aligned}|s|^2&=\frac{1-e^{-i2\pi ryq_{max}/N}}{1-e^{-i2\pi ry/N}}\frac{1-e^{i2\pi ryq_{max}/N}}{1-e^{i2\pi ry/N}}=\frac{2-\left(e^{i2\pi ryq_{max}/N}+e^{-i2\pi ryq_{max}/N}\right)}{2-\left(e^{i2\pi ry/N}+e^{-i2\pi ry/N}\right)}\\&=\frac{1-\cos\left(2\pi ryq_{max}/N\right)}{1-\cos\left(2\pi ry/N\right)}\\&=\frac{\sin^2\left(\pi ryq_{max}/N\right)}{\sin^2\left(\pi ry/N\right)} .\end{aligned} \tag{15.34}$$

CARDY: A lot of math went by there. Please explain what happened.

ALICE: Sure. In the first line of Eq. (15.34), we multiplied the geometric series sum by its complex conjugate and then multiplied out the numerator and denominator, remembering that $e^{i\theta}e^{-i\theta}=1$. In the second line, we used Euler's relation in Eq. (15.25) to replace the exponentials with cosines. Finally, in the last line we used the trig relationship $1-\cos\theta=2\sin^2\theta/2$ in both the numerator and the denominator.

Try It 15.7

Go through the algebra and trigonometry in Eq. (15.34) by multiplying out the numerators and the denominators.

ALICE: Remember that the Shor algorithm allows us to find the repetition period r of a function, where r is an integer. In Eq. (15.34), the sine functions will be largest (and the probability will be largest) when the arguments are $\pi/2$ times an odd integer. Since q is, by design, an integer, we want yr/N to be an integer. However, we can't guarantee that yr/N will be exactly an integer. But we do expect y to be close to N/r. So, we introduce a new variable $\delta = y - N/r$ and anticipate that the highest probability will occur when δ is close to 0. We make that substitution in the last line of Eq. (15.34) and use another trig relationship, $\sin(\theta + n\pi) = \pm \sin\theta$. Since we are squaring the sines in Eq. (15.34), it doesn't matter whether we have plus or minus. Recalling the normalization factors from Eq. (15.32), we end up with our desired result:

$$P(\delta) = \frac{1}{q_{\max}N} \frac{\sin^2(\pi r \delta q_{\max}/N)}{\sin^2(\pi r \delta/N)}.$$

(15.35)

Try It 15.8

Make the suggested substitutions and check the results in Eq. (15.35).

Recall from Chapter 13 that once we have found $y = N/r$ from Eq. (15.35), we can solve for the repetition period r, and that once we have r, we can find the prime factors of N—just what the Shor algorithm is designed to do.

CARDY: That was a bit of work, but I appreciate seeing how we got to the result that is critical for the Shor algorithm.

BOB: I always understand a procedure better when I work out a simple example. Here is one from Zubairy (2020). Let's return to the Fourier transform and apply it to the state function in the Shor circuit in Figure 13.11. In particular, let's look at the system state $|\psi_3\rangle$ after we apply the quantum Fourier transform to the upper-register states. The result, which is an entangled state, is expressed in Eq. (13.18) repeated here:

$$|\psi_3\rangle = \frac{1}{N} \sum_x \sum_y \underbrace{U_F(x,y)\,|y\rangle}_{\text{upper register}} \underbrace{|f(x)\rangle}_{\text{lower register}}$$

$$= \frac{1}{N} \sum_x \sum_y \underbrace{e^{i2\pi yx/N}\,|y\rangle}_{\text{upper register}} \underbrace{|f(x)\rangle}_{\text{lower register}}.$$

(15.36)

In Eq. (15.36), yx means ordinary multiplication. Note that when we sum over x for a fixed y, the argument of the exponential function changes in steps of $i2\pi y/N$.

To see what is going on, let's write this out for the case of $N = 8$ (a three-qubit system). For typographical simplicity, we will label the states $|f(x)\rangle$ with the *decimal* values of the arguments; for example, we will denote $|f(x = 010)\rangle$ as $|f(2)\rangle$. Here are the first few terms in the sums in Eq. (15.36) with $y = 0$, 1, and 2:

$y = 0$

$$\tfrac{1}{8}|0\rangle\,[|f(0)\rangle + |f(1)\rangle \ldots + |f(7)\rangle]$$
(15.37)

$y = 1$

$$\tfrac{1}{8}|1\rangle\left[e^{i0}|f(0)\rangle + e^{i\pi/4}|f(1)\rangle + e^{i\pi/2}|f(2)\rangle \ldots + e^{i\pi 7/4}|f(7)\rangle\right]$$
(15.38)

$y = 2$

$$\tfrac{1}{8}|2\rangle\left[e^{i0}|f(0)\rangle + e^{i\pi/2}|f(1)\rangle + e^{i\pi}|f(2)\rangle + \ldots + e^{i\pi 7/2}|f(7)\rangle\right].$$
(15.39)

Let's put these amplitudes and the other amplitudes into a table.

Table 15.1 lists the arguments θ of the phase factors $e^{i\theta}$ (amplitudes) of the states in Eq. (15.36). I have not simplified fractions because the unsimplified forms make the pattern more obvious. For example, for the state $|3\rangle$ in the left column, the arguments of the phase factors for the $|f(j)\rangle$ state in the top row are integer multiples of $3\pi/4$ (starting with 0): $0\pi/4$, $3\pi/4$, $6\pi/4$, $9\pi/4$, etc. Note that for a particular row in the table, the argument increases as you progress across the row by the entry in the $|f(1)\rangle$ column. I have also included an arrow, which gives a graphical representation of the phase factor with the assumption that the arrow associated with $e^{i0} = 1$ points to the right along the horizontal direction. The arrows each rotate counterclockwise by the $|f(1)\rangle$ angle from the previous arrow's direction. You can add the amplitudes by adding up the arrows graphically, as shown below.

We now invoke the periodicity of the function $f(x) = a^x \bmod N$. For our simple case, let's assume that $r = 2$ for some combination of a and N. In that case, we have

$$f(0) = f(2) = f(4) = f(6)$$
$$f(1) = f(3) = f(5) = f(7).$$
(15.40)

As an example, we can use the results in Eq. (15.40) to rewrite Eq. (15.38) for $y = 1$ as

Table 15.1 The arguments θ of the phase factors $e^{i\theta}$ (amplitudes) for the states in Eq. (15.36), along with the arrow representations of those phase factors. For example, $e^{i\pi/4}$ is represented by ↗ and $e^{i\pi/2}$ is represented by ↑.

$\lvert y \rangle$	$\lvert f(0) \rangle$	$\lvert f(1) \rangle$	$\lvert f(2) \rangle$	$\lvert f(3) \rangle$	$\lvert f(4) \rangle$	$f(5)$	$\lvert f(6) \rangle$	$\lvert f(7) \rangle$
$\lvert 0 \rangle$	0 →	0 →	0 →	0 →	0 →	0 →	0 →	0 →
$\lvert 1 \rangle$	0 →	$\pi/4$ ↗	$2\pi/4$ ↑	$3\pi/4$ ↖	$4\pi/4$ ←	$5\pi/4$ ↙	$6\pi/4$ ↓	$7\pi/4$ ↘
$\lvert 2 \rangle$	0 →	$\pi/2$ ↗	$2\pi/2$ ←	$3\pi/2$ ↓	$4\pi/2$ →	$5\pi/2$ ↑	$6\pi/2$ ←	$7\pi/2$ ↓
$\lvert 3 \rangle$	0 →	$3\pi/4$ ↖	$6\pi/4$ ↓	$9\pi/4$ ↗	$12\pi/4$ ←	$15\pi/4$ ↘	$18\pi/4$ ↑	$21\pi/4$ ↙
$\lvert 4 \rangle$	0 →	$4\pi/4$ ←	$8\pi/4$ →	$12\pi/4$ ←	$16\pi/4$ →	$20\pi/4$ ←	$24\pi/4$ →	$28\pi/4$ ←
$\lvert 5 \rangle$	0 →	$5\pi/4$ ↙	$10\pi/4$ ↑	$15\pi/4$ ↘	$20\pi/4$ ←	$25\pi/4$ ↗	$30\pi/4$ ↓	$35\pi/4$ ↖
$\lvert 6 \rangle$	0 →	$6\pi/4$ ↓	$12\pi/4$ ←	$18\pi/4$ ↑	$24\pi/4$ →	$30\pi/4$ ↓	$36\pi/4$ ←	$42\pi/4$ ↑
$\lvert 7 \rangle$	0 →	$7\pi/4$ ↘	$14\pi/4$ ↓	$21\pi/4$ ↙	$28\pi/4$ ←	$35\pi/4$ ↖	$42\pi/4$ ↑	$44\pi/4$ ↗

$$\tfrac{1}{8} \lvert 1 \rangle \left[\left\{ e^{i0} + e^{i\pi/2} + e^{i\pi} + e^{i3\pi/2} \right\} \lvert f(0) \rangle + \left\{ e^{i\pi/4} + e^{i\pi 3/4} + e^{i\pi 5/4} + e^{i\pi 7/4} \right\} \lvert f(1) \rangle \right]$$
$$(15.41)$$

using the complex exponentials with phases or, using the arrows to indicate phases, as

$$\tfrac{1}{8} \lvert 1 \rangle \left[\{ → + ↑ + ← + ↓ \} \lvert f(0) \rangle + \{ ↗ + ↖ + ↙ + ↘ \} \lvert f(1) \rangle \right] .$$
$$(15.42)$$

If you evaluate the quantities inside the curly braces, you find that they give 0; the amplitude for measuring the $\lvert 1 \rangle$ state in the upper register is 0. You can do that addition either by adding up the complex numbers associated with the complex exponentials or you can add up the arrows using the usual "tail-to-tip" rule. If the arrows form a closed figure when added tail-to-tip as shown in Figure 15.3, the result is zero. In particular, for the arrows in Eq. (15.41) we get the results shown in the figure. This result is said to exhibit destructive interference among the state vectors. The net effect is that they add to 0.

CARDY: I guess I could use a spreadsheet to calculate the exponentials, but is there an easier way to see that the sum of the numerical value of the exponentials gives zero?

BOB: Yes, there is. You can use the rule for the sum of exponents, $e^x e^y = e^{x+y}$, and $e^{i\pi} = -1$ to see that the terms inside the curly braces cancel pairwise. For example, in the second set of curly braces in Eq. (15.41), we find $e^{i\pi 7/4} = e^{i\pi} e^{i\pi 3/4} = -e^{i\pi 3/4}$.

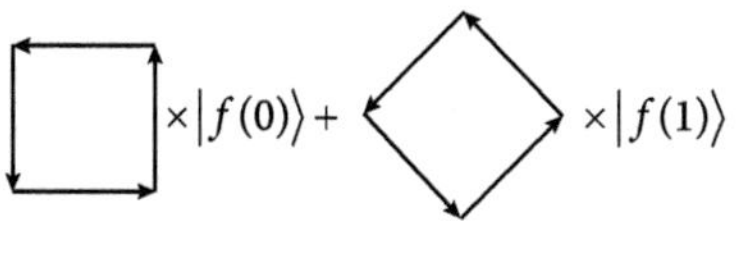

Fig. 15.3 The addition of the vectors inside the curly braces in Eq. (15.41).

Continuing with our example, if you work through the details, you find that the amplitudes for the upper register states $|1\rangle$, $|2\rangle$, $|3\rangle$, $|5\rangle$, $|6\rangle$, and $|7\rangle$ are all 0 (destructive interference), meaning that the probability of getting those states as results of measurements on the upper register states is 0. Only the states $|0\rangle$ and $|4\rangle$ have probabilities that are not zero, and in this case both have probabilities equal to 50%. For those states, the arrows will tend to line up in the same direction to give a large amplitude, often called constructive interference.

Let's see how the addition works for states $|0\rangle$ and $|4\rangle$. For the state $|0\rangle$ we have:

$$\tfrac{1}{8}|0\rangle\left[\{e^{i0}+e^{i2\pi}+e^{i4\pi}+e^{i6\pi}\}|f(0)\rangle+\{e^{i\pi}+e^{i3\pi}+e^{i5\pi}+e^{i7\pi}\}|f(1)\rangle\right]$$
$$=\tfrac{1}{8}|0\rangle\left[\{\rightarrow+\rightarrow+\rightarrow+\rightarrow\}|f(0)\rangle+\{\rightarrow+\rightarrow+\rightarrow+\rightarrow\}|f(1)\rangle\right]$$
$$=|0\rangle\left[\tfrac{1}{2}|f(0)\rangle+\tfrac{1}{2}|f(1)\rangle\right]. \tag{15.43}$$

For the state $|4\rangle$, we find

$$\tfrac{1}{8}|4\rangle\left[\{e^{i0}+e^{i2\pi}+e^{i4\pi}+e^{i6\pi}\}|f(0)\rangle+\{e^{i\pi}+e^{i3\pi}+e^{i5\pi}+e^{i7\pi}\}|f(1)\rangle\right]$$
$$=\tfrac{1}{8}|4\rangle\left[\{\rightarrow+\rightarrow+\rightarrow+\rightarrow\}|f(0)\rangle+\{\leftarrow+\leftarrow+\leftarrow+\leftarrow\}|f(1)\rangle\right]$$
$$=|4\rangle\left[\tfrac{1}{2}|f(0)\rangle-\tfrac{1}{2}|f(1)\rangle\right]. \tag{15.44}$$

I should note that the amplitudes work out to be zero or non-zero only if N is evenly divided by r. In more general cases, some of the amplitudes will add to small numbers, but not quite equal to zero. In those cases, the arrows will almost form a closed figure, but not quite. For those states, the probabilities for observing those states upon measurement will be small, but not zero. The resulting amplitude is given by the length of the vector from the tail of the first vector to the tip of the final vector, indicated by the short red arrows in Figure 15.4. Those small amplitudes mean that the algorithm will not give the desired result exactly, but the probability for getting the correct result is close to 1.

BOB: Now, we reverse the question to match what the Shor algorithm accomplishes: How do we find the repetition period r, if we get $|0\rangle$ or $|4\rangle$ in the upper-register measurements? In general, if $f(a) = f(a+r)$ and if N is evenly divisible by r, the only y states with non-zero amplitudes are

$$|0\rangle,\ |N/r\rangle,\ |2N/r\rangle,\ldots,|(r-1)\,N/r\rangle. \tag{15.45}$$

Since the $|0\rangle$ state doesn't give us any useful information, we try $N/r = 4$ in our example and find $r = 2$, the desired period. If N is not evenly divided by r, then we need to use the methods embodied in Eq. (15.35).

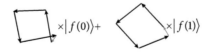

Fig. 15.4 The sum of the arrows may not form a closed circuit.

Let's review what we have done. The Shor algorithm works by using the quantum Fourier transform to find the repetition period of the function $f(x) = a^x \bmod N$. We know from number theory that given a, N, and the period, we can find the prime factors of N. The Shor algorithm cleverly makes use of superposition and entanglement to produce system states that give high probabilities to be detected in a measurement if the state label y satisfies $y \approx N/r$ and small otherwise. For the high-probability states, the amplitudes add together constructively with the sum close to 1. For the low-probability states, the amplitudes add together destructively with the sum close to 0. This is often called quantum interference.

15.4   Relative Phases in Quantum States

ALICE: In Chapter 8, we showed that the relative sign between the amplitudes in a quantum superposition state leads to observable results, that is, to different measurement outcomes. In particular, if the superposition state is prepared in one set of basis states, but measured with respect to a different set of basis states, the measurement outcomes can tell us the relative sign between the superposition amplitudes in the original basis states. We will now revisit this using complex exponentials.

Let's start with a general phase difference between the amplitudes in superposition states. We start with a superposition state for a single qubit, and for the sake of simplicity we assume the amplitudes have a magnitude $1/\sqrt{2}$:

$$|s\rangle = \frac{1}{\sqrt{2}} \left[|0\rangle + e^{i\phi} |1\rangle \right]. \tag{15.46}$$

CARDY: Why did you put the complex exponential in front of the $|1\rangle$ basis state and not the $|0\rangle$ basis state?

ALICE: That is a convention. In the more general case, we could have a complex exponential for both amplitudes $a_0 = |a_0| e^{i\phi_0}$ and $a_1 = |a_1| e^{i\phi_1}$ as shown in the following example:

$$\begin{aligned}
|s\rangle &= \frac{1}{\sqrt{2}} \left[a_0 |0\rangle + a_1 |1\rangle \right] \\
&= \frac{1}{\sqrt{2}} \left[e^{i\phi_0} |a_0| |0\rangle + e^{i\phi_1} |a_1| |1\rangle \right] \\
&= \frac{e^{i\phi_0}}{\sqrt{2}} \left[|a_0| |0\rangle + e^{i(\phi_1-\phi_0)} |a_1| |1\rangle \right] \\
&= \frac{e^{i\phi_0}}{\sqrt{2}} \left[|a_0| |0\rangle + e^{i\phi} |a_1| |1\rangle \right], \tag{15.47}
\end{aligned}$$

where $\phi = \phi_1 - \phi_0$. The complex exponential multiplying the overall state does not change the probabilities calculated from the state (just like an overall minus sign does not give a distinct quantum state). The coefficient of $|1\rangle$ contains a term that is the relative phase between $|0\rangle$ and $|1\rangle$, which, as we shall see, can result in measurement outcomes that depend on ϕ. The take-away message is that only the relative phase between the two amplitudes matters, not the global value of the phase of the two amplitudes. As I mentioned, by convention we assign the complex exponential representing that phase to the $|1\rangle$ state.

Try It 15.9

Work through each line of Eq. (15.47) and explain both the manipulation of the complex exponentials and the Dirac state vector notation.

ALICE: Now I will show you that we can find the relative phase ϕ by making measurements with a different set of basis states, à la Chapter 8. Let's use the state in Eq. (15.46) as an example. The method and notation will match the ones we used in Chapter 8, where we expressed the original state in terms of the basis states $|B_0\rangle$ and $|B_1\rangle$ that will be used by the measurement device. For our example, I'll use a simple transformation:

$$|0\rangle = \frac{1}{\sqrt{2}}\left[|B_0\rangle + |B_1\rangle\right]$$

$$|1\rangle = \frac{1}{\sqrt{2}}\left[|B_0\rangle - |B_1\rangle\right]. \tag{15.48}$$

We then use Eq. (15.48) in Eq. (15.46), regrouping terms for the B basis states:

$$|s\rangle = \frac{1}{\sqrt{2}}\left[\frac{1}{\sqrt{2}}\left(|B_0\rangle + |B_1\rangle\right) + \frac{e^{i\phi}}{\sqrt{2}}\left(|B_0\rangle - |B_1\rangle\right)\right]$$

$$= \frac{1}{2}\left[\left(1 + e^{i\phi}\right)|B_0\rangle + \left(1 - e^{i\phi}\right)|B_1\rangle\right]. \tag{15.49}$$

To find the probabilities for the measurement outcomes, we calculate, as usual, the absolute-value-squared of the state vector amplitudes. The probability for the measurement outcome B_0 is

$$P(B_0) = \frac{1}{4}\left(1 + e^{i\phi}\right)^*\left(1 + e^{i\phi}\right)$$

$$= \frac{1}{4}\left(1 + e^{-i\phi}\right)\left(1 + e^{i\phi}\right) = \frac{1}{4}\left(1 + e^{i\phi} + e^{-i\phi} + 1\right)$$

$$= \frac{1}{2}\left(1 + \cos\phi\right), \tag{15.50}$$

where we used the "reverse" Euler formula in Eq. (15.25).

> **Try It 15.10**
>
> Use the same type of calculation to show that $P(B_1) = \frac{1}{2}(1 - \cos \phi)$. Do the measurement probabilities add up to 1?

ALICE: The important take-home message is that measurements carried out with basis states different from those in which the original state was prepared allow us to find the relative phase between the two parts of the qubit state. Or conversely, different relative phases lead to different measurement outcomes. We learned more about the state by making measurements relative to a different basis.

For QIS and QC applications, we would like to have a quantum gate that can change the relative phases in a quantum state because that change will be observable. In fact, there are many phase operators of various kinds in QC, but we will stick to a relatively simple but powerful example for single-qubit systems. Our example is built using the Z operator, which we met in Chapter 4, and the complex exponential we have just become familiar with. The phase operator is

$$U_Z(\phi) = e^{-i\phi Z/2}. \tag{15.51}$$

CARDY: Whoa! What does it mean to have an operator in an exponential function? I feel we are getting into some rather strange mathematical territory.

ALICE: I agree this is likely not familiar to most of our readers. We won't go into the details, but because the Z operator has the nice property that $ZZ = 2I$ (twice the identity operator), the phase operator can be written as

$$e^{-i\phi Z/2} = I\cos(\phi/2) - iZ\sin(\phi/2)$$

$$= \cos(\phi/2)\begin{pmatrix} 1 & 0 \\ 0 & 1 \end{pmatrix} - i\sin(\phi/2)\begin{pmatrix} 1 & 0 \\ 0 & -1 \end{pmatrix}$$

$$= \begin{pmatrix} \cos(\phi/2) - i\sin(\phi/2) & 0 \\ 0 & \cos(\phi/2) + i\sin(\phi/2) \end{pmatrix}$$

$$= \begin{pmatrix} e^{-i\phi/2} & 0 \\ 0 & e^{i\phi/2} \end{pmatrix} = e^{-i\phi/2}\begin{pmatrix} 1 & 0 \\ 0 & e^{i\phi} \end{pmatrix}. \tag{15.52}$$

We should apply this in a simple case to make sure it does what we want. To make the math easier to read, we will use the matrix form of Z and the column vector representation of the state vector:

$$U_Z(\phi)(a_0\,|0\rangle + a_1\,|1\rangle) \Rightarrow e^{-i\phi/2}\begin{pmatrix} 1 & 0 \\ 0 & e^{i\phi} \end{pmatrix}\begin{pmatrix} a_0 \\ a_1 \end{pmatrix}$$

$$= e^{-i\phi/2}\begin{pmatrix} a_0 \\ a_1 e^{i\phi} \end{pmatrix} \Rightarrow e^{-i\phi/2}\left(a_0\,|0\rangle + a_1 e^{i\phi}\,|1\rangle\right). \tag{15.53}$$

So, we see that the operator creates a relative phase between the two state amplitudes and produces an overall phase factor in front of the matrix, which we can ignore because it disappears when we calculate probabilities—just what we want. The ability to adjust phase differences as well as amplitudes in superposition states gives us powerful tools to build quantum algorithms. A method called phase estimation shows up in the quantum Fourier transform and in quantum algorithms for solving systems of linear equations.

CARDY: I don't see how phases are related to systems of linear equations. I learned about linear equations in Algebra II and there wasn't a phase in sight.

ALICE: The connection between phases and the solution of linear equations is not obvious. To see that connection, let's take a little mathematical detour by writing the linear equations in terms of a matrix and column vectors. For example, the system of equations

$$5x + 6y = 3$$
$$7x + 2y = 4 \tag{15.54}$$

can be written as

$$\begin{pmatrix} 5 & 6 \\ 7 & 2 \end{pmatrix} \begin{pmatrix} x \\ y \end{pmatrix} = \begin{pmatrix} 3 \\ 4 \end{pmatrix} \quad \text{or with } A = \begin{pmatrix} 5 & 6 \\ 7 & 2 \end{pmatrix},$$
$$A \begin{pmatrix} x \\ y \end{pmatrix} = \begin{pmatrix} 3 \\ 4 \end{pmatrix}. \tag{15.55}$$

To solve the equations, we find the eigenvalues λ_1 and λ_2 of the matrix A and use those eigenvalues to build the inverse of the matrix, which allows us to solve for x and y. Multiplying both sides of the last line in Eq. (15.55) by A^{-1} and using $A^{-1}A = I$, we get

$$A^{-1} \left[A \begin{pmatrix} x \\ y \end{pmatrix} \right] = A^{-1} \begin{pmatrix} 3 \\ 4 \end{pmatrix}$$
$$\begin{pmatrix} x \\ y \end{pmatrix} = A^{-1} \begin{pmatrix} 3 \\ 4 \end{pmatrix}. \tag{15.56}$$

The phase estimation comes in by noting that the eigenvalues of any unitary matrix U can be found from

$$U \left| \lambda \right\rangle = e^{2\pi i \lambda} \left| \lambda \right\rangle, \tag{15.57}$$

where $\left| \lambda \right\rangle$ is the eigenstate associated with the eigenvalue λ of the matrix U. For more details on how this process makes use of quantum phase estimation, see Further Reading: Qiskit, sections 3.6 and 4.1.1.

15.5 The Bloch Sphere

BOB: Now that we know something about complex numbers in quantum mechanics, we need a representation different from our customary two-dimensional state space diagram. The angle of the state vector in a two-dimensional state space diagram shows only the different state amplitudes of the superposition state; it does not show information about the relative phases we just discussed.

The most common representation that includes the relative phase is called the Bloch sphere, named after Felix Bloch, a Nobel Laureate in physics, and a pioneer in the field of nuclear magnetic resonance. Those techniques eventually led to magnetic resonance imaging (MRI), one of the most important medical imaging techniques available to us today. Bloch was concerned with the nuclei of atoms and their magnetic and spin properties. In fact, the nucleus of a hydrogen atom (a proton) is an example of a qubit because it has spin-1/2. If you read more about qubits and their quantum states, you will eventually come across this pictorial representation: the Bloch sphere.

ALICE: The Bloch sphere may look familiar if you think about the geography of the Earth. To describe a place on the surface of the planet, we usually specify a latitude and longitude. For example, New York City is at $40.7128°$ N (latitude) and $74.0060°$ W (longitude). For the Bloch sphere, latitude corresponds to θ_B and longitude to ϕ. The angle θ_B is called the "co-latitude" because it is the complement of the angle used to describe latitude on Earth.

A quantum state whose Bloch vector tip is at $90°$ co-latitude on the Bloch sphere is a superposition state where the probabilities associated with the measurement outcomes $|0\rangle$ and $|1\rangle$ are equal (50/50). Other angles between $0°$ and $90°$ and between $90°$ and $180°$ are superposition states with unequal probabilities. A Bloch vector pointing to the "north pole" at $0°$ co-latitude represents a state that is not a superposition but is simply $|0\rangle$.

CARDY: So, the south pole of the Bloch sphere is $180°$ and must be purely $|1\rangle$. Instead of the axes of the two states being the traditional state-space orthogonal horizontal and vertical axes, they are along the line joining the north pole and the south pole.

ALICE: Yes, the Bloch sphere representation twists the basis state vectors so the corresponding Bloch vectors point in opposite directions. That, in a sense, opens up the diagram, so we can display information about the phase difference between the amplitudes. I should point out that in many cases, we use spin-down and spin-up basis states instead of $|0\rangle$ and $|1\rangle$. By convention, the spin-up state $|\uparrow\rangle = |0\rangle_z$ points upward along the z axis (towards the north pole of the Bloch sphere) and the spin-down state $|\downarrow\rangle = |1\rangle_z$ points towards the south pole.

CARDY: What about longitude?

ALICE: Longitude in the Bloch sphere diagram represents the relative phase of the complex exponentials, which we discussed in the previous section. In the Bloch sphere representation, the longitudinal angle ϕ is measured counterclockwise from the x axis (looking down on the north pole). So, a superposition state Bloch vector lying in the Bloch diagram xz plane has no phase difference between the basis state amplitudes.

BOB: Let's work with the Bloch sphere by expressing a qubit state in terms of the computational basis states and a complex exponential as part of the amplitude of the $|1\rangle$ state:

$$|\psi\rangle = \cos(\theta_B/2)\,|0\rangle + e^{i\phi}\sin(\theta_B/2)\,|1\rangle, \tag{15.58}$$

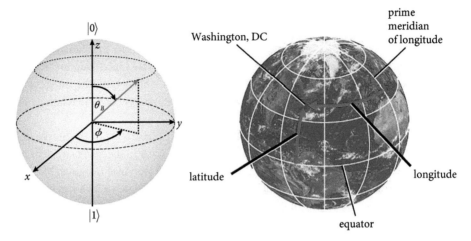

Fig. 15.5 On the left, the Bloch sphere for single-qubit states. The red vector represents the qubit state with a *Bloch vector*. The *co-latitude* angle θ_B is related to the magnitudes of the state vector amplitudes, while the *longitude* angle ϕ represents the phase difference between the amplitudes. The central dashed oval indicates the Bloch sphere's equator. The upper dashed oval is the constant co-latitude curve (parallel to the equator) corresponding to the co-latitude of the tip of the state vector. On the right, a picture of latitude and longitude lines on the surface of Earth.
Source: The image, by an unknown author, is licensed under CC BY.

where we introduced $\cos(\theta_B/2) = a_0$ and $e^{i\phi}\sin(\theta_B/2) = a_1$ for the basis state amplitudes. This way of writing the state leads to the geometric picture depicted in Figure 15.5. Note that $\theta_B/2$ is the standard *state space* angle between the $|0\rangle$ basis state vector and the qubit state vector $|\psi\rangle$. When $\theta_B = \pi$, the Bloch vector points toward the south pole and $|\psi\rangle = |1\rangle$ as Cardy pointed out.

Try It 15.11

Draw the Bloch vector on a Bloch sphere diagram with $\theta_B = 60°$ and $\phi = 30°$. Repeat with $\theta_B = 60°$ and $\phi = 90°$. Hint: $\cos 60° = 0.5$ and $\sin 30° = 0.5$. Then use Eq. (15.58) to express the state as a superposition of the computational basis states.

BOB: In Eq. (15.58), the complex exponential phase term, as usual, is part of the coefficient of the $|1\rangle$ basis state. States that have the same angle θ_B have the same magnitudes of the amplitudes but different phases as specified by ϕ. For example, all state vectors that lie in the equatorial plane (for which $\theta_B = 90°$) are superposition states that give equal probabilities for spin-up and spin-down but differ in phase between the two basis states. Another way of saying this is that changing the phase difference moves the Bloch vector tip around a circle of constant co-latitude, as shown in Figure 15.6.

In Figure 15.6, we look down on the north pole of the Bloch sphere. The angle ϕ is the longitude angle shown in Figure 15.5. This diagram is almost the same as the visualization that Richard Feynman used in his book *QED*, which explains the interaction between photons and

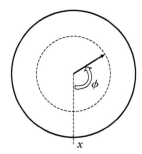

Fig. 15.6 Looking down on the north pole of the Bloch sphere. The angle ϕ is the angle for the phase difference between the two basis state amplitudes. The angle is measured counterclockwise from the Bloch sphere x axis. The dashed circle is a circle of constant co-latitude.

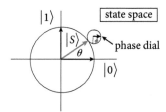

Fig. 15.7 A state space diagram showing a state space vector $|S\rangle$ and the associated phase dial, indicating the phase difference ϕ between the amplitudes of the components of the state space vector along the basis states.

matter. Rather than using a Bloch sphere to keep track of the quantum state phase, Feynman used a "phase dial" (like a clock face). You could also attach the phase dial to state vectors in our usual two-dimensional state space, as shown in Figure 15.7. The position of the clock hand indicates the relative phase between the amplitudes of the basis states in a superposition state.

CARDY: The Bloch sphere is a neat picture, but what is it good for?

BOB: The Bloch sphere representation is a nice way of visualizing how quantum gates act on single-qubit quantum states. The gates can affect both the magnitudes of the amplitudes and their phase difference. In fact, there are some nice online animated visualizations of how various single-qubit gates affect the Bloch vector (see Further Reading). That said, the Bloch sphere representation is generally not helpful if we have systems of more than one qubit.

In addition to providing a nice visualization of single-qubit states, the Bloch sphere diagram helps us understand what is called the U_B (for "universal") single-qubit gate (the subscript B reminds us that the gate operates on a Bloch vector). U_B is universal in the sense that it can operate on a Bloch vector to produce any other Bloch vector. In particular, $U_B(\alpha, \beta, \phi)$ operates on a Bloch sphere vector with phase ϕ and adds α to the phase and β to the co-latitude. For a single-qubit state, the U_B gate in the computational basis is represented by a matrix:

$$U_B(\alpha, \beta, \varphi) \Rightarrow \begin{pmatrix} \cos(\beta/2) & -e^{-i\phi}\sin(\beta/2) \\ e^{+i(\phi+\alpha)}\sin(\beta/2) & e^{i\alpha}\cos(\beta/2) \end{pmatrix}. \tag{15.59}$$

Note that we have assumed that ϕ appears in the complex exponential as shown in Eq. (15.58).

When U_B operates on the computational basis states (which have $\phi = 0$), it generates superposition states:

$$U_B(\alpha, \beta, 0)\,|0\rangle = \cos(\beta/2)\,|0\rangle + e^{i\alpha}\sin(\beta/2)\,|1\rangle$$
$$U_B(\alpha, \beta, 0)\,|1\rangle = -\sin(\beta/2)\,|0\rangle + e^{i\alpha}\cos(\beta/2)\,|1\rangle. \tag{15.60}$$

Try It 15.12

Use the column vector forms of the basis states $|0\rangle$ and $|1\rangle$ and Eq. (15.59) to check the results in Eq. (15.60). Verify that the resulting states are normalized states.

CARDY: That is cool! The universal gate operator looks just like a generalization of the Hadamard gate we used before. Is that right?

BOB: Absolutely. In fact, the connection is

$$U_B(\pi, \pi/2, \pi) = \frac{1}{\sqrt{2}}\begin{bmatrix} 1 & 1 \\ 1 & -1 \end{bmatrix} = H. \tag{15.61}$$

The U_B operator is widely used in quantum algorithms because of its ability to set both amplitudes and relative phases in a superposition state. There are also special cases of U_B, the S operator and the T operator, that appear in many QC and QIS applications:

$$S \Rightarrow \begin{pmatrix} 1 & 0 \\ 0 & e^{i\pi/2} \end{pmatrix}$$
$$T \Rightarrow \begin{pmatrix} 1 & 0 \\ 0 & e^{i\pi/4} \end{pmatrix}. \tag{15.62}$$

Try It 15.13

What values of α, β, and ϕ in Eq. (15.59) lead to the operator matrices in Eq. (15.62)?

ALICE: We can derive Eq. (15.59) by using the z axis rotation operator in Eq. (15.53), which changes the phase, and the operator for rotations about the Bloch y axis. The operator for a rotation by an angle θ about the Bloch sphere y axis is

$$R_y(\theta) \Rightarrow \begin{pmatrix} \cos(\theta/2) & -\sin(\theta/2) \\ \sin(\theta/2) & \cos(\theta/2) \end{pmatrix}, \tag{15.63}$$

which you should recognize from the operator for changing basis states in Chapter 8. Note that this is the form only for Bloch vectors that lie in the Bloch xz plane.

Try It 15.14

Challenge: You can use Eqs. (15.53) and (15.63) to derive the U_B matrix in Eq.(15.59). First you operate on the Bloch sphere state in Eq. (15.58) with $R_z(-\phi)$. This brings the state longitude to $0°$. You then apply the matrix $R_y(\beta)$ to add β to the co-latitude and finish by applying a longitude rotation $R_z(\alpha + \phi)$ to make the relative phase $\alpha + \phi$. This procedure will lead to $U_B(\alpha, \beta, \phi) = R_z(\alpha + \phi)R_y(\beta)R_z(-\phi)$.

The Bloch sphere also provides a nice visualization of different paths by which a state can evolve from one position to another, for example, from spin-up to spin-down. Paths can be chosen to minimize phase disruptions or amplitude disruptions to decrease errors, for example.

It is worth mentioning that there is a construction for polarization states for light similar to the Bloch sphere. That construction is called a Poincaré sphere. The mathematician and physicist Henri Poincaré developed this representation in 1892, long before the development of quantum mechanics. There is also a way to use matrices and column vectors to represent states of polarized light and how those states change when they interact with devices such as polarizing beam splitters. The method is called the Jones calculus after its inventor R. C. Jones, who developed the analysis in 1941. See the Further Reading section for more information.

 CHAPTER SUMMARY

- To go further in QC with qubit systems that have more degrees of freedom, complex numbers must be used to satisfy the requirements of orthogonality and normalization within a change of basis. They also allow for mining information about the relative phases of quantum states.

- The complex exponential function is a fundamental building block for many quantum operators.

- The quantum Fourier transform makes use of complex exponentials to find the period of periodic functions.

- A Bloch sphere representation shows the interplay between the amplitudes in different basis states and relative phases for a qubit state. It can be used as a visual tool for thinking about how quantum gates affect a single-qubit state.

FURTHER READING

Complex numbers in quantum mechanics

Ricardo Karam, "Why are complex numbers needed in quantum mechanics? Some answers for the introductory level." *American Journal of Physics* **88**, 39 (2020). The author explores the arguments put forth in physics textbooks about why quantum mechanics requires complex numbers.

M. Suhail Zubairy, *Quantum Mechanics for Beginners* (Oxford University Press, Oxford, 2020). The quantum Fourier transform example is on pages 248–251.

Theodore A. Corcovilos, "A simple game simulating measurements of qubits." *American Journal of Physics* **86**, 510 (2018). A game using 20-sided dice and the Bloch sphere representation of single-qubit states nicely illustrate many features of quantum measurements carried out on qubits.

Richard P. Feynman, *QED: The Strange Theory of Light and Matter* (Princeton University Press, Princeton, NJ, 1985). Feynman uses the "phase dial" ("phase clock") picture to explain a wide variety of quantum phenomena.

Bloch Sphere

A Bloch Sphere animation is available at https://www.st-andrews.ac.uk/physics/quvis/simulations_html5/sims/blochsphere/blochsphere.html.

Sliders allow you to change θ_B and ϕ to see how the Bloch vector changes.

For other simulations, search online for "Bloch sphere animations."

Poincaré sphere—the optical equivalent of the Bloch sphere

"Polarization (waves)—Poincaré sphere," Wikipedia, https://en.wikipedia.org/wiki/Polarization_(waves).

"Jones calculus," Wikipedia, https://en.wikipedia.org/wiki/Jones_calculus. This article includes a discussion of the parallels between Jones vectors and quantum states for polarized light. The Jones calculus uses matrices and column vectors that are quite similar to those used in quantum mechanics of qubit systems.

Qiskit

IBM Qiskit, accessible at https://qiskit.org/, provides many tutorials on all aspects of QC.

16 Present and Future QIS and QC

It's tough to make predictions, especially about the future.

Apparently an old Danish proverb, but also attributed to Yogi Berra and Niels Bohr

16.1 QC and QIS Overview

ALICE: We want to congratulate you, Cardy and our dedicated readers, on your persistence through a lot of challenging material. You should be proud of your work. Both Bob and I hope that you will continue your studies of QIS and QC. In this chapter, we suggest several "next steps" you might take to deepen your understanding of QIS and QC. We will also talk about programming quantum computers and developments such as the quantum internet. A quick survey of current QC and QIS technology will give you a sense of the broad range of skills deployed in those fields. We will also rub the quantum crystal ball to suggest areas of QIS and QC to keep an eye on over the next few years, and consider some ethical issues that are important in QIS and QC.

CARDY: I do appreciate the work you and Bob put into the presentation. I feel I have learned a lot. But I sense there is a lot more to understand.

ALICE: QIS and QC are still in their beginning stages of development. Scientific journals are publishing new results in practically every issue. Governments, especially in the US, Europe, and China, are pouring research funds into QIS and QC. Major technology companies like Microsoft, Google, and IBM are also active in these areas. As of this writing (2021), Google, IonQ, and IBM have QCs with 20–50 qubits. There is obviously both intellectual excitement in these developments and many possibilities for rewarding careers. QC and QIS are truly multidisciplinary fields requiring experts in physics, computer science, mathematics, engineering, materials science, and electronics technology; and of course in business management to get the resources to make all this work and to put QC and QIS to use in projects that will benefit the world.

BOB: Let's get an overview of QC before discussing future directions for the field. A good place to start is the diagram of the components of a quantum computer system shown in Figure 16.1. It starts with software and ends with hardware.

The A level in the diagram includes the development of quantum algorithms and quantum programming languages that users can employ to implement algorithms without worrying about the details of the qubit hardware. This is similar to the situation with classical computers, where we use spreadsheets, word processors, and programming languages such as Python, R,

Quantum Computing: From Alice to Bob. Alice Flarend and Bob Hilborn, Oxford University Press.
© Alice Flarend and Robert C. Hilborn (2022). DOI: 10.1093/oso/9780192857972.003.0016

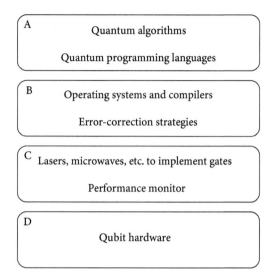

Fig. 16.1 A diagram of a quantum computer system.

or C# that are (we hope) understandable to humans, without needing to know anything about the actual computer hardware.

Layer B refers to the operating systems that coordinate the operations of the computers' physical components and the translation of the code written in the programming language to the code that the computer hardware itself understands. For classical computers, we have operating systems such as Unix, Linux, Windows, and iOS, which translate the high-level language commands into what the hardware understands: assembly language or machine language. In quantum computing, layer B includes translating the code into qubit gate operations. We saw several examples of that translation in Chapters 12 and 13.

Layer C contains hardware that is set up more-or-less permanently to organize gate operations and to carry out measurements. This hardware is called quantum firmware and is used to improve stability and to reduce various qubit errors. There is a tremendous amount of engineering work required to develop firmware. That layer also carries out the error-correction protocols like those we mentioned in Chapter 12.

The bottom layer D is the actual qubit hardware—the physical machinery to prepare, process, and measure the qubits. More about that in section 16.3.

The take-away message is that making a working (and useful) QC requires many kinds of skills and many different ways of thinking. Most QC users will probably focus on the top layer. However, a quantum algorithm or a QC program is useless if there is no machine on which to implement the program. To construct a QC, we need people to build the components, physically install the electrical wires, etc. That means there are many QC and QIS employment opportunities in the trades, too.

ALICE: A way to gain experience with QCs is to use a classical computer to simulate a QC. The basic idea is that the gates acting on quantum states can be represented, as we have shown again and again in this book, by matrices acting on column and row vectors. Most classical programming languages handle those calculations easily. The only tricky part is simulating quantum measurements. That can be done by using computer-generated random numbers to produce measurement outcomes consistent with the probabilities encoded in the quantum states. Of course, this method does require some knowledge of computer programming, but it is not overly complicated. Further Reading has citations to papers that illustrate how these simulations work. However, the simulations become very time-consuming as the number of qubits grows, and they will not be able to replace a QC when large numbers of qubits are required. After all, if they could, we would not need to build QCs! Nevertheless, the simulations will help test algorithms and build your intuition about how QCs operate.

16.2   Programming Quantum Computers

CARDY: I want to ask about programming quantum computers—layer A in your diagram about quantum systems. You really haven't said anything specific about that so far.

BOB: One way to think about programming QCs is to break the process into two steps. First, we develop an algorithm that makes use of superposition states, entanglement, quantum gates, and quantum measurements. As we saw in previous chapters, representing those algorithms as quantum circuits is something like writing flow charts for classical computer programming. In those circuits and flow charts, we describe the logic of the process we want to implement without worrying about the computer programming code needed to send the appropriate instructions to the computer. The second step is writing code that can be sent to the QC to implement the algorithm.

For the short-term future, it is highly likely that the commands you send to a quantum computer will be specific for each device because you need to communicate directly with that device's hardware. Eventually, higher-level quantum programming languages will be developed akin to Python or C#; so, the typical user will not need to worry about how to communicate with the specific underlying hardware. That is certainly the situation in classical computers today, but within living memory, there was a time when a programmer needed to be aware of how memory was set up in a computer and which part of the central processing unit did what. I remember working with some of the first mini-computers in the 1970s. You had to flip a set of mechanical toggle switches to get the machine up and running. Computer engineers quickly figured out how to build in those instructions using integrated circuits and more-or-less permanent computer memory units.

Figure 16.2 shows a simple quantum circuit, which we will use as an example of QC programming. The circuit contains the first two steps we just mentioned. You will recognize the familiar elements in this circuit: two qubits, their initial states, a Hadamard gate, a CNOT gate, and two measurement devices. We also included two classical bits that are used to record the outputs of the measurement devices. Those outputs are passed along to a classical computer for further processing.

CARDY: I remember! H and CNOT produce an entangled superposition state.

BOB: Exactly right! I am impressed that you have become so fluent in using those quantum terms. You are definitely catching on to the world of quantum computing. You should also notice that on the far right of Figure16.2, the circuit contains computational basis state

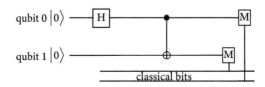

Fig. 16.2 The circuit diagram for the quantum programming example. H is a Hadamard gate; the Ms are measurement devices. Two classical bits, used to record the outcome of the measurements, are shown at the bottom of the diagram.

measurements. The measurement outcomes are recorded by a classical computer. Below is an example of how we translate that circuit into programming code using the syntax from IBM's Quantum Experience website. The code sets up a quantum circuit with the two qubits. Then the code adds a Hadamard gate (H gate) followed by a CNOT gate with qubit 0 in Figure 16.2 acting as the control and qubit 1 acting as the target. The code also adds two classical bits to capture the measurement outcomes. The quantum circuit program is then run many times, in this case for a total of 1000 iterations. Statements following # are comments to tell you what is happening in various parts of the code, a common practice among programmers.

```
# Specify a quantum circuit consisting of two qubits.
circuit = QuantumCircuit (2,2)
# Include a Hadamard gate (h) acting on qubit 0
circuit.h(0)
# Include a CNOT (CX) gate with control qubit 0 and target qubit 1
circuit.cx (0,1)
# Include measurements based on the computational basis states
circuit.measure( [0,1],[0,1])
# assign the program to a quantum processor and execute the steps 1000
times to get a set of measurement results
myJob = execute(circuit, quantum processor, shots=1000)
# get the results
result = myJob.result()
# get the counts for each of the two possible measurement outcomes
counts = result.get_counts(circuit)
# print the results for further analysis
print("\nTotal count for 00 and 11 are:",counts)
# END
```

ALICE: Of course, the details of the code syntax will vary from one programming setup to another, just like the differences among Python, C#, Mathematica, MATLAB, and other programming environments for classical computers. Eventually, many of the individual gate operations will be pulled together into higher-level programming structures. As a simple example, the combination of the Hadamard gate and a CNOT gate is so common, it is likely to be put together into something like a BuildSuperposition function.

CARDY: I had some programming in high school and I think that the quantum programming code looks a lot like classical computer code where variables are defined and assigned.

I also see steps that repeat the calculation for a specified number of shots and print statements to see the results.

ALICE: You are correct. The primary differences are in setting up the quantum circuit with qubits, gates, and measurement devices and in handling multi-qubit states, which as we know can be a bit trickly. Of course, the hard work is in developing the algorithm and its corresponding quantum circuit.

While you are waiting to have your own QC and want to use more than a QC simulator, it is now possible to access remotely quantum computers operated by companies such as IBM, IonQ, and D-Wave. For example, Amazon AWS has set up an interface called Braket that allows access to other companies' QCs. These interfaces allow you to send instructions to the QC using special functions built into standard programming languages like Python and C#. The companies also provide lots of good background information on quantum computing that should be accessible now that you have finished this book. The math requirements are a bit higher than those used in this book, but they are not out of reach. The Further Reading section contains links to these resources.

16.3 Physics of Qubits

BOB: In this section, we'll describe the types of physical qubits being deployed in today's (as of 2021) quantum computers. The discussion will be primarily qualitative because understanding the details requires substantial background in physics, materials science, and engineering. With the rapidly developing technology in QC, what we write here will most likely be out of date within a year or two. Moreover, as we mentioned before, most QC users will not need to know those details, any more than the typical classical computer user needs to understand the details of the semiconductor devices used in their laptops and smart phones. Of course, the details are important if you want to build a QC. In fact, there are many fascinating and challenging issues to be addressed to build fully functioning quantum computers and, as we mentioned before, lots of great jobs in the labs and businesses that are designing and building those machines.

Before looking at specific systems, let's talk about what is needed in a QC qubit. To me, it is amazing that we can controllably manipulate individual atoms or other atomic-like particles. Transistors on silicon wafers in classical computers are small, but the atomic world is even smaller. Of course, a qubit must be a quantum object whose state can be manipulated into superposition states and changed via gates by means of external, controllable stimuli. These stimuli can be laser light or other electromagnetic fields applied at different places and times in the circuit. These quantum objects must be available in large enough quantities and have nearly identical properties to make a useful system. As mentioned earlier, the largest general-purpose QC systems at this time have about 50 qubits, but that number is likely to increase dramatically over the next few years.

The qubits must be able to be measured. In most practical QCs, the measurement bases stay fixed, and we apply various gates to transform the states of the qubits. As we saw in Chapter 8, using different basis states is a pathway to extracting the maximum information from a quantum system. Importantly, the objects must be robustly immune to unwanted outside influences to limit changes in those states. While we can use error-correction protocols as we discussed in Chapter 12, there are limits to the fixes, and it is more efficient, when possible, to reduce the effects due to undesirable outside influences.

The two types of qubits currently used in general-purpose quantum computers are superconducting circuits and trapped atomic ions. In both these systems, the qubits are acted upon and measured through interactions with electromagnetic fields, making use of the tremendous advances in lasers and microwave devices developed over the past 50 years or so.

Superconducting Qubits

BOB: Let's start with qubits made from superconducting materials. Superconductors are materials that allow electrical currents to flow through them without electrical resistance. That means that if you get the current started, it will flow forever. Superconductivity comes about through quantum effects that cause the materials' charge carriers, usually electrons, to pair up (in so-called Cooper pairs) in such a way that electrical resistance disappears. However, with the materials available today, the materials must be cooled to very low temperatures (typically 1 K—one degree above absolute zero—or lower) and isolated from their warmer surrounding environment to operate properly as superconducting qubits.

Superconducting qubits use superconducting materials separated by a thin insulating layer. Such a structure is called a Josephson junction device, named after Nobel Laureate Brian Josephson, who predicted their behavior theoretically in the 1960s. Normally, the insulator would stop the electrons from flowing from one material to another, but thanks to another quantum phenomenon called tunnelling the Cooper pairs can penetrate that layer. When the Josephson junction is placed in an external electrical circuit, the combination has a preferred oscillation frequency, usually in the microwave frequency range. Measuring the oscillation frequency allows you to observe properties of a qubit's quantum state. Conversely, by having the qubit interact with externally controlled microwaves you can manipulate the quantum state and perform gate operations. There are many different flavors of superconducting qubits, those used in several types of QCs are called transmons.

CARDY: Why are they called transmons?

BOB: Physicists like to name atomic-sized or smaller objects with words that end with "on." You are familiar with photon, electron, proton, neutron, and ion. "Transmon" is an abbreviation for "transmission-line shunted plasma oscillation," which is quite a mouthful. In simple terms, a transmon is a circuit built from a Josephson junction device, and other electrical components, including a region that traps microwaves, much like your microwave oven, but on a much smaller size scale, typically a few millimeters on a side.

Although a transmon is much larger than an electron or ion, its electrical current behavior can be a described as a quantum two-state system with different amounts of current in the two states. Measurement, control, and coupling between the transmons is performed by observing microwaves emitted by the transmons or shining microwaves on them to shift them from one quantum state to another.

Since transmons are manufactured devices, we can't guarantee that they will all be exactly the same, and the QC must be designed to tolerate slight differences among the qubits. However, the technology of Josephson devices is well established, and we know a lot about how to control and measure microwaves thanks to decades of work in microwave communications devices and their use in both commercial and basic science applications.

Trapped Ions

BOB: Several companies have developed quantum computers using atomic ions as the qubits. An ion is an atom or molecule that has lost or gained one or more electrons, leaving it with a net electrical charge. That charge makes it relatively easy to use electric and magnetic fields to push the ions around and to trap them individually in separate spatial regions. In current QCs, the total area occupied by the trapped ions is about 0.1 mm × 0.4 mm, which is small compared to a human, but very large compared to a typical atom's diameter—about 0.1 nm $(10^{-10}$m$)$.

Using techniques that have been developed to a high level of sophistication over the past five or six decades, you can operate on those states with light from lasers. You can create superposition states and entangled states and you can make measurements of various kinds.

Ion qubits don't need super-cold temperatures, and all the qubits are the exactly the same. Since the qubits are manipulated by laser light and not wires, every qubit in the system can interact with any other qubit in the system, giving what is called complete connectivity. The main challenge is getting the laser beams that control the qubit states connected uniformly to all the qubits and synchronized in a controlled way.

The company IonQ uses ytterbium (Yb) positively charged ions in its 32-qubit quantum computer. Yb is used because its atomic energy states are convenient for manipulation by laser light, and it is relatively easy to trap the ions in specific spatial locations with electric fields. The collective motion of the ions in the trap region provides another quantum degree of freedom; so, each ion contributes an energy qubit and is part of a motion qubit. Because Yb has been used in many sensitive atomic clock experiments, the technology for manipulating Yb ions and their states is well understood.

ALICE: Let's now look at some other physical systems that are being developed for use as qubits in QIS and QC applications.

Nitrogen Vacancies in Diamond

ALICE: Another interesting qubit is a "vacancy" in diamond crystals next to an "impurity" atom substituting for one of the carbon atoms. Diamond is a crystalline form of carbon. A local gap in the crystal structure is called a vacancy. See Figure 16.3. Suppose a nitrogen atom replaces one of the carbon atoms in the crystal next to the vacancy. The nitrogen atom plus vacancy system has energy states that lead to the absorption and emission of visible light; so, the vacancy plus nitrogen atom is called a *color center*. Those energy states are easily manipulated with light from lasers. Furthermore, the electrons' spin can be manipulated with magnetic fields (via their magnetic dipole moments), giving us another "handle" to operate on the qubit states. As of this writing, an Australian company, Quantum Brilliance, is building a

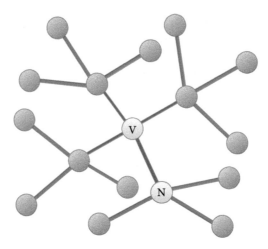

Fig. 16.3 A schematic drawing of carbon atoms (gray) in a diamond lattice, a vacancy V (blue), and a substitute nitrogen atom N (yellow). The colored lines represent chemical bonds. Only a few of the bonds are shown. The nitrogen atom and the vacancy constitute a quantum object that can be used as a qubit.

QC based on these nitrogen vacancies in diamond. Another qubit possibility is to use the spins associated with the nuclei of the atoms in a nitrogen vacancy. This possibility has already been implemented in laboratory situations.

In addition to nitrogen substitutions, scientists are also exploring the use of tin atoms in diamond. The tin atom version has the advantage that it is relatively easy to build into nanophotonic structures that combine the vacancies with material shapes that trap microwaves. In addition, the tin version can operate at somewhat higher temperatures than the nitrogen vacancies, thus simplifying the cooling systems needed to build the QC. Although tin vacancies have yet to be used in QC and QIS devices, their advantages seem likely to drive the research and engineering needed to make that happen.

Photon Qubits

CARDY: Why don't we use photons for qubits in QCs? We talked about how to use their polarization states for quantum cryptography. My impression is that photons are easy to produce and easy to detect.

ALICE: There are two problems with using photons as qubits in a QC. First, photons travel fast—at the speed of light. So, they might travel through the QC faster than we can act on them with gates. You can, however, use long optical fibers to provide time delays if needed. The second problem is that we need to have the photons "made to order" and ready to be manipulated at specific times. In most common photon sources (flashlights, LEDs, candle flames, lasers, and the like), the photons are produced randomly in time. There are some laboratory devices that can produce photons in pairs. Then the detection of one of the photons (the "herald") tells us that the other photon is on its way. But those devices produce the pairs at random times. However, there are other laboratory techniques that produce single photons at controllable times. One example is placing atoms placed between pairs of mirrors ("optical cavities"). That setup can produce large interactions between the atoms and photons in the cavity in such a way that a photon can be emitted upon demand. This technique is still at the laboratory development stage and has not yet been used in a practical QC.

Another strategy is to sort the photons into time bins (rather than polarization modes or path modes). Two different time bins are equivalent to a two-state qubit. Three-state qubits ("qutrits"), four-state qubits ("ququads"), and so on are also relatively easy to produce this way. However, QC work with photons is just in its developmental stage and it remains to be seen if a reliable QC with a large enough number of qubits can be constructed using photons. The company PsiQuantum is working on a photon QC. See Further Reading for the company's URL.

Neutral Atom Qubits

BOB: Neutral (no net electric charge) atoms are another possible set of qubits. With so-called laser-cooling techniques, it is possible to trap atoms in a vacuum in regular spatial arrays set up by intersecting light beams. The interactions between the atoms can be controlled relatively easily and the atoms' states can be manipulated and measured by other light beams. The techniques for manipulating and measuring the trapped atoms are highly developed and are similar to the trapped-ion methods mentioned previously. My guess is that QCs based on trapped neutral atoms will be operational over the next few years.

Quantum Dots

BOB: Another quantum technology, one utilizing quantum dots, also looks promising for QCs.

CARDY: What are quantum dots? How can they be used as a QC?

BOB: A quantum dot is a material in which electrons are confined to a small (few nm = few $\times 10^{-9}$ m) region. This is larger than an atom, but it behaves in some ways like a single atom. "Dot" simply means that the confinement region is small, and when observed, it looks like a dot. The colors of light absorbed and emitted by the confined electron depend on the size of the dot rather than the chemical composition. By way of contrast, the different colors that appear in fireworks are due to different chemical compounds. Larger quantum dots tend to emit and absorb red light (longer wavelengths), while smaller dots absorb and emit blue light (shorter wavelengths). See Figure 16.4.

There are two common ways to produce quantum dots: (1) put small particles, usually semi-conductor materials, in liquid solutions, and (2) use electric and magnetic fields to trap the electrons near the surface of some material. The latter method allows greater control over the position of the trap region, and many quantum dots can be created near one another, leading to a compact set of qubits. By the way, quantum dots may also be used in solar cells to enhance the efficiency of the cells in converting solar light energy into electrical energy.

You can buy solutions of quantum dots that emit a range of colors when stimulated by ultra-violet light. You can even make the solutions at home. Quantum dots available in commercial

2–10 nm

Fig. 16.4 Quantum dots fluoresce (emit light) when illuminated by ultraviolet light. The smaller dots on the left emit violet and blue light, while the larger ones on the right emit yellow and red light.

solutions are often small clusters of cadmium, selenium, or other semiconductor materials. But because the dots are in solution, they are not easily controllable for QC purposes.

The quantum dots that are being explored for possible use in QCs are fabricated in thin layers of solid semiconductors and other materials with the electrons trapped by applied electric and magnetic fields. These are sometimes called "gate confined" states. Here, gate does not mean logic gates but rather materials to which electric voltages are applied to produce the electric fields that trap the charged particles, just like a gate in a fence can be used to help corral horses. Quantum dot qubits have the advantage that they can be fabricated in dense arrays using techniques similar to those widely used in integrated-circuit construction. However, many technical issues, such as uniformity and removal of disturbances from other charges in the system, need to be addressed before quantum dots can be widely used in QCs.

Topological Qubits

ALICE: There is another possibility for qubits that is very different from the others. Scientists have been searching for many years for materials that could be used to produce "topological" qubit states. These kinds of quantum states depend on the topological properties (edges, holes, etc.) of the device. Since the states depend on these spread-out properties of the device, it is believed that they will be less susceptible to local effects such as material impurities and stray electric and magnetic fields. Defects in semiconductor nanowires might provide the appropriate devices, but much basic research remains to be done.

In all these current and future qubits, improving the properties of the materials may reduce the need for error correction and sophisticated quantum control systems that make QCs quite complex. Research in materials science will play a large role in making otherwise fragile qubits more robust.

16.4   Quantum Information Science and the Quantum Internet

ALICE: Throughout this book we have used QIS and QC as if they were friendly siblings, always being careful to give them equal time. Indeed, they are closely related, but in some sense, QIS is more general because it includes both the transmission of information and computations carried out on that information.

Just as QC can be viewed as a generalization of classical computing, QIS can be viewed as a generalization of classical information science. Both QIS and classical information science are concerned with the overall theory of information, not just computer algorithms. QIS, of course, takes advantage of the constraints and opportunities provided by quantum mechanics. That often leads to a deep level of abstraction that makes the concepts difficult to fathom.

When I first learned about information theory, it seems to have little overlap with what I worry about when sending information via email, Twitter, or a postcard. I finally realized that information theory is really not about information in the everyday sense, but more about the abstract structure of what is being sent back and forth between Bob and me, for example, and the advantages and disadvantages of various kinds of information structures.

Let's add learning more about information theory and QIS to our list of topics to study to understand the broader principles that underly QC.

Quantum Internet

ALICE: One area in which QIS is likely to make a big splash within the next few years is the so-called quantum internet. This does not mean faster streaming services for your favorite cat videos. Today's internet does a great job in producing the fast transfer of information from point A to point B, regardless of where A and B are. However, we know that "bad actors" can often access that information and use it for malevolent purposes. A quantum internet has the potential to provide super-high-security data transmission over long distances. Polarized photons appear to be the practical choice for the qubits because they can travel extended distances at high speeds over special polarization-maintaining optical fibers, thus allowing their quantum states to be undisturbed as they travel.

There are two approaches to quantum-enabled secure communications. The first is called post-quantum cryptography, which makes use of standard public cryptography keys that are believed to be sufficiently resistant to current quantum algorithms. The second uses quantum mechanics to distribute the cryptographic key and is called quantum key distribution (QKD). We described a simple version of QKD in Chapter 8. Quantum state teleportation (Chapter 10) is part of QKD as well. Commercial quantum communication packages are now available that operate over city-size distances. See Further Reading.

For longer-distance communications, devices called repeaters are needed because the probability that a photon will be lost due to absorption or scattering increases with the length of the optical fiber. Such repeaters are used in classical transmission lines to amplify the signal, which is inevitably degraded by various "losses" in the electrical cables, or, nowadays, in optical fibers.

If we are using photons and their quantum states to transmit information, how do we build a quantum repeater? That would require detecting and then rebuilding the quantum state. The infamous no-cloning theorem tells us that is not possible in general. If that idea jumped into your mind before you read it, then you are thinking quantum mechanically!

To get around the no-cloning theorem problem, Bob and I can make use of a combination of quantum state teleportation and entanglement swapping (Chapter 10). We each entangle a pair of photons at our ends of the communication channel and we each send one of our photons towards the repeater unit. When the repeater successfully receives a photon from either Bob or me, it stores it in quantum memory and waits for a photon from the other person. When the second photon is received, the repeater performs a Bell state measurement on the two photons and the entanglement is swapped (as we saw in Chapter 10) between the two memory photons. They are then launched on their respective ways toward the receivers.

CARDY: What is a quantum memory? Is that like memory in a classical computer?

ALICE: There are similarities. A classical computer memory stores a pattern of 0s and 1s and allows us to read out that pattern later. A quantum memory stores a quantum state and allows us to produce a qubit described by that quantum state. Again, quantum memories are just beginning to be developed, but the idea is that a photon qubit could interact with other qubits (perhaps trapped atoms or ions). Those other qubits cannot be easily transported but they can store the state temporarily until they are commanded to emit a photon whose state is the same as that of the incoming photon.

Although a fully functioning quantum repeater has yet to be built, research labs are testing various repeater components. In addition, some researchers are using classical computer simulations of quantum repeater networks. Such simulations will help engineers come up with a design protocol for the repeater and allow them to study the complex dynamics of a repeater system. For more information about the quantum internet, please see the references in the Further Reading section at the end of the chapter.

16.5   Quantum Computing and Machine Learning

ALICE: Two of the most rapidly growing areas in computer science are machine learning and artificial intelligence. In simple terms, machine learning means training a computer to recognize patterns—either spatial patterns or temporal patterns or a combination of the two. More generally, machine learning is a procedure by which a computer can extract information from a database. The training consists of giving the computer some input data and the desired output and having the software work out connections that allow the computer to recognize similar patterns in input data on which it has not been trained. For example, you might give the computer several images of different kinds of dogs. The computer analyzes those images and comes up with some rules that allow it to produce an output "dog" when similar images are presented to it later. Much of machine learning involves solving systems of linear equations. We mentioned in Chapter 15 that there are quantum algorithms that can carry out those solutions more efficiently than classical computers.

Related to image recognition is image processing. As a concrete example, cameras in satellites in Earth orbit send images that are useful for weather prediction, monitoring environmental changes (e.g. erosion and glacier deterioration), and military reconnaissance. If we could put fast image-processing computers on satellites, we could reduce substantially the data-transmission and reception requirements (Fan et al., 2017). Once QCs become more portable, image processing is likely to be one of the first space applications of QCs. Until then, quantum machine learning is likely to focus on solving problems related to quantum mechanics, such as quantum simulation.

Artificial intelligence (widely known as AI) is similar to machine learning. The computer is trained to sense its environment and learn what happens when it responds to that environment in various ways as it tries to achieve some specified goal. For example, AI allows a computer to learn to be competitive in strategic games such as go and chess, to control an autonomously operating automobile, to provide intelligent routing of messages in content delivery networks, and to run sophisticated military simulations. With AI, the computer is supposed to be able to learn all these things without human intervention.

What is the role of quantum computing in machine learning and AI? The complete answer is not yet in, but current work focuses on using quantum computer algorithms to speed up mathematical operations that are part of classical machine learning and AI. These operations include solving systems of linear equations, using what is called principal-component analysis to focus on the major features of large data sets, and finding eigenvectors and eigenvalues of matrices. See Chapter 13 of *Programming Quantum Computers* (Further Reading) for more examples of the use of quantum computers in machine learning and AI.

16.6 Alternative Quantum Computer Architectures

ALICE: Thinking ahead about what is coming in QC, I should mention that there are different ways of designing QCs. So far, we have described QCs based on quantum gates interacting with qubits. Quantum gate architecture is much like the architecture in classical computers. But there are other ways qubits can be manipulated and measured for computational purposes. These alternative structures may turn out to be faster or more easily implemented than the gated-based structures that have been used so far. Moreover, alternative architectures may help circumvent noise, loss, and decoherence effects.

As one alternative, we have already mentioned briefly in Chapter 4 and earlier in this chapter the possibility of using quantum systems with more than two states. It turns out there are some advantages of using these multi-state systems in terms of energy efficiency, coding flexibility, and a reduction in susceptibility to noise and other decoherence effects. But so far, it is qubit (two-state) technology and algorithms based on qubit manipulation that have dominated QIS and QC. Developing technologies to produce qutrits in specified states and getting them involved in entangled states is still in the development stage. But we believe this is an aspect of QIS and QC that may take off in the next decade or so.

Quantum Annealing

BOB: A quite different architecture uses qubits arranged in arrays with various connections between the qubits. Instead of using gate operations on one or two qubits, these systems have the qubits undergo a type of time evolution called quantum annealing.

CARDY: I heard about annealing in my high school shop class. We annealed some pieces of glass and some metal welding joints.

BOB: Quantum annealing is similar. You computationally "heat up" a system of qubits and then let them "cool" slowly. The slow cooling allows the system, be it blown glass, a metal weld, or the qubits, to relax slowly into its lowest-energy state. The opposite process—quenching—cools the system quickly, freezing in defects and resulting in a state of higher energy.

In quantum annealing, you prepare the qubits individually in some state and then use controllable coupling between the qubits to form an entangled state. You next use external fields, often magnetic fields, to control the speed of relaxation to the lowest-energy state. To make quantum annealing useful for computation, you need to formulate your problem in terms of minimizing the system's energy or, more generally, in optimizing some property of the system. That turns out to be possible in many cases, including search problems, factoring problems similar to those addressed by the Shor algorithm, and molecular configuration problems.

The Canadian company D-Wave has been selling quantum annealing devices based on superconducting qubits. The advantage of this method is that it is relatively easy to include many qubits. However, you usually have only a limited number of connections among the qubits. The company's website (see Further Reading) has useful information about the quantum annealing architecture and the types of problems that can be formulated to take advantage of that technique. Some authors characterize quantum annealing as an example of "hardware acceleration," in which the specially designed qubit circuitry carries out a fixed task efficiently but lacks the flexibility of a general-purpose computer processor.

Measurement-Based Quantum Computing on Cluster States

ALICE: Yet another architecture uses clusters of qubits and implements quantum algorithms by making state-dependent measurements on a sequence of clusters. For example, you might have your qubits set up in an array as shown in Figure 16.5. The array can be three-dimensional. You divide the qubits into clusters (the shaded groups shown in the figure) where the size and shape of the clusters have been chosen to implement a particular quantum algorithm. After preparing each cluster in an entangled state, you then make a measurement of a single qubit in cluster 1. The result of that measurement affects what you do with cluster 2. Eventually the

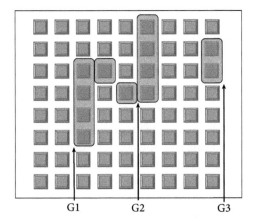

G1   G2   G3

Fig. 16.5 General scheme of a quantum computer with cluster states. The qubit clusters G1, G2, and G3 (the beveled rectangles with gray shading in the background) are measured one after the other. The results of earlier measurements determine the measurement bases of later ones.

measurement outcomes give you the sought-for answer. The whole computation is nothing but a series of single-qubit measurements. Alas, as we know, the measurements destroy the entangled states; so, they can be used only once. To run the algorithm again (e.g. as part of a computational loop), you need to reset the entangled states.

16.7 Ethical Issues

BOB: Ethical issues might not be foremost in your mind if you are interested in QC and QIS. Your thoughts probably tend to focus on qubits, superposition, entanglement, and measurement. Rightfully so. But because information technology has become so integrated with our lives, both professionally and personally, we know that its benefits are accompanied by side-effects—sometimes serious and dangerous side-effects. We are all familiar with identity theft, malicious hacking, ransomware, distribution of misinformation . . . the list goes on. We should be worried that QIS and QC might enhance the capabilities to break security codes, to increase the gathering of private information, and to deploy more sophisticated ways to disrupt the enemy's (competitor's) information technology systems.

These issues are already present in our society. The use of QIS and QC will likely accelerate their rate of growth, making each more pervasive and more difficult to deal with. For example,

- QC and QIS may make it easier to access your private information unless you are using a sophisticated quantum encryption algorithm.
- They may accelerate machine learning and AI in ways that we as a society are not prepared to deal with.
- Hackers and information technology disrupters may gain enhanced capabilities with the use of QCs.
- Quantum computational chemistry may lead to a plethora of new drugs, which might overwhelm our drug-testing capabilities.
- If new drugs are developed by QC methods, who gets access to these new drugs? Just those in the country whose scientists designed the drugs?
- QC chemistry may also lead to new ways of manipulating genetic codes—those of plants, animals, and humans—by finding more efficient and targeted gene-editing techniques such as CRISPR.

Other ethical issues are more focused on social justice, rather than technology. We should be asking hard questions such as the following:

- Will QC and QIS widen the gap between the haves and the have-nots, both within a country and among countries? Will only those companies and countries that have access to QCs be able to enjoy their benefits?
- How do we assure diversity, equity, and inclusion in the QC and QIS workforce?
- How do we adapt our education systems to meet the new workforce needs generated by QC and QIS?

Both Alice and I encourage you to take these issues seriously and to talk about them with friends, colleagues, and those who make technology and educational policy.

16.8   Preparing for Your Quantum Career

ALICE: This is good time to give Cardy some advice about next steps.

CARDY: Thanks, Alice. I am interested in hearing about the math, computer science, and physics I will need to know if I continue with my studies of QIS and QC.

ALICE: Although there are many specific jobs that need to be done in building and operating a QC, each requiring specialized knowledge, there is some common background needed by almost everyone who wants to build or use QCs and other quantum information-processing devices.

In 2020, the US White House Office of Science and Technology Policy released a report *Key Concepts for Future QIS Learnings* (see Further Reading). The report was the product of a conference attended by secondary school and university educators, industry researchers, and representatives from educational and professional organizations. Here is the list of key concepts from the report, with brief explanatory material:

1. **Quantum information science** (QIS) exploits quantum principles to transform how information is acquired, encoded, manipulated, and applied. Quantum information science encompasses quantum computing, quantum communication, and quantum sensing, and spurs other advances in science and technology.

2. **A quantum state** is a mathematical representation of a physical system, such as an atom, and provides the basis for processing quantum information.

3. **Quantum applications** are designed to carefully manipulate fragile quantum systems without observation to increase the probability that the final **measurement** will provide the intended result.

4. **The quantum bit, or qubit**, is the fundamental unit of quantum information, and is encoded in a physical system, such as polarization states of light, energy states of an atom, or spin states of an electron.

5. **Entanglement**, a correlation among multiple qubits, is a key property of quantum systems necessary for obtaining a quantum advantage in most QIS applications.

6. For quantum information applications to be successfully completed, fragile quantum states must be preserved, or kept **coherent**.

7. **Quantum computers**, which use qubits and quantum operations, will solve certain complex computational problems more efficiently than classical computers.

8. **Quantum communication** uses entanglement sent over a transmission channel, such as optical fiber, to transfer quantum information between different locations.

9. **Quantum sensing** uses quantum states to detect and measure physical properties with the highest precision allowed by quantum mechanics.

We have covered all the concepts in this book, with the exception of quantum sensing. That area requires more detailed knowledge of physics, chemistry, materials science, and engineering.

Let's talk about the mathematical knowledge needed for your future work in QIS and QC. In this book, we have kept the math requirements to a minimum, but we also showed that we need extended mathematical reasoning to understand how quantum algorithms work. To dig deeper into QC and QIS, you will need to be on good speaking terms with complex variables and, by implication, functions of complex variables. You should certainly learn more about linear algebra, which is the math underlying right and left vectors, state space, quantum gates (operators), and so on. In linear algebra, the kind of state space we have been using is called Hilbert space, after the famous mathematician David Hilbert. You should also learn more about eigenvalues and eigenstates—concepts and language which permeate the applications of linear algebra in physics and engineering. You don't necessarily need a full course in linear algebra because a lot of information about linear algebra is available on the web. However, a good course or a good textbook will provide a systematic overview, which will clarify the mathematical context of those parts of linear algebra needed in QC and QIS.

BOB: It is also important to learn more about computer science since fundamentals such as the theory of computation, algorithms, data structures, computer security, and cryptography apply equally well to QC and to classical computing.

If you are interested in building and operating actual quantum computers, you will need to learn more about physical qubits such as polarized photons, atoms and molecules with spin, trapped ions, and superconducting circuits—the devices used currently in QC and QIS. Courses in this area are just beginning to be offered by colleges and universities.

We should also mention that many of the jobs in QIS and QC are "engineering" positions: developing and building devices based on the underlying quantum and materials principles, including electrical, cryogenics, vacuum, and nanofabrication systems. These can be exciting and well-paying positions with the opportunities to work on many kinds of QIS and QC systems (see Asfaw et al., 2021).

Brief biographical sketches of 52 women scientists and engineers working in QIS and QC are available in Quantum Daily (2021, see Further Reading).

16.9 Grand Finale

ALICE: Bob and I appreciate your persistence and devotion to learning more about QC and QIS. We confidently predict that over the next few decades, the advances in QC and QIS will dramatically alter almost every area of science and technology. We believe that you, devoted readers, are now well positioned to understand those advances and, if you so decide, to join the QC and QIS workforce that will make this happen. Whether this is your first journey into the world of quantum or your last, you have chosen to be part of what is now being called the *second quantum revolution*. Just as the first quantum revolution in the 20th century brought us lasers, semiconductors, classical computers, solar cells, nuclear power, radiation treatments, and MRI in medicine, the second quantum revolution has the promise of changing lives in dramatic and unexpected ways. We wish you the best as you continue your QC and QIS journeys.

By the way, April 14 is World Quantum Day. Mark your calendars! And look for local activities that you might join.

BOB: When Alice and I began writing this book, we were at first worried that it might not be possible to develop a presentation that is both accessible to a broad audience and faithful to the concepts of quantum mechanics and their applications in quantum algorithms. We trust that we were reasonably successful in achieving those goals. We also hope that you have learned as much about QIS and QC as we did in writing the book.

We anticipate that some of our readers will contribute to the development of new quantum algorithms, quantum programming languages, and the building of actual QCs. The ground is fertile, and the seeds are ready to be planted; we just need lots of people to tend the plants and harvest the fruits of our labors.

Let's wrap up with some quantum humor. I know that sounds oxymoronic, especially to those not versed in quantum, but here goes:

There are three types of people in this world:

those who understand quantum computing;

those who do not understand quantum computing;

and those who simultaneously do and do not understand quantum computing.

 CHAPTER SUMMARY

- Quantum computer programming does not (yet) have universal programming languages akin to Python and C#. Currently, each elementary gate needs to be specified instead of choosing a complex operation such as creating a state of superposition. There are websites offering access to quantum computers with interfaces that use standard programming languages.

- Further work in QIS and QC requires understanding many diverse fields. Obviously, quantum physics, complex variables, linear algebra, and other mathematical tools are needed. In addition, an understanding of the nature of information and how it is shared is an important underlying foundation for manipulating the quantum world.

- Exciting applications of QC to machine learning, artificial intelligence, and a quantum internet are currently being explored. These will most likely be a pairing of classical and quantum computing to increase the efficiency and therefore the power of these tools.

 FURTHER READING

Open-Access Resources

The journal *Physical Review X Quantum* is an open-access journal. Most of the articles are moderately technical, but we have listed several under the heading QIS and QC Overview that can be accessible to more casual readers. The journal also has occasional tutorials on basic quantum topics.

www.arXiv.org is an online repository of preprint articles, freely accessible.

Wikipedia has many articles dealing with QC and QIS principles and techniques. https://en.wikipedia.org/wiki/Main_Page

QIS and QC Overview

National Academies of Sciences, Engineering, and Medicine, 2019. *Quantum Computing: Progress and Prospects*. (The National Academies Press, Washington, DC, 2019). Free download available at https://www.nap.edu/read/25196/chapter/1

Ivan H. Deutsch, "Harnessing the Power of the Second Quantum Revolution," *Physical Review X Quantum* **1**, 020202 (2020). A broad overview of QIS and QC and their relationship to classical computing.

Harrison Ball, Michael J. Biercuk, and Michael R. Hush, "Quantum firmware and the quantum computing stack," *Physics Today* **74**, 29 (2021). An expanded discussion of the organization of quantum computing development as illustrated in our Figure 16.1.

The US federal website https://www.quantum.gov/ has an extensive collection of reports on all aspects of QC and QIS.

Yuri Alexee et al., "Quantum Computer Systems for Scientific Discovery," *Physical Review X Quantum* **2**, 017001 (2021). A wide-ranging article about current and future QCs. Should be accessible to anyone who has finished this book.

Programming Quantum Computers

Eric R. Johnston, Nic Harrigan, and Mercedes Gimeno-Segovia, *Programming Quantum Computers* (O'Reilly, Sebastol. CA, 2019). Aimed at experienced computer programmers.

"Cloud-based quantum computing," Wikipedia. An overview of what is available for online access to quantum computers.

Companies and labs in the United Kingdom have developed an open-source universal quantum operating system that will enable programmers to write programs for quantum computers with different hardware structures: https://thequantumdaily.com/2021/09/23/uk-companies-tout-major-step-toward-universal-quantum-operating-system/.

Google cirq https://quantumai.google/cirq. Python programming framework and library for designing, manipulating, and optimizing quantum circuits. Run circuits on a QC simulator or actual Google QC. Tutorials are available but they assume familiarity with quantum computing at about the level of the textbook by Michael Nielsen and Isaac Chuang, *Quantum Computation and Quantum Information* (Cambridge University Press, Cambridge, 2009).

IBM Qiskit https://qiskit.org/ provides access to IBM quantum computers and simulators as well as many tutorials on all aspects of QC.

AWS Braket https://aws.amazon.com/braket/ Provides access to QCs at several different companies. Access must be purchased, but scientists at universities can get credits (grants).

The system also provides Jupyter notebooks that come installed with the Amazon Braket Software Development Kit (SDK).

Microsoft Azure Quantum https://azure.microsoft.com/en-us/services/quantum/ As of this writing, Microsoft does not yet have its own quantum computer. However, the system provides access to other companies' QCs. Q#, Microsoft's open-source programming language, can be used to develop your own quantum algorithms.

Sandia Labs scientists have developed an assembly language for their trapped-ion computer. "Coding for Qubits: How to Program in Quantum Computer Assembly Language" https://spectrum.ieee.org/tech-talk/computing/software/qscout-sandia-open-source-quantum-computer-and-jaqal-quantum-assembly-language.amp.html.

Jed Brody and Gavin Guzman, "Calculating spin correlations with a quantum computer," *American Journal of Physics* **89**, 35–40 (2021). This article describes how to use the online accessible IBM quantum computers to calculate the average value of "spin correlations" for two-qubit states. The authors assume that the reader is familiar with Pauli matrices, complex numbers, and rotation matrices. We have touched on all those topics but putting them together may be somewhat challenging for our readers. Nevertheless, our dedicated readers should be able to skim the article to get the basic idea of how to program and use a quantum computer to produce measurements of quantum properties of a system.

Physics of Qubits

Superconducting qubits:

The following website has good introductory-level information on using superconducting qubits with microwave resonators.

https://www.jst.go.jp/erato/nakamura_mqm/english/project_overview/p01.html

Trapped ions:

Ion Q: https://ionq.com

The website of the company IonQ has a nice introduction to QC with trapped ions.

Nitrogen-vacancies in diamond:

V. Sewani, H Vallabhapurapu, Y. Yang, et al., "Coherent control of NV⁻ centers in diamond in quantum teaching lab." *American Journal of Physics* **88**, 1156–1169 (2020). This article provides a nice introduction to the energy states associated with a nitrogen-vacancy in diamond and gives details of an experimental setup to prepare and measure the appropriate quantum states.

Evangelia Takou and Sophia E. Economou, "Tin Qubits Give Diamond a New Shine." *Physics* **14**, 105 (2021). (www.physics.aps.org). Using tin atoms in place of nitrogen atoms in diamonds has several advantages.

Neutral atom qubits:

David S. Weiss and Mark Saffman, "Quantum computing with neutral atoms." *Physics Today* **70** (7), 44 (2017).

Semiconductor spins:

Lieven M. K. Vandersypen and Mark Eriksson, "Quantum computing with semiconductor spins." *Physics Today* **72** (8), 38 (2019).

Photons as qubits for QC:

Peter B. R. Nisbet-Jones, Jerome Dilley, Annemarie Holleczek, Oliver Barter, and Axel Kuhn, "Photonic, Qubits, Qutrits, and Ququads Accurately Prepared and Delivered on Demand." arXiv:1203.5614v2 (2012).

PsiQuantum (Palo Alto, CA) is developing a photonic QC: https://psiquantum.com/.

Nathalie P. de Leon et al., "Materials challenges and opportunities for quantum computing hardware." *Science* **372**, 253 (2021). Discusses the materials science research needed to improve superconducting qubits, gate-defined quantum dots, color centers, ion traps, and topological qubits in semiconductor nanowires.

Examples of Quantum Computing Companies

ionQ, https://ionq.com, trapped-ion quantum computer.

Honeywell Quantum Solutions, https://www.honeywell.com/us/en/company/quantum, trapped-ion quantum computer.

Rigetti Quantum Computer, https://www.rigetti.com/, superconducting circuit QC.

IBM Qiskit, https://qiskit.org/, provides access to IBM superconducting circuit QCs and simulators as well as many tutorials on all aspects of QC.

Google, https://quantumai.google/, superconducting qubit gate-based computer system.

Microsoft Azure Quantum, https://azure.microsoft.com/en-us/solutions/quantum-computing/.

Microsoft is attempting to build a quantum computer based on Majorana (topological) qubits. Microsoft has available a quantum development kit for the Q# quantum programming language and Azure Quantum projects.

D-Wave, https://www.dwavesys.com/, a quantum annealing system rather than a general-purpose quantum computer.

Quantum Internet

Marcos Curty, Koji Azuma, and Hoi-Kwong Lo, "A quantum leap in security." *Physics Today* **74**, 37 (2021). How the use of one-photon and two-photon interference can lead to enhanced cryptographic protocols.

R. Valivarthi et al., "Teleportation Systems Towards a Quantum Internet." arXiv 2007.11157v2 (2020).

Commercially available quantum information networks: https://www.aliroquantum.com/

QuISP is a simulation package for quantum repeater networks: https://aqua.sfc.wide.ad.jp/quisp_website/

A nice survey of current Quantum Key Distribution systems: https://www.nature.com/articles/s41534-019-0221-4#Tab1

The US National Security Agency currently does not recommend the use of QKD systems: https://www.nsa.gov/Cybersecurity/Quantum-Key-Distribution-QKD-and-Quantum-Cryptography-QC/

Quantum Computing, Machine Learning, and AI

"Machine Learning," Wikipedia.

"Artificial Intelligence," Wikipedia.

Xanadu: https://pennylane.ai/. Lots of information about quantum machine learning.

E. Johnson, N. Harrigan and M. Gimeno-Segovia *Programming Quantum Computers* (O'Reilly Media Inc, Sebastopol, CA 2019). Chapter 13: "Quantum Machine Learning."

"Quantum Image Processing," Wikipedia.

F. Fan, A. M. Iliyasu, and P.Q, Le, "Quantum image processing: A review of advances in its security technologies." *International Journal of Quantum Information* **15** (3): 1730001–44 (2017). doi:10.1142/S0219749917300017 (open access).

Alternative QC Architectures

Ehud Altman et al., "Quantum Simulators: Architectures and Opportunities." *Physical Review X Quantum* **2**, 017003 (2021). A wide-ranging survey of current and future QC qubits and architectures. Generally accessible descriptions. No math.

Qutrits:

Qutrits can be used in quantum teleportation: https://www.scientificamerican.com/article/qutrit-experiments-are-a-first-in-quantum-teleportation/

Quantum annealing:

"D-Wave Systems," Wikipedia.

Measurement-based QCs:

Robert Raussendorf, Daniel E. Browne, and Hans J. Briegel, "Measurement-based quantum computing on cluster states." *Physical Review A* **68**, 022312 (2003).

Sara Bartolucci et al., "Fusion-based quantum computation." arXiv:2101.09310 (2021). Describes a QC architecture based on entangling measurements and resource states.

David Hunger, "Quantum logic at a distance." *Science* **371**, 576 (2021). Describes work demonstrating that QCs can be scaled by breaking up the system into smaller sets of neutral atom qubits connected by optical fibers.

M. Pompili et al., "Realization of a multimode quantum network of remote solid-state qubits." *Science* **372**, 259 (2021). Uses entanglement among three nitrogen-vacancies in diamond, including one memory qubit.

Quantum charge-coupled device architecture:

J. M. Pino et al., "Demonstration of the QCCD trapped-ion quantum computer architecture." arXiv:2003.01293v2 (2020). The paper describes a variation on a trapped-ion computer in which the ions can be transported back and forth so that in principle all qubits can be interconnected. The Honeywell company is building a quantum computer using this architecture: https://www.honeywell.com/us/en/company/quantum.

Ethical Issues

Ronald de Wolf, "The Potential Impact of Quantum Computers on Society." arXiv: 1712.05380 (2017). An essay focusing on cryptography, optimization, and simulation of quantum systems.

Cathy O'Neil, *Weapons of Math Destruction* (Broadway Books, New York, 2017). A candid look at the downsides of machine learning and big data.

Quantum Education

The report *Key Concepts for Future QIS Learners* is available at https://qis-learners.research.illinois.edu/.

David Matthews, "How to get started in quantum computing." *Nature* **591**, 166–167 (2021). Available online at https://www.nature.com/articles/d41586-021-00533-x.

Abraham Asfaw et al., "Building a Quantum Engineering Program." arXiv:2108.01311 (2021). The title is a bit misleading; in addition to designing quantum engineering curricula, the authors write about a wide range of QIS and QC careers and the topics you should study to become involved in those careers.

Qubit-by-qubit: https://www.qubitbyqubit.org/ The organization offers a year-long high school course on QC and QIS. It also runs summer workshops (1–2 weeks and 4 weeks) for both high school and college students.

Quantum.country: https://quantum.country/ A series of "essays" about QC and QIS. The essay "Quantum computing for the very curious" introduces quantum states and quantum gates but assumes some familiarity with linear algebra and complex numbers. The essay might serve as a review of our Chapters 1–5 and 9. The pace is fast, but the writing is lucid and informative.

Quantum for all: K-12 quantum education materials: https://quantumforall.org/.

National Q-12 Education Partnership: https://q12education.org/. Has links to many quantum educational materials. https://q12education.org/learning-materials.

PhETs: Freely accessible simulations on a variety of STEM topics. Search for "quantum phenomena." https://phet.colorado.edu/.

ACEQM: Online materials for an introductory quantum mechanics course: https://www.physport.org/curricula/ACEQM/ (instructors only).

Qubitekk: www.qubitekk.com, sells equipment for quantum education and research using photons. Single-photon detectors, entangled photon sources, and so on.

EPiQC: Enabling Practical Scale Quantum Computing https://www.epiqc.cs.uchicago.edu/. The site includes QC and QIS educational materials at various levels.

D. Candela, "Undergraduate computational physics projects on quantum computing." *American Journal of Physics* **83**, 688 (2015). Shows how to simulate a quantum computer to illustrate gate operations on qubit states and to simulate the Grover and Shor algorithms. The article's exercises provide nice extensions of the methods described in this book.

Mark Beck, *Quantum Mechanics: Theory and Experiment* (Oxford University Press, Oxford, 2012). Chapters 1–8 and 17 provide a somewhat higher-level coverage of this book's Chapters 1–11. An excellent resource for physics and engineering majors.

People in QIS and QC

https://q12education.org/ hosts short bios of a diverse set of QC and QIS workers:

https://q12education.org/about/careers.

The Quantum Daily, https://thequantuminsider.com/, has several articles about people in QC and QIS careers. Short bios of 52 women scientists and engineers working in QC and QIS: https://thequantuminsider.com/2021/08/24/52-wonder-women-working-in-industry-as-quantum-scientists-engineers/.

Appendix Quantum Toolkit

Philosophy is written in this grand book–I mean the universe–which stands continually open to our gaze, but it cannot be understood unless one first learns to comprehend the language and interpret the characters in which it is written. It is written in the language of mathematics.

Galileo, *The Assayer* (1623)

In this Appendix, we provide a summary, without proofs or derivations, of quantum mechanics expressions that occur frequently in QIS and QC. We also collect terminology dealing with matrices and vectors.

Quantum States

Superposition state—a linear combination of basis states weighted by state amplitudes. Example: a superposition of computational basis states $|0\rangle$ and $|1\rangle$

$$|\psi\rangle = a_0 |0\rangle + a_1 |1\rangle , \qquad\qquad (A.1)$$

where a_0 and a_1 are the state amplitudes.

Normalization condition $a_0^2 + a_1^2 = 1$ (for real amplitudes) and $a_0^* a_0 + a_1^* a_1 = 1$ (for complex amplitudes).

Probability interpretation (Born rule): The probability of getting $|0\rangle$ as a measurement outcome for the state $|\psi\rangle$ is $|a_0|^2$ while $|a_1|^2$ is the probability of observing $|1\rangle$.

Right and left vectors (column and row vectors)

See Figure A.1.

Right vectors and column vectors

$$|\psi\rangle = a_0 |A_0\rangle + a_1 |A_1\rangle \Rightarrow \begin{pmatrix} a_0 \\ a_1 \end{pmatrix}. \qquad\qquad (A.2)$$

Left vectors and row vectors

$$\langle\psi| = a_0^* \langle A_0| + a_1^* \langle A_1| \Rightarrow \begin{pmatrix} a_0^* & a_1^* \end{pmatrix}. \qquad\qquad (A.3)$$

State Products

There are several kinds of state vector products: inner, tensor, outer, Kronecker, exterior.

Inner (scalar, dot) product of two state vectors is given by the Dirac bracket $\langle s_0 | s_1 \rangle$, the projection of $|s_1\rangle$ along $|s_0\rangle$. If both state vectors are normalized, $\langle s_0 | s_1 \rangle = \cos \theta_{01}$, where θ_{01}

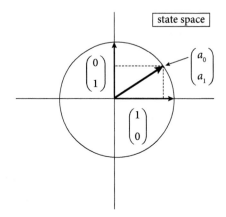

Fig. A.1 The state space vector tip has coordinates a_0 and a_1 ($|a_0|$ and $|a_1|$ if the amplitudes are complex numbers). The two basis states are shown in column vector form. The circle has a radius equal to 1.

is the state space angle between the two state vectors. In terms of state amplitudes, we write

$$|A\rangle = a_o |0\rangle + a_1 |1\rangle$$
$$|B\rangle = b_o |0\rangle + b_1 |1\rangle \tag{A.4}$$

and the inner product becomes

$$\langle B| A\rangle = b_0^* a_0 + b_1^* a_1. \tag{A.5}$$

$\langle A|B\rangle = 0$ means the state vectors $|A\rangle$ and $|B\rangle$ are orthogonal. If $\langle A \mid A\rangle = 1 = \langle B \mid B\rangle$, the vectors are normalized. If all three relations hold, the vectors are said to be *orthonormal*.

Tensor Product—used to form basis vectors for the state space formed as a product of two subspaces. $|\psi_A\rangle$ lives in the A subspace, $|\psi_B\rangle$ in the B subspace.

$$|\psi_A\rangle = a_0 |A_0\rangle + a_1 |A_1\rangle$$
$$|\psi_B\rangle = b_0 |B_0\rangle + b_1 |B_1\rangle. \tag{A.6}$$

Two right vectors

$$|\psi_A\rangle |\psi_B\rangle \Rightarrow \begin{pmatrix} a_0 \\ a_1 \end{pmatrix} \begin{pmatrix} b_0 \\ b_1 \end{pmatrix} = \begin{pmatrix} a_0 \begin{pmatrix} b_0 \\ b_1 \end{pmatrix} \\ a_1 \begin{pmatrix} b_0 \\ b_1 \end{pmatrix} \end{pmatrix} = \begin{pmatrix} a_0 b_0 \\ a_0 b_1 \\ a_1 b_0 \\ a_1 b_1 \end{pmatrix}. \tag{A.7}$$

Two left vectors

$$\langle \psi_A| \langle \psi_B| \Rightarrow \begin{pmatrix} a_0^* & a_1^* \end{pmatrix} \begin{pmatrix} b_0^* & b_1^* \end{pmatrix} = \begin{pmatrix} a_0^* \begin{pmatrix} b_0^* & b_1^* \end{pmatrix} & a_1^* \begin{pmatrix} b_0^* & b_1^* \end{pmatrix} \end{pmatrix}$$
$$= \begin{pmatrix} a_0^* b_0^* & a_0^* b_1^* & a_1^* b_0^* & a_1^* b_1^* \end{pmatrix}. \tag{A.8}$$

The *outer product* produces a matrix from two column vectors (not necessarily of the same dimension). Example, with each column vector with two rows:

$$|\psi_A\rangle \otimes_{\text{outer}} |\psi_B\rangle \Rightarrow \begin{pmatrix} a_0 \\ a_1 \end{pmatrix} \otimes_{\text{outer}} \begin{pmatrix} b_0 \\ b_1 \end{pmatrix} = \begin{pmatrix} a_0 \\ a_1 \end{pmatrix} \begin{pmatrix} b_0 & b_1 \end{pmatrix} = \begin{pmatrix} a_0 b_1 & a_0 b_1 \\ a_1 b_0 & a_1 b_1 \end{pmatrix}. \tag{A.9}$$

The *Kronecker product* is the "vectorization" of the outer product. That is, the product matrix entries are written as a column vector:

$$\begin{pmatrix} a_0 \\ a_1 \end{pmatrix} \otimes_{\text{Kronecker}} \begin{pmatrix} b_0 \\ b_1 \end{pmatrix} = \begin{pmatrix} a_0 b_1 \\ a_0 b_1 \\ a_1 b_0 \\ a_1 b_1 \end{pmatrix}. \tag{A.10}$$

The *exterior product* is a generalization of the vector cross product used in physics and engineering, $\vec{C} \times \vec{D} = \vec{E}$. The magnitude of $\vec{E}$ is $|C||D|\sin\theta$, where θ is the angle between the two vectors. The exterior product of state vectors is not used in ordinary quantum mechanics.

Changing Basis States

See Figure A.2.
The relationships among the basis states:

$$|A_0\rangle = \cos\theta_{AB} |B_0\rangle + \sin\theta_{AB} |B_1\rangle$$
$$|A_1\rangle = -\sin\theta_{AB} |B_0\rangle + \cos\theta_{AB} |B_1\rangle \tag{A.11}$$

θ_{AB} is the *state space* angle between the $|A_0\rangle$ basis direction and the $|B_0\rangle$ basis direction. In matrix and column vector form:

$$\begin{pmatrix} a_0 \\ a_1 \end{pmatrix} = \begin{pmatrix} \cos\theta_{AB} & \sin\theta_{AB} \\ -\sin\theta_{AB} & \cos\theta_{AB} \end{pmatrix} \begin{pmatrix} b_0 \\ b_1 \end{pmatrix}. \tag{A.12}$$

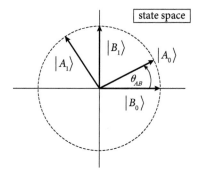

Fig. A.2 A state space diagram of two sets of basis states.

Entangled States

An entangled state is a state for two (or more) qubits that cannot be expressed as a simple product state $|\psi_A\rangle\,|\psi_B\rangle$ of individual qubit states. Example:

$$|AB\rangle = \frac{1}{\sqrt{2}}\,|0_A\rangle\,|0_B\rangle + \frac{1}{\sqrt{2}}\,|1_A\rangle\,|1_B\rangle. \tag{A.13}$$

The subscripts are the qubit labels.

Operators

An operator C acts on a state vector to produce a new state vector $C\,|\psi\rangle = |\phi\rangle$.
Right-left vector (outer product) form of an operator $C = |\phi\rangle\,\langle\psi|$.

Matrix Form of an Operator

Most operators can be represented using both the left-right vector (outer product) form and matrices.
    Example: the Pauli X operator:
    State vector form in the computational basis:

$$X = |0\rangle\,\langle 1| + |1\rangle\,\langle 0|. \tag{A.14}$$

The *matrix form* in that basis is formed by "sandwiching" X between the various basis states:

$$\begin{aligned}
X_{00} &= \langle 0|\,X\,|0\rangle = \langle 0|0\rangle\,\langle 1|0\rangle + \langle 0|1\rangle\,\langle 0|0\rangle = 0 \\
X_{01} &= \langle 0|\,X\,|1\rangle = \langle 0|0\rangle\,\langle 1|1\rangle + \langle 0|1\rangle\,\langle 0|1\rangle = 1 \\
X_{10} &= \langle 1|\,X\,|0\rangle = \langle 1|0\rangle\,\langle 1|0\rangle + \langle 1|1\rangle\,\langle 0|0\rangle = 1 \\
X_{11} &= \langle 1|\,X\,|1\rangle = \langle 1|0\rangle\,\langle 1|1\rangle + \langle 1|1\rangle\,\langle 0|1\rangle = 0.
\end{aligned} \tag{A.15}$$

The matrix form of X using these basis states is

$$X = \begin{pmatrix} 0 & 1 \\ 1 & 0 \end{pmatrix}. \tag{A.16}$$

This method works for any operator and any state basis set.

Projection Operators

$$P_{B_0} = |B_0\rangle\,\langle B_0| \tag{A.17}$$

projects a state $|\psi\rangle$ along the $|B_0\rangle$ direction in state space

$$P_{B_0}\,|\psi\rangle = |B_0\rangle\,\langle B_0|\psi\rangle \tag{A.18}$$

The resulting vector has length $\langle B_0|\,\psi\rangle$.

Identity operator I is defined as $\mathrm{I}\,|\psi\rangle = |\psi\rangle$ for all state vectors. The identity operator can be written as a sum of projection operators:

$$\mathrm{I} = |a_0\rangle\,\langle a_0| + |a_1\rangle\,\langle a_1| + \ldots, \tag{A.19}$$

where the sum includes all the projection operators that make up a "complete" set of states. (*Complete* means that any quantum state for the system can be expressed as a sum of the basis states, with appropriate amplitudes.) For a single-qubit state there are only two terms corresponding to the two basis states: $\mathrm{I} = |a_0\rangle\,\langle a_0| + |a_1\rangle\,\langle a_1|$.

Eigenvalues and Eigenvectors

An operator M has associated with it a set of states that satisfy

$$M\,|\psi_1\rangle = \lambda_1\,|\psi_1\rangle$$
$$M\,|\psi_2\rangle = \lambda_2\,|\psi_2\rangle \tag{A.20}$$

for a qubit system with two states. λ_1 and λ_2 are numbers called the *eigenvalues*. $|\psi_1\rangle$ and $|\psi_2\rangle$ are the *eigenstates* (or *eigenvectors*) of the operator M.

Average Values (Expectation Values)

If the system state is $|\psi\rangle = \sum_j c_j\,|\psi_j\rangle$, where $|\psi_j\rangle$ are the eigenstates of M, the average value of the quantity represented by operator M is given by

$$\langle\psi|\,M\,|\psi\rangle = \sum_j \lambda_j\,|c_j|^2. \tag{A.21}$$

If M represents a measurable property of the qubit, then M's eigenvectors are the corresponding *measurement basis states*.

Matrices and Tensors

In strict mathematical terms, a matrix is simply an array of entries, usually numbers, variables, or even other matrices. Tensors are also arrays of entries, but a tensor comes along with one or more transformation rules that tell us how the entries change, say, under rotations or translations of coordinates or the change of basis states in quantum mechanics, which, as we have seen, can often be represented by rotations in state space. In this language, scalars, vectors, and "tensors" (two-dimensional arrays) are rank 0, rank 1, and rank 2 tensors, which transform differently under rotations of coordinate systems. In many cases, the transformation rules can be expressed as a matrix operating on the matrix representation of the tensor.

Matrices can often be used to represent tensors. That leads to somewhat confusing language in which matrix and tensor terminology get mixed.

In what follows, we restrict ourselves to operations on square $n \times n$ matrices.

The *entries in a matrix* (the so-called *matrix elements*) will be labeled by subscripts indicating the row-column location of the entry within the matrix. For example, F_{23} indicates the entry in the second row and third column of the matrix F.

Transpose F^T of a matrix F means swap the corresponding elements across the upper-left to lower-right (main or primary or principal) diagonal

$$\begin{pmatrix} F_{11} & F_{12} \\ F_{21} & F_{22} \end{pmatrix}^T = \begin{pmatrix} F_{11} & F_{21} \\ F_{12} & F_{22} \end{pmatrix}. \tag{A.22}$$

Hermitian conjugate $F^\dagger$ means transpose (swap the corresponding elements across the upper-left to lower-right diagonal) and take the complex conjugate of the elements:

$$\begin{pmatrix} F_{11} & F_{12} \\ F_{21} & F_{22} \end{pmatrix}^\dagger = \begin{pmatrix} F_{11}^* & F_{21}^* \\ F_{12}^* & F_{22}^* \end{pmatrix}. \tag{A.23}$$

Inverse: F^{-1} defined such that $F^{-1}F = I$.

Orthogonal matrix: the matrix elements are real and $F^{-1} = F^T$, which is equivalent to $F^T F = I$.

The rows of an orthogonal matrix constitute an orthonormal set of left (row) vectors. The columns contain an orthonormal set of right (column) vectors.

Unitary operators and matrices have $U^{-1} = U^\dagger$, that is, the inverse is equal to the Hermitian conjugate:

$$U^\dagger U = U^{-1}U = I. \tag{A.24}$$

The eigenvalues of a unitary operator satisfy $|\lambda_u|^2 = 1$. The eigenvectors and eigenvalues for a unitary operator (matrix) can be written as

$$U|\psi_u\rangle = \lambda_u|\psi_u\rangle = e^{ia_u}|\psi_u\rangle, \tag{A.25}$$

where λ_u is the eigenvalue associated with the state $|\psi_u\rangle$ and a_u is a real number.

Matrix Products

There are several types of matrix products.

Matrix product: This works only if the number of columns in one matrix is equal to the number of rows in the other. For 2×2 matrices:

$$\begin{pmatrix} a_{11} & a_{12} \\ a_{21} & a_{22} \end{pmatrix} \begin{pmatrix} b_{11} & b_{12} \\ b_{21} & b_{22} \end{pmatrix} = \begin{pmatrix} a_{11}b_{11} + a_{12}b_{21} & a_{11}b_{12} + a_{12}b_{22} \\ a_{21}b_{11} + a_{22}b_{21} & a_{21}b_{12} + a_{22}b_{22} \end{pmatrix}. \tag{A.26}$$

The matrix product of a $p \times m$ matrix with a $m \times q$ matrix is a $p \times q$ matrix.

In quantum mechanics, the matrix product gives the matrix representation of operators acting successively within a state space. Normally, the numbers of rows and columns match the dimensionality of the state space.

Matrix times a vector: A matrix multiplying a vector yields another vector (equivalent to an operator acting on a state vector)

$$\begin{pmatrix} F_{11} & F_{12} \\ F_{21} & F_{22} \end{pmatrix} \begin{pmatrix} a_0 \\ a_1 \end{pmatrix} = \begin{pmatrix} F_{11}a_0 + F_{12}a_1 \\ F_{21}a_0 + F_{22}a_0 \end{pmatrix}. \tag{A.27}$$

If the matrix is orthogonal (or more generally, unitary), the length of the newly produced vector is the same as that of the vector being multiplied. For quantum mechanics, we say that orthogonal (unitary) matrices maintain normalization.

Outer product of matrices F and G:

$$F \otimes G = \begin{pmatrix} F_{11} \otimes G & F_{12} \otimes G \\ F_{21} \otimes G & F_{22} \otimes G \end{pmatrix} = \begin{pmatrix} F_{11}G_{11} & F_{11}G_{12} & F_{12}G_{11} & F_{12}G_{12} \\ F_{11}G_{21} & F_{11}G_{22} & F_{12}G_{21} & F_{12}G_{22} \\ F_{21}G_{11} & F_{21}G_{12} & F_{22}G_{11} & F_{22}G_{12} \\ F_{21}G_{21} & F_{21}G_{22} & F_{22}G_{21} & F_{22}G_{22} \end{pmatrix}. \tag{A.28}$$

For a multi-qubit system in quantum mechanics, the outer product is used to build matrix representations of operators from those that act on the qubit subspaces. In most cases the outer product is the same as the *Kronecker product*. However, the Kronecker product gives a product of matrices of any size, not necessarily the same, which is useful if the tensor product space is built from subspaces with different dimensions.

Index